The Conceptual Revolution in Geography

For Ken Davies, who gave me the opportunity

The Conceptual Revolution in Geography

edited by
Wayne K. D. Davies, B.Sc., Ph.D.
Associate Professor of Geography, University of Calgary, Canada

Rowman and Littlefield
Totowa, New Jersey 07512

First published in the United States 1972
by Rowman and Littlefield, Totowa, New Jersey

ISBN 0 87471 107 X

Printed and bound in Great Britain by
Hazell Watson and Viney Ltd, Aylesbury, Bucks

Contents

6   *Contents*

Preface

Recent years have seen a fundamental re-examination of the methodology of geographical enquiry in which the re-alignment of geography within the mainstream of modern science and the increasing contact with other disciplines probably represent the most influential trends. Although the broad objectives of geographical endeavour remain the same, these changes have had a profound effect upon the organization of geographical material. Indeed, the formal study of this organization or methodology has now become a basic field of geographical interest. The result is that introductory geography courses can no longer be concerned only with case studies or with the history of the discipline. Moreover, all geographers, of whatever specialization, have by necessity been paying more attention to the derivation and application of their concepts, rather than claiming academic distinctiveness by an empirical justification based on their subject matter. This collection of essays attempts to emphasize the primacy of methodology in the geographical debate during the past decade by reprinting the text of twenty papers derived from the English language literature. The decision to produce a collection of essays rather than an integrated methodological statement was not taken lightly, especially in view of the criticism made of similar projects in other spheres of geography. However, it was felt that the volume would provide college and university students with convenient access to the original arguments developed by the authors, rather than with paraphrased summaries into which problems of distortion and bias would inevitably intrude. In this way, the recent methodological debate may be more easily appreciated and understood.

The papers were selected according to two principles: either they represent a comprehensive statement of a particularly methodological (not statistical) point; or they represent one of the first statements on a specific issue which has continued to be the subject of geographical discussion since this collection was finally completed in 1969. Although it must be stressed that the essays were independent pieces of work, their overall comprehension was thought to be enhanced by grouping them into four sections, and by providing an introductory discussion to the theme of each. These sections are:

  I. *Geography and the Role of Ideas*
 II. *Geography and the Methods of Modern Science*
III. *Geography and the Systems Approach*
 IV. *Geography and Behaviour*

Acknowledgments

The publication of this collection of essays would not have been possible without the full co-operation of the following authors, to whom I wish to express my appreciation:

E. A. Ackerman, D. Amedeo, B. J. L. Berry, J. M. Blaut, L. A. Brown, I. Burton, R. J. Chorley, L. Curry, S. Gale, R. G. Golledge, D. Grigg, P. W. Lewis, D. Lowenthal, I. M. Matley, G. Olsson, D. R. Stoddart, J. Wolpert.

I should also like to thank the editors of the following journals and collected papers, in which the selected essays were first published, for their permission to reprint the articles in this volume:

Annals of the Association of American Geographers
The Canadian Geographer
Geografiska Annaler
Geography
Northern Geographical Essays (Newcastle-upon-Tyne)
Papers and Proceedings of the Regional Science Association
The Professional Geographer
Tijdschrift voor Economische en Sociale Geografie
U.S. Geological Survey Professional Paper Series

We are grateful for their permission to reproduce copyright material.

GENERAL NOTE   American spelling has been retained in papers reproduced from American sources.

W. K. D. Davies

The Conceptual Revolution in Geography

Whatever we see could be other than it is.
Whatever we can describe at all could be other than it is.
There is no *a priori* order of things.

L. Wittgenstein, *Tractatus Logico-Philosophicus*

Science ultimately depends upon man's perception of order in the universe, the individual disciplines being distinguished not by the particular objects they study but by the questions they ask and by the integrating concepts, propositions and perspectives that their workers use. There is no ultimate all-embracing truth; St. Paul's imagined paradise,[1] a place in which we have complete understanding, is now regarded as an anachronistic ideal. Instead of such an epistemological monism,[2] most scientists today adhere to concepts of epistemological dualism. Thus, 'whatever knowledge we have of real objects is indirect or representative, the datum whereby . . . [we] . . . know any such object is not identical with the object known.'[3] Recognition and acceptance of these views makes it easier to appreciate the fact that the major scientific advances in knowledge are made, not by more precise observations, but by the development of new ways of looking at things.[4] Certainly this does not negate the value of technical refinement, for such sophistication is closely geared to, and represents a fundamental pre-requisite for, the growth of an objective and substantive discipline. What it does do is to place technical ability in its true perspective; it is but one weapon in the whole methodological and procedural armoury of each discipline. Hence one must consider it a travesty to include all the changes in attitudes and objectives that have characterized geographic research in the last decade under the simple title of a 'quantitative' revolution. More fundamental developments which have transformed the face of geographical enquiry are observed in the process.

One consequence of the more rational scientific methodology embraced by many geographers in the last few years has been to lead the discipline closer to the contemporary attitudes and developments in other disciplines. No longer do geographers consider their discipline to be methodologically unique:[5] they are concerned with similar problems to other scientists. More-over, these currents have re-awakened the possibility of cross-fertilization among the sciences. An earlier period of cross-fertilization in the nineteenth century was responsible for the development of the institutionalized and professional discipline we know today, but since then geographers seem to have spent much of their time in academic isolation from the mainstreams of scientific thought. Apparently self-imposed (one must be careful here because the methodological process has not been accurately charted)[6] the development was probably controlled by two features. The first was a retreat from the environmentally based cause-effect doctrine, the second by an advance into a dominantly empirical phase, an attitude that may be summarized by that well worn jingle, attributed to James Fairgrieve, 'geography is learnt through the soles of one's boots,' an aphorism still indelibly imprinted on the feet of most contemporary geographers. Yet not all workers accepted the rampant empiricism of the early and middle twentieth century. As in all disciplines the progress of the subject was marked by numerous eddies of interest and cross currents of method. However, although the theorists[7] continued to develop their models and concepts and to point out alternative ideas of progress, they were not utilized by the majority of academic geographers. The conventional wisdom lay in a one-sided empiricism that rejected such rationalist viewpoints.

These developments are not, however, unique to geography. Similar developments took place in most of the other observational sciences. All have some history of introversion and concern about their methodological distinctiveness. Thus, to take only two examples, sociology[8] went through its own phase of empiricism, particularly in the 1920's and 1930's, whilst psychology[9] in the United States still wrestles with its over-emphasis on 'behaviourism', an over-zealous restriction of its subject matter to observable activities, to the exclusion of consciousness.

Yet again we must be careful. Though it may be fashionable in some quarters to be intolerant of the empiricist attitudes, one must recognize that not all is on the debit side. We have already intimated that there is no single path to knowledge and, however one looks at the situation, empiricism has provided a solid backlog of knowledge to aid our interpretation of the geographical patterns in the world. Moreover, objectivity and precision of statement replaced earlier speculative judgements, aiding the development of controlled observation and research design. It has also constrained theorists to keep their studies in touch with the real world and to relate their ideas to the realm of observation. Despite these merits one must, on the other hand, criticize the mere accumulation of evidence. It seems impossible to escape the conclusion that the acceptance of a primarily empirical philosophy presents great dangers. Research can easily degenerate into a mere collection of uncodified and trivial facts, and the discipline may become fascinated by a concern for the particulars, rather than the generalities dealt with by science. An idiographic rather than a nomothetic attitude is engendered. Moreover, it is usually recognized that the empiricist trend makes it more difficult to extract spatial principles and generalizations from the mass of factual evidence, whilst it is all too easy to replace the naïve attitudes of subjectivity held by many of the nineteenth century rationalists with another naïveté. In this case it is the acceptance of epistemological monism rather than epistemological dualism, in other words the failure to recognize the assumptions behind all empirical observations. Fortunately, many geographers are aware of these problems and have re-emphasized the need for theoretical and conceptual schemes as well as for factual evidence.[10] The approaches reinforce one another, leading to the abandonment of the extremes of single-minded empiricism and subjective rationalism[11] in favour of a more coherent approach to knowledge.

The new frontiers

In many ways it must be counted as rather symbolic that these new outlooks are permeating geographical teaching and research at a time when man's major exploratory interest lies outside the earth system. Exploration was traditionally one of the supports of geographical enquiry; indeed, the discipline occupied its greatest relative position among the sciences in the so-called 'Golden Age of Discovery' spanning the fifteenth to the nineteenth centuries. In this era geographers actively aided exploration by their technical ability – cataloguing and collating the variation of the earth's surface in map as well as in verbal form. Today the specialization of technical effort required for such a technological triumph as space exploration takes one outside the realm of data collection and processing, and geographers are inevitably excluded from the frontiers of exploration. Moreover, even the compilation of facts about the earth's surface and the establishment of precise locations is mainly in the hands of government departments and specialized agencies – a situation that has contributed to the explosive growth of information about the world in which we live. Indeed, a situation of information overload has developed, the scale of the problem being nicely demonstrated by Haggett and Chorley[12] in their observation that the size of the problem facing Humboldt and Ritter was about one thousand times as small as that facing contemporary geographers. Even allowing for exaggeration there is no room for complacency, for it will not be long before such a situation will occur. It will be drastically aided by information derived by remote sensors[13] and by the growth of computerized data banks in the natural and social sciences, a feature emphasized by Gaits[14] when he pointed out that a locational reference will characterize all sorts of human and physical data in these systems. Confronted by these situations it is perhaps not so surprising that the technical concern of geographers has moved away from the stage of wholesale data collection to data manipulation, and to the search for new ways of looking at and presenting this information. Computerization and quantification are the most obvious tools that can cope with our increased spatial knowledge, but again it must be stressed that they are only part of the process, for every observable feature of the earth's surface cannot be used at the same time. Certainly high speed electronic calculators aid the geographer's ability to deal with more facts, and together with statistical sophistication will contribute substantially to the factual basis of the discipline. However, manipulation is not enough; techniques are the common property of all scientists. If a discipline is to play a distinctive and useful role in the scientific pantheon it must possess a distinctive *raison d'être* and a distinctive corpus of concepts. Geography, as the contents of this collection of essays set out to show, is increasingly acquiring a new conceptual apparatus, in addition to its earlier concerns. These developments are, in the long run, more fundamental than high speed calculators or remote sensors. Indeed, the increased attention paid to conceptualization represents the realization that this is the means whereby knowledge is more easily codified, and deeper insights into the possible spatial variations are thereby provided. Without a body of concepts to act as a check list of current problems the geographic viewpoint can easily degenerate into an encyclopedic list of facts and regularities. Such an approach, although adequate one hundred years ago, would hardly be enough to ensure the continued recognition of geography as a distinctive body of academic opinion. The reason is not that the attitude of the geographer has changed, because a basic interest in searching for and explaining spatial distribution, organizations and processes still exists. Rather it is because science has become more sophisticated and people have their own more varied travel experiences. In addition they are subjected to a daily diet of data about the world by newspapers, journals, television and films. Hence geography has lost its unique role in mediating information obtained about various parts of the earth. One must also remember that many of the problems of former years – in particular the construction of maps and charts and the creation of an objective locational framework upon which to describe the position of places and countries have been largely solved – at least for geographic purposes. (This is not true for other purposes; the requirements of jet and space travel require ever more rigorous locational accuracy, and sophisticated electronic equipment is provided to deal with the task.) This success has meant that the research frontiers of other years are now part of the heritage of common knowledge or are part of a specialized, separate discipline such as cartography. However, although geographers may have lost their role as the only people concerned with the provision of accurate information about the spatial differences and regularities of the world, the distinctiveness of the discipline has not been submerged. Indeed, as Stoddart has recently shown, there has been an acceleration in the amount of geographical literature.[15] This situation is explained by the fact that geographers still remain the only group of people who are fundamentally interested in establishing an organized and coherent discipline to study and explain spatial regularities and distributions, though the growth of regional science[16] and perhaps ekistics[17] represent allied, but currently more theoretically orientated disciplines. This is not to deny, of course, the role of other scientists in providing and processing information on a spatial basis; geologists, botanists, sociologists and economists, in particular, fulfil such functions. But it must be stressed that in these cases the description and explanation of the spatial variations are not the basic objectives of the enquiry, the information being derived as the by-product of some other organizational theme. At its most basic level, therefore, it is the organization of the material used that is all important, not the material itself. It is this that gives modern geography its distinctive role, though one must admit that some ecologists and regional scientists would adopt similar positions. In creating this distinctive role, methodological and conceptual advances are critical; not only are they important steps in the provision of a continuing specialized role for geography among all the sciences, but they also provide a rational guide to future and contemporary investigations.

Despite this concern with conceptual evidence one must not forget that the last decade has seen more traditional approaches to the problem of information abundance. In particular mention must be made of the *Geography Abstract Series*[18] created by Professor Clayton now at University of East Anglia and the continuing provision of an updated bibliography of the 'International List of Geographic Serials'[19] by Professor Chauncy Harris at Chicago. These represent valuable additions to the individual reviews of the literature in the specialist journals[20] and to the annual lists of new geographical publications.[21] All are essential to the task of information retrieval, and decrease the search time taken for individual studies on any problem. But this is not all. Like most research orientated disciplines, the typical eighteen month time lag between the acceptance of a research paper and publication means that the journals now have a limited role in quickly disseminating the results of any research. The result has been an outpouring of research papers and discussion papers from the major university departments and, particularly from the United States, a series of reports aimed specifically at providing reviews of the field as well as dealing with the professional status of the discipline.[22] Yet despite the existence of these bibliographic guides and reports it is still difficult to appreciate the scale of development in any field. Hence it is not surprising that a group of academics have started a series aimed at providing authoritative and contemporary reviews of the major lines of enquiry. This is appropriately titled *Progress in Geography* and is under the general editorship of Board, Chorley, Haggett and Stoddart.[23]

In the light of all these developments it is clearly presumptuous to claim that the collection of essays in this book represents either the most up-to-date literature in particular fields or a complete check list of all the methodological articles written in the last decade. In addition no attempts are made to encompass the accelerating range of educational material dealing with the introduction of these new methods to schools and colleges.[24] Instead the collection is designed to provide examples of what are considered to be the more significant methodological trends that have appeared in the English language journals since 1961. The temptation to enlarge the scope of the work by including key articles from foreign language journals was resisted. Not only does this deserve a book in itself, but it is felt that the methodological variations in interpretation that have historically characterized different language and culture systems would create more problems than it would solve, as Matley's article (Chapter 4) demonstrates. Finally it must be emphasized that this book specifically excludes examples of the technical and statistical progress made, for instance, by Dacey in the analysis of spatial patterns[25] or Tobler in map transformations.[26] Two recently published collections of essays: 'Spatial Analysis' edited by Berry and Marble and 'Quantitative Geography' edited by Marble and Garrison, adequately cover this ground.[27]

Dispute will inevitably rage over the choice of essays, but it must be borne in mind that any choice involves exclusion, and a deliberate attempt has been made to keep the number of essays as small as possible in order to avoid the bulky, repetitive and usually expensive volume that characterizes most collections of essays. Perhaps a more fundamental problem revolves around the fact that some of these essays have been superseded by larger works. Again this would seem to be inevitable in view of the eight-year time-lag since some of them were written, though this is, incidentally, a satisfying feature in the progress of the discipline.[28] However, it is to be hoped that this small collection is representative enough to demonstrate, not by second-hand reviews, but by the original articles, the current methodological status of geography. No longer is the discipline purely concerned with empirical problems. It is increasingly developing an interest in epistemological questions and is acquiring the technical and conceptual apparatus of modern science in its search for order in the spatial system of the earth. Moreover it is worth noting the inter-disciplinary nature of footnotes to the papers in this collection. They show that geographers are aware of developments in other disciplines. No longer can geographers be considered to be introverted and concerned only with their own individual problems; they are aware that such positions would lead to the persistence of error[29] and to another period of methodological isolation.

Although each of the essays in this collection represents an independent piece of work, and should be considered in this light, many common themes and linkages will be recognized. In an effort to focus attention on some of these common links, the papers have been grouped into four sections. The first section 'Geography and the Role of Ideas' emphasizes the need to scrutinize the way we look at and interpret evidence; frequently this cannot be divorced from the socio-economic and cultural system of which we are a part. The second section, 'Geography and the Methods of Modern Science', shows how geographers are using modern scientific methodology to compile and analyse evidence in their attempt to build up a distinctive corpus of knowledge about spatial regularities and differences. This approach frequently involves the abstraction of information from its spatial setting and this may obscure its relevance to other facts that are not part of the analysis. Hence the third section, 'Geography and the Systems Approach' shows how geographers can bring together individual pieces of evidence (especially those that would seem to represent disparate studies), as part of a common framework of scientific knowledge. The need for geographers to study functional relationships and processes as part of their study of the world ecosystem is also stressed in this section.

The final section of this collection of essays, Geography and Behaviour, emphasizes the current trend towards behavioural studies. Geographers have been fairly slow to recognize the value of interdisciplinary research in this field but are starting to reformulate their ideas about spatial behaviour. In the past spatial studies of behaviour have adopted the rationality assumptions of the economist and only recently have these theories been challenged. A valuable by-product of this trend has been the integration of many geographical studies of physical and human distributions. Frequently these studies depend upon the perception of the physical potentialities of any site or area and the approach provides an added justification for the increasing concern with the role of ideas in geographical enquiry. Certainly these developments mean that the ghost of the continuing debate between the dichotomy of determinism and possibilism has finally been laid, and the energies of geographers have been translated into more useful and productive lines of enquiry.

Although it has proved possible to provide a justification for this four-fold division, these comments should not be allowed to obscure the overall unity of this collection of essays. The similarity of methodological approach would seem to be more important than the differences in factual content. However, at the risk of destroying this unity it was thought useful to group the essays into a series of sections – purely to emphasize some of the more fundamental approaches and association of ideas. Each section is preceded by a small introduction designed to simplify the arguments developed in more detail in the individual papers and also to attempt, by means of a small bibliography, to lead to the most pertinent additional essays dealing with the various themes. It is to be hoped that these commentaries will link rather than interrupt the continuity of the ideas and concepts and will provide a useful guide to the developing methodology of geographic enquiry.

References

1   'For now we see through a glass, darkly; but then face to face: now I know in part; but then I shall know even as also I am known.' Corinthians I, Chapter 13, verse 11.

2   MCEWAN, WILLIAM P. *The Problem of Social Scientific Knowledge*, Totowa, New Jersey: Bedminster Press (1963), p. 7.

3   LOVEJOY, ARTHUR C. *Revolt against Dualism*, La Salle, Illinois: Open Court Publishing Co. (1930).

4   Among the many examples in science one of the best known is W. Heisenberg's re-ordering of the existing knowledge about atomic structure. See HOFFMAN, B., *The Strange Story of the Quantum*, Harmondsworth (1963).

5   An attitude attacked by SCHAEFER, F., 'Exceptionalism in Geography: A Methodological Examination'. *Annals of the Association of American Geographers*, Vol. 43 (1953), pp. 226–49.

6   Several books have recently appeared dealing with the development of geography but most of them are content to deal with the history of the discipline:

a.   FREEMAN, T. W. *One Hundred Years of Geography*, London: Duckworth (1961).

b.   DICKINSON, R. E. *The Makers of Modern Geography*, London: Routledge and Kegan Paul (1969).

7   Two of the best examples are:
a.   CHRISTALLER, W. *Die Zentralen Orte in Suddeutschland,* Jena (1933).
b.   LOSCH, A. *The Economics of Location,* New Haven: Yale University Press (1954).
8   MCKINNEY, J. C. 'Methodology, Procedures and Techniques in Sociology', in H. Becker and A. Boskoff (Eds.) *Modern Sociological Theory,* London: Holt, Rinehart and Winston (1966).
9   KOESTLER, A. *The Ghost in the Machine,* London: Hutchinson (1967), p. 5 et seq.
10   BUNGE, W. *Theoretical Geography,* Gleerups, Lund, Sweden (1963).
11   Approaches to knowledge characterized by speculative theories and ideals that were verbally and subjectively expressed, rather than being formulated as verifiable propositions.
12   CHORLEY, R. J. and HAGGETT, P. *Models in Geography,* London: Methuen (1967), p. 31.
13*a.*   MOORE, E. G. and WELLER, B. S. 'Urban Data Collection by Remote Sensor', *Journal of the American Institute of Planners,* Vol. 35, no. 1 (1969), pp. 35–43.
   b.   A new interdisciplinary journal has recently appeared to deal with these issues: *Remote Sensing of the Environment,* University of Kansas, Lawrence, Vol. 1 (1969).
14*a.*   GAITS, G. M. 'A Co-ordinate Reference System for Planners,' in the Contribution by the Ministry of Housing and Local Government to: *Third annual Conference of the Regional Studies Association* (1967).
   b.   One of the latest pleas for the need for data banks is provided by: LLOYD, P.E. and DICKEN, P. E. *Town Planning Review,* Vol. 38, No. 4 (Jan 1968).
15   STODDART, D. R. 'The Growth and Structure of Geography', *Transactions of the Institute of British Geographers,* Vol. 41 (1967), pp. 1–19.
16   Regional science has been largely inspired by the work of Professor W. Isard at the University of Pennsylvania. Its exponents have been drawn from many social science disciplines and have been drawn together by a common concern for the theoretical aspects of spatial regularities – particularly with the mathematical description of those regularities. Although the Regional Science Association has branches in many parts of the world the major journals are still:
a.   *Papers and Proceedings of Regional Science Association,* Vol. 1 (1954) – Vol. 24 (1969) (continuing).
b.   *Journal of the Regional Sciences,* Vol. 1 (1958)–Vol. 9 (1969) (continuing).
17   To a greater extent than regional science, *ekistics,* the science of human settlements, is largely the work of one man, Constantine A. Doxiadis. Based in Athens, Doxiadis and his associates have produced a monthly journal called *Ekistics* and a substantive work of the same title: DOXIADIS, C. A. *Ekistics,* London: Hutchinson (1968).
18   *Geographical Abstracts 1966–present,* University of East Anglia, Norwich, UK. Currently consisting of four series appearing six times a year the reviews are preceded by 'Geomorphological Abstracts 1960–65'.
19*a.*   HARRIS, C. D. and FELLMAN, J. D. 'International List of Geographical Serials', *Department of Geography Research Series,* Chicago University, No. 63 (1960).
   b.   MULLINS, L. S. 'New Periodicals of Geographical Interest', *Geographical Review,* Vol. 59 (1969), pp. 147–51, 290–4.
20   For example: STEEL, R. W. 'A review of I.B.G. Publications 1946–60', *Transactions of the Institute of British Geographers,* Vol. 29 (1961), p. 129–47.
   MCDONALD, J. R., 'Publication Trends in a Major French Journal', *Annals of the Association of American Geographers,* Vol. 55 (1965), p. 125–39.
   FAIRCHILD, W. B. 'Adventures in Longevity', *Geographical Review,* Vol. 56 (1966), pp. 1–11.

21 The major bibliographic sources are: *Referativny Zhurnal, Current Geographical Publications, New Geographical Literature and Maps, Bibliographie Géographique Internationale.*

22*a.* Some of the first examples of this trend in the United States and Britain may be provided by the following series from the Departments of Geography in the following universities:

i. U.S.A.

University of Chicago Research Papers in Geography, Nos. 1–117 (cont.)

Northwestern University Studies in Geography, No. 1 (1952) (cont.)

Michigan Interuniversity Community of Mathematical Geographers, Vol. 1 (1963) (cont.)

University of British Columbia Geographical Series, No. 1 (1963) (cont.)

ii. U.K.

London School of Economics Research Paper Series.

Southampton University Research Papers.

Hull University Occasional Paper Series.

Durham University Occasional Papers and Research Series, No. 1 (1957).

The list is not a comprehensive guide but is merely an indication of the longest standing series.

  b. Associations of American Geographers: Commission on College Geography

    i. *Geography in Undergraduate Liberal Education* (1965).

    ii. *A Basic Geographical Library* (1966).

    iii. *Geographic Manpower – a Report on Manpower in American Geography* (1966).

    iv. *New Approaches in Introductory College Geography Courses* (1967).

    v. *Introductory Geography – Viewpoints and Themes.*

    vi. *Undergraduate Major Programs in American Geography* (1968).

    vii. *A Survey Course: Energy and Mass Budget at the Surface of the Earth* (1968).

    viii. *A Systems Analytic Approach to Economic Geography* (1968).

    ix. *Computer Assisted Instruction in Geography* (1969).

23 BOARD, C., CHORLEY, R., HAGGETT, P., STODDART, D. *Progress in Geography*, Vol. 1, London: (1969), Arnold (1969).

24 Examples of these new methods are:

  a. American High School Project – See GRAVES, N. J. 'The High School Project of the Association of American Geographers', *Geography*, Vol. 53 (1968), pp. 33 and 68.

  b. WALFORD, REX *Games in Geography*, London: Longmans (1969).

25 DACEY, M. F. has produced over a score of articles dealing with spatial patterns. Two examples are:

  a. 'A probability model for central place locations', *Annals of the Association of American Geographers*, Vol. 56 (1966), pp. 550–68.

  b. 'Some properties of order distance for random point distributions', *Geografiska Annaler 49B* (1967), pp. 25–32.

26 TOBLER, W. For example:

  a. 'Geographic area and map projections', *Geographical Review*, Vol. 53 (1963), pp. 59–78.

  b. 'Medieval distortions: the projections of ancient maps', *Annals of the Association of American Geographers*, Vol. 56 (1966), pp. 351–60.

27*a.* BERRY, B. J. L. and MARBLE, D. *Spatial Analysis. A Reader*, Prentice Hall (1968).

  b. GARRISON, W. L. and MARBLE, D. F. (Eds.) 'Quantitative Geography', Parts I and II *Northwestern Studies in Geography*, Nos. 13 and 14, Evanston, Chicago (1967).

28 Two of the most productive research interests, namely model building and perception studies have been the subject of the following books:

28 *a.* CHORLEY, R. J. and HAGGETT, P. op. cit., note 13.

   b. LOWENTHAL, D. (Ed.) 'Environmental Perception and Behaviour', *Department of Geography, University of Chicago Research Paper* No. 109, Chicago (1967).

29   JAMES, PRESTON E. 'On the Origin and Persistence of Error in Geography', *Annals of the Association of American Geographers,* Vol. 57 (1967).

I Geography and the Role of Ideas

'The trouble isn't what people don't know.
It's what they do know, that isn't so.'

Will Rogers

For most of this century geographers have pursued their discipline primarily in empirical terms. This methodological emphasis tends, however, to obscure the fact that interpretations of real world patterns are highly coloured by the point of view as well as the training of the observer. The five essays in this section provide examples of work recently carried out in an attempt to redress this balance. By emphasizing the need to understand the meaning of particular concepts, and by tracing the impact of ideas and imagination on geographical enquiry, the essays draw attention to certain epistemological questions and illuminate the dangers of a discipline based purely on the empirical approach.

In many ways this group of five essays goes a long way towards placing the conceptual turmoil of the last decade in perspective. They show that rapid shifts of methodological emphasis as well as periods of relative stagnation characterize the history of any discipline. Today, at the end of a decade of debate, it is worth remembering that geography is merely experiencing the consequences of one such revolution and with it the attendant birth-pains associated with the growth of a new methodological orthodoxy. Yet it must not be thought that this will inevitably lead to a new methodology as dominating or as self-centred as the Hettner-Hartshorne philosophy formulated in an earlier period.[1] The debate is likely to continue because of the growing interest among geographers in the theory of knowledge. This could ensure that all concepts and ideas will be subject to constant scrutiny and discussion, with the methodological developments taking place in allied disciplines providing the necessary guide lines.

The comparative lack of interest in epistemological questions in the past has meant that a great deal of misinterpretation about fundamental issues exists in the literature. The first two papers of this section provide examples of the confusion that can result. In Chapter 1 'Theory, Science and Geography', Davies attempts to resolve the paradoxical fact that the quantitative frontiersman of the early 1960's always stressed the need to make geography 'scientific', whereas most geographers had always felt themselves to be members of a scientific discipline. In this example the paradox was resolved by recourse to the changing meaning of the term 'science'. Blaut in 'Space and Process' (Chapter 2), also deals with the perspective of certain ideas in relation to a particular term, in this case 'space', and he again clarifies the ambiguity associated with it by similar semantic means.

These two papers emphasize one of the important principles that can be formulated from a study of the ideas and concepts currently in use, or historically used in geography. This principle is that the new interpretations made at any point in time (or space for that matter) are not timeless. New perspectives are always being developed to question the adequacy of the current conventional wisdom and older ideas are frequently being rendered obsolete. Indeed, it seems as if the turmoil observed within individual disciplines during periods of change is part of the price paid for academic progress. If this is the case then we would be advised not to place too many barriers in the way of the development of new perspectives. Otherwise the new concepts may be stifled and this may lead to sterility in ideas and fossilization of viewpoints – no matter how productive the initial set of concepts proved to be. A vivid example of this point is provided by the third essay of this section: Chapter 3, 'Darwin's Impact on Geography', by D. R. Stoddart. His discussion of the impact of mid-nineteenth century biological thinking on geography shows that the fertility of Darwin's original ideas led to a series of important developments in geography, but these ideas were eventually translated into a single-minded devotion to the course of evolution. At the same time important concepts relating to random processes were ignored and abandoned.

These introductory remarks may seem to imply that a consensus is bound to have a restrictive effect upon any discipline, muffling new ideas and, through familiarity, breeding indifference to the unexplored quarters of older systems of thought. It is worth remembering that this is far from the truth, that some sort of academic consensus is needed at various points in time, if only to organize and to preserve the continuity of knowledge. Two different discussions of this sort of consensus are provided. The first, by Lowenthal (Chapter 4, 'Geography, Experience, and Imagination'), really represents the keystone of this section, for the essay explores the factors that lie behind any consensus, demonstrating the relationship of the multitude of individual views to the general view, as well as showing the extent to which images and experience condition the opinions that are held. The second paper by Matley (Chapter 5, 'The Marxist Approach to the Geographical Environment') provides a case study of one particular consensus, a consensus conditioned by the political philosophy of the state in a totalitarian setting. Even here, it is worth noting one of the points made above, namely, that the consensus is not timeless. Marxist geographers have experienced certain shifts in the methodological interpretation of the term 'environment' and in the extent to which it conditions society.

Whilst these comments demonstrate the general unity of the essay in this section, each essay does demonstrate other important principles that are best discussed individually.

Chapter 1, 'Theory, Science and Geography', by Davies, represents an overview of this whole collection because it draws upon many of the themes more rigorously developed by these essays. The basic objective of the article is to present the case for a geography that follows modern scientific lines of enquiry, and stresses the need for the development of theories and precision in observation and analysis. Two major points emerge from a brief review of nineteenth century scientific methodology. The first is that a methodological orientation based solely on induction observes the true nature of scientific discovery, because it ignores the role of deduction and the flash of inspiration that usually lies at the heart of major scientific discoveries. The second point is that the rejection of the determinist philosophy in the early years of this century seems to have led to the abandonment of much of the contemporary scientific method associated with this philosophy. The result was that most geographers were content to describe the differences between areas in a subjective way and on a formalized basis. These developments led to the doctrine of the unique case and to the enshrinement of a particular type of region, the compage, as the core of the discipline.

Both points add emphasis to an argument already referred to, namely, that the methodological attitudes of past workers in any discipline should not determine its future development, however illustrious these workers might have been. An evolving methodology is a vital necessity, and links with other disciplines must be maintained if the subject is not to become sterile, self-centred and subjected to perpetuating its own errors of interpretation.[2]

A great deal of attention is also paid in the essay to the often repeated assertion that a single methodology cannot embrace the general and the unique. In retrospect this may represent a rather exaggerated view if applied to the systematic fields of geography, though it is still an important principle if applied to traditional regional geography. Thus, Lewis (Chapter 7), points out that the recognition of two unique cases does not preclude the search for similarities. Moreover it might be noted that generalizations provide important guides for studies of unique cases, while sets of uniquely orientated studies provide the raw material for future generalizations. Yet these are relatively minor qualifications, for the nomothetic approach is upheld as being the most significant methodology because it is the most productive and most intellectually satisfying. Thus Davies hopes that with an accepted body of theory and modern scientific methods, geographers, like other scientists, can go straight to the heart of the problem, and geographic work will become ever cumulative, in a way that is impossible in an art. If this is achieved, the discipline will become more and more open to the possibility of cross-fertilization and new ideas. An example of this is provided by the replacement of the doctrine of the inevitable, completely mechanical trend of the Newtonian world by the probable trends of quantum mechanics. Hence, Davies concludes, the keys to future progress lie in the development of an organized and productive methodology.

Blaut's small but very pertinent paper, 'Space and Process' (Chapter 2), represents a second contribution to the clarification of concepts used in geography. In the course of an investigation into the difference between the two meanings of the term, that is 'relative' as opposed to 'absolute' space, he observes that it is intellectually unsound for certain geographers to continue to refer to their discipline as a 'science of space', because geographers only deal with a particular part of space (cf. the interests of fundamental particle physicists and astronomers'). In any case all too often the term relates to concepts of absolute space, ideas that are outmoded in terms of the general consensus of scientific knowledge.

After outlining the difference between the two concepts (absolute space relating to a distinct physical and eminently real entity, and relative space referring to the relationship between events or parts of events), Blaut traces the use of each concept in terms of geographical and scientific enquiry. He demonstrates that ideas of absolute space were central to the scientific system of Newton and to the philosophic beliefs of Kant. However, although the Newtonian system has been overthrown by new physical theories, the geographical literature still provides examples of the use of the term in this sense, as for example when space is introduced as a causal force. Yet of even greater methodological significance is the influence that the philosophic system of Kant had on geographical enquiry. Kant's three-fold division of reality into objects, spatial arrangements and temporal arrangements, pro-vided the basis of the methodology propounded by Hettner and Hartshorne. This division, Blaut maintains, reduces the role of the student of spatial distribution and processes to the role of a cataloguer or of a cartographer, and only leaves him with the task of providing the introduction to the scienti-fic study of objects. Anything else that is left, as for instance in the labelling of mathematical space or in the measurement of distances, is the common property of all sciences. Yet this does not signal the demise of the distinctive body of knowledge called geography, for these criticisms rest on the assump-tion of an absolute space, an assumption of dubious value in view of the methodological limitations of the concepts demonstrated in all branches of scientific study. Thus, instead of dealing with concepts of absolute space, it is more important to investigate the potentialities of relative space. For Blaut, this is the more useful and productive concept because, by definition, relative space cannot be divorced from relative time. Hence he concludes that the focus of inquiry in geography should be the study of these two interrelated notions, namely process.

Blaut concludes, however, on a note of depression, for despite his thorough study of three cases he maintains that he is unable to find any 'process' concept that can qualify as a basic organizing concept for the disci-pline. Yet it would be surprising that one such basic concept could be found in view of the wide variety of spatial processes that could be investigated. Indeed, at this stage it is worth mentioning Berry's point (Chapter 11) that geographers do not claim to study the universe: they are concerned only with those processes operating in the world-wide ecosystem of which man is the dominant part. Acceptance of this limitation would seem to partially resolve Blaut's problem. In any case there is no reason to fully accept Blaut's point about the painfully confining nature of the study of simultaneity in any system, whether the system is a farming or an urban one. Certainly it is true at present that geographers only investigate the events or particles existing in a system at any point in time. However, the events that did not occur in the past or may be likely to occur in the future could be incorporated into the system and studied at the same time: several studies already point the way to this possibility.[3]

In an exceptionally well documented article, Lowenthal (Chapter 4), sets himself the task of exploring the much neglected theory of geographical knowledge, instead of dealing with the methods or meaning of geography or with the analysis of particular concepts. His general theme can best be summarized by one of his concluding remarks, namely that 'every image and idea about the world is compounded of personal experience, learning and memory', and he maintains that this is as true for the discipline of geography as for the layman's perception of the milieu.

Avoiding any academic pretentiousness or claim to uniqueness, Lowenthal explains that geographers deal with the milieu on the same sort of scale and in similar categories to those usually dealt with by the layman, although the demands of a discipline make it, among other things, more 'narrowly focused, orderly and theoretical'. He shows that, despite certain exceptions, man has reached some sort of general agreement with the character of the world, and this world view replaces the regional fragments that constituted knowledge in the past. Yet certain dangers exist in this situation. The very coincidence of ideas that constitute this world view means that the consensus *may* be treated as the only worthwhile view of reality. Instead, Lowenthal maintains that we should be humble and realize that there are many alternative views of any situation (cf. the quotation from Wittgenstein on p. 9), a point neatly illustrated by his reference to Kohler's observation that the physics of the grasshopper world would be different from the physics developed by man. Yet such humility does not rest only on tolerance, it also acts as a check on any tendency to cherish the consensus as the only guide to knowledge. Certainly this guide provides a valuable function by organizing existing lines of enquiry but, as Davies has pointed out in Chapter 1, there is a danger that the guide may become too restrictive and smother new ideas. Perhaps, therefore, heretics should be welcomed, not only because their challenge to orthodoxy provokes fresh debate upon the efficacy of any viewpoint but also as Lowenthal implies, because some of the heretical views of any point in time are quite likely to provide part of the new orthodoxy of another generation.

Lowenthal's essay also provides a welcome reminder that all knowledge has a subjective as well as an objective component, and that it is difficult to consider any kind of ultimate truth. Geographers, just like other academics faced with the complexity of the real world, observe, select, classify, and analyse according to particular purposes and circumstances. It is salutary experience to realize that all of them, by the very nature of their individual perspectives, warp the world in their own particular way. Yet this does not mean that the essay represents a defence of subjectivity: nothing could be further from the truth. Rather it is a literate plea for the need to consider the limitations of any knowledge we might have. Indeed, the bulk of the essay consists of a careful discussion of the nature of individual world views, focusing particular attention upon the limited nature of these private milieus, and the extent to which they vary among groups as well as individuals. Particularly important is the discussion of the way in which these attitudes transcend objective reality, and can be influenced by past memory or stereotypes. Thus the essay demonstrates the limitations as well as the value of the knowledge achieved to date.

Although Stoddart's paper 'Darwin's impact on geography' (Chapter 3), represents a departure from the wider epistemological concerns of Lowenthal, or the narrower clarification of the particular concepts studied by Davies and Blaut, it emphasizes many of the points already raised. Indeed, Stoddart's study of the intellectual impact of biological thinking on geography provides one of the few comprehensive case studies of the derivation and the application of individual geographic ideas and concepts. Particularly worth noting is the fact that many of the ideas attributed solely to Darwin were already being discussed in the mid-nineteenth century and that Darwin's work exercised a synthesizing function. Moreover it must be stressed that, despite the general consensus of opinion, Darwin was not just concerned with the course of evolution: rather it was with the mechanism of change. This provides, once again, an illustration of the difference between the currently accepted interpretation of Darwin's work and its real substantive content.

The bulk of Stoddart's essay consists of an examination of the influence of biological thinking organized in four themes: time and evolution, organization and ecology, struggle and selection, and finally randomness and chance. Stoddart shows that the first three were taken up with enthusiasm. Applied to geographical examples they led to new understanding. In particular they provided unifying principles that subsumed the greatly increased quantities of data being derived about the world and thereby released geography from being condemned to a destiny of inventory and catalogue.

The first of these themes, time and evolution, generated the greatest initial success. Darwin's own study of coral reefs led the way in the study of landforms, but the climax of this evolutionary approach was represented by Davisian geomorphology, a trend that emphasized 'stage' and incorporated many biological analogies such as 'youth', 'adolescence' and 'old age' into its terminology. Eventually this led to a narrow and restrictive focusing of the discipline around denudation chronology. Similar developments occurred in ecology and pedology, but the limitations of such lines of development seem to have been recognized earlier in these fields. Less successful, perhaps, was the application of these ideas to the social sciences, though in view of Matley's thesis (Chapter 5) it is worth noting that the ideas made a substantial contribution to the development of the Marxist view of world society.

The second and third themes: organization and ecology (cf. Section III), struggle and selection, derive largely from Darwin's long voyage in the *Beagle* and his observations about the complexity of inter-relationships between living things and their close relationships with the environment. In geography, one of the most important 'influences' of these themes was the realization that man was not at the centre of life, but was a part of it, a revolution in thought that is comparable in impact to the Copernican super-seding of the medieval interpretation of the relationships between the plants. These developments made it possible to accept the organic unity of the world (or any region) as a unifying theme for the spatial description of man and his environment. Expressed in subjective free-will terms it accommodated the possibilism of Vidal de la Blache and, in rigid mechanical form, led to the extreme determinism that Stoddart maintains was one of the last fields of operation of the Newtonian world view.

It is the fourth theme, however, that seems to be a crucial one for modern methodology, because it represents an example of the way in which a critically influential idea can be obscured by the prevailing consensus and the inadequate development of associated fields. Although Darwin carefully distinguished between the course of evolution and the way it was effected, he was unable to explain the basic variation of the mechanism for change. Eventually he abandoned the distinction, a feature no doubt partially due to the pressure of contemporary opinion, though it must be remembered that Darwin lacked the intellectual equipment upon which random variations could be precisely formulated and incorporated into his basic thesis. Not for seventy years after the publication of his original work would the complete set of statistical and genetical tools be forthcoming in the work of Mendel and Fisher. By that time geography had set off on a self-centred evolutionary tack.

All the essays in this section have shown that geography does not exist in a cultural vacuum. Its ideas and concepts are influenced by the broader spectrum of scientific philosophy and, as Stoddart in particular has shown, the discipline has progressed by accepting some ideas yet rejecting others. How-ever, this process of acceptance and rejection is influenced in certain parts of the world by the prevailing state philosophy, and Matley's essay 'The Marxist Approach to the Geographical Environment' (Chapter 5) draws attention to the extent of this political control.

Matley shows that the Marxist view of the nature of the environment and the extent of its relationship with the development of society is at variance with the general consensus among Western geographers. However, in both intellectual camps one must not forget Lowenthal's point about the influence of stereotypes on opinion.[4] Stoddart has already provided one example in his discussion of Darwinism, but Matley draws attention to another – the tendency among Marxist geographers to regard Western geographers as determinists. Part of this attitude is probably explained by the prevalence of this particular philosophy in the Western World when the Marxists came to power in Russia, but it may also be attributed to the criticisms made by Marxists about the extent of exploitation in Western bourgeois society. Most Western geographers would reject the former point, of course, for it is based on a misunderstanding of progress of Western schools of geography. However, the second point may have some validity, although it is not the place here to enter into a discussion of all the issues involved. It is enough to be able to draw attention to the possible need to temper the Western 'world view' with an alternative philosophy. In other words, we must be careful not to develop a one-sided, Western-dominated methodological approach. Stereotypes may also exist in the views held by the Western world about Marxists. Indeed one of the most valuable points of Matley's paper is that it shows that Marxist geographers have not maintained the unflinching methodological stance that many Western geographers suggest. Soviet geography has undergone several shifts of opinion about the nature of the environment and its relationship with society. In part this is probably due to the peripheral position occupied by geography in Soviet society, and to the major tenets of Marxism. After all, Marx and Engels devote little space in their studies to any consideration of the discipline of geography, but one must also remember that Soviet policy is often pragmatic, however rigid and doctrinaire the official stance seems to be. Despite these mitigating points, it must not be thought that Soviet geographers arrive at a general consensus in the same way as their counterparts in the Western world. Official party pronouncements lay down the general policy format on certain issues though the official view is usually preceded by internal debate.

The first policy statement on the nature of the environment and its influence on society was, according to Matley, rather late in appearance. It was not until 1938 that Stalin codified existing opinion and laid down the official view in which environment and society were regarded as separate entities. The general thesis was that Marxist laws of society accounted for the development of man and society, and this development could only be influenced by the environment. One of the consequences of this view was that geography developed in a segregated manner: the physical and economic branches were separated. It is, perhaps, interesting to note that this sort of dichotomy characterized much of the work of Western geographers in the same period, however much lip-service was paid to an over-riding unity. Moreover it may be suggested that the Soviet segregation of the two branches lies in the acceptance, among Marxists, of the pre-Darwin view of the separation of man and his environment. Darwinism, or rather the stereotyped attitudes of Darwinism held in the early twentieth century, seemed to have pointed the way to a determinism alien to Marxist ideas.

The party pronouncement of 1938 is not, however, the end of the story. Soviet geography has also experienced a conceptual revolution in the 1960's[5]. After a series of important attacks on the official policy, particularly by V. Anuchin, a major policy statement by Illyichev in 1963 rejected the dichotomy between the two branches of geography and supported the concept of a unified geography. (It may be noted that a gradual consensus seems to have emerged in certain sectors *before* the official seal of approval was given.) Today the concept of the 'environment' is regarded in a broader context, for it includes more than that rigid, unchanging part of the earth external to man. Now it also incorporates that part of nature transformed by man.

Hence it opens the way to a two-way interaction between society and environment, instead of the rigid separation of the concepts or of proposing a deterministic relationship between them – the only alternatives that seemed possible in earlier periods.

Although one might conclude these few words of introduction by observing that more points of comparison can be made between the Western and Marxist interpretations of geography, it must be stressed that it is unlikely that this will lead to any world consensus about the nature of geography. The Marxists still rigidly preserve their independence by rejecting the Western approach and claiming a Marxist line of reasoning for all their theorie and methodology. However, perhaps it is noteworthy that, despite the differences in historical development, both major schools of geography experienced a conceptual revolution in the 1960's. In both cases the revolution was associated with the consequences of the separation of the discipline: in the Western world the isolation of geography from the mainstream of scientific thought; in the Marxist world the division of the discipline into two separate parts, a division intensified by the great rift that developed between the natural and social sciences. It is proposed that adequate appreciation of the nature of these methodological revolutions and the factors that have contributed to these rather rapid changes cannot be made without reference to epistemological questions. Only in this way will a balanced perspective of the state of the discipline be derived.

References

1   SCHAEFER, F. 'Exceptionalism in Geography', *Annals of the Association of American Geographers,* Vol. 43 (1953), pp. 226–49.

2   JAMES, P. E. 'On the Origin and Persistance of Error in Geography', *Annals of the Association of American Geographers,* Vol. 57 (1967), pp. 1–24.

3   See CURRY, L. (Chapter 16). For an empirical approach to the same problems see:

a.   MEINIG, D. W. 'A Comparative Historical Geography of Two Railnets', *Annals of the Association of American Geographers,* Vol. 52 (1962), pp. 394–413.

b.   DAVIES, W. K. D. 'Latent Migration and Space Preferences', *Professional Geographer,* Vol. 16 (Sept. 1966), pp. 300–304.

4   LOWENTHAL, D. (Chapter 4). See also:

a.   BUCHANAN, W. and CANTRIL, H. *How Nations See One Another,* Urbana: University of Illinois Press (1953).

b.   SECORD, P. F. *et al.* 'The Negro Stereotype and Perceptual Association', *Journal of Abnormal and Social Psychology,* Vol. 53 (1956), pp. 78–83.

5   See also:

a.   HOOSON, J. 'Methodological Clashes in Moscow', *Annals of the Association of American Geographers,* Vol. 52 (1962), p. 469–75.

b.   Comment on Matley's article:
CHAPPELL, J. E. 'Marxism and Environmentalism', *Annals of the Association of American Geographers,* Vol. 57 (1967), pp. 203–207.

c.   Soviet Geography: Review and Translation, Vol. 1 (1960) – continuing. Published by American Geographical Society.

Selected bibliography for further reading

BOOKS

National Academy of Sciences—National Research Council: *Report of the Ad Hoc Committee on Geography, Earth Sciences Division* (Publication No. 1277), *The Science of Geography,* Washington D.C. (1965).

ARDREY, R. *The Territorial Imperative,* New York: Atheneum (1966).

BRODBECK, M. *Readings in the Philosophy of the Social Sciences,* London: Macmillan (1968).

CHORLEY, R. J. and HAGGETT, P. *Frontiers in Geographical Teaching,* London: Methuen (1965).

CHORLEY, R. J. and HAGGETT, P. (Eds.), *Models in Geography,* London: Methuen (1968).

DRAY, W. *Laws and Explanation in History,* London: Oxford University Press (1957).

GLACKEN, C. J. *Traces on the Rhodian Shore: Nature and Culture in Western Thought from ancient times to the end of the eighteenth century,* University of California Press (1967).

HALL, E. T. *The Hidden Dimension,* Doubleday, New York (1966).

HARVEY, D. *Explanation in Geography,* Arnold (1969).

HOWARD, I. P. and TEMPLETON, W. B. *Human Spatial Orientation,* New York (1966).

JAMMER, M. *Concepts of Space,* Cambridge (Mass.): Harvard University Press (1954).

KATES, R. W. 'Hazard and Choice Perception in Flood Plain Management' – *Research Paper No. 78,* Department of Geography, University of Chicago (1962).

KOESTLER, A. *The Ghost in the Machine,* London: Hutchinson (1967).

KÖRNER, S. *Experience and Theory,* London (1966).

KÖRNER, S. *Conceptual Thinking,* Cambridge (Mass.) (1955).

LAMBERT, W. E. and KLINEBERG, D. *Children's Views of Foreign Peoples: A Cross-National Study,* New York: Appleton Century Crofts (1967).

LYNCH, K. *The Image of the City,* Cambridge (Mass.) (1960).

NAGEL, E. *The Structure of Science,* London: Routledge and Kegan Paul (1961).

OSER J. *The Evolution of Economic Thought,* Harcourt, Brace and World, New York (1963).

PIAGET, J. and INHELDER, B. *The Child's Conception of Space,* London (1956).

POPPER, K. R. *The Logic of Scientific Discovery,* London: Hutchinson (1959).

ROBINSON, J. *Economic Philosophy,* London: Watts (1962).

SAARINEN, T. F. 'Perception of the Drought Hazard on the Great Plains', *Research Paper No. 106,* Department of Geography, University of Chicago (1966).

SEGALL, M. H., CAMBELL, D. T. and HERSKOVITS, M. J. *The Influence of Culture on Visual Perception,* Indianapolis (1966).

TOULMIN, S. and GOODFIELD, J. *The Discovery of Time,* London: Hutchinson (1965).

WRIGHT, J. K. *Human Nature in Geography,* Cambridge (Mass.): Harvard University Press (1966).

ARTICLES

BERGMANN, G. 'Ideology', *Ethics,* Vol. 51, (April 1951), pp. 205–218.

BRODBECK, M. 'Explanation, Prediction and Imperfect Knowledge' in H. Feigl and G. Maxwell: *Minnesota Studies in Philosophy of Science,* No. 3, Minnesota (1962).

BROMBERGER, S. 'A Theory about the Theory of theory and about the theory of theories', *Delaware Seminar on Philosophy of Science,* Vol. 2, (1963), pp. 79–106.

BROOKFIELD, H. C. 'On the Environment as Perceived' in C. Board, R. J. Chorley, D. Stoddart and P. Haggett: *Progress in Geography,* Vol. 1, Arnold (1969).

CHAPMAN, J. D. 'The Status of Geography', *Canadian Geographer,* Vol. 3 (1966), pp. 133–44.

DOWNS, R. M. 'The Role of Perception in Modern Geography, *Seminar Papers,* Series A, No. 11, University of Bristol (1967).

GOULD, P. 'On Mental Maps', *Michigan Inter-University Committee of Mathematical Geographers,* No. 9 (1966).

GOULD, P. 'Problems of Space Preference Measures and Relationships', *Geographical Analysis,* Vol. 1 (1969), pp. 31–4.

GOULD, P. and WHITE, R. R. 'The Mental Maps of British School Leavers', *Regional Studies,* Vol. 2 (1968), pp. 161–82.

HARRISON, J. D. 'An Annotated Bibliography on Environmental Perception with Emphasis on Urban Areas', *Exchange Bibliographies,* Council of Planning Librarians, Monticello, Illinois, 61856.

HARTSHORNE, R. 'The Concept of Geography as a Science of Space', *Annals of the Association of American Geographers,* Vol. 45 (1955), pp. 205–44.

HEATHCOTE, R. L. 'Drought in Australia: A Problem of Perception', *Geographical Review,* Vol. 59 (April 1969), pp. 175–94.

HOOSON, D. 'Phases in the 20th century Development of Russian and American Geography' *Proceedings of the Association of American Geographers,* Vol. 1 (1969) pp. 66–69.

INKELER, A., HAUFMANN, E. and BEIER, H. 'Modal Personality and Adjustment to Soviet Socio-Political Systems' in B. Kaplan (Ed.): *Studying Personality Cross-Culturally,* Evanston, Illinois: Row, Peterson and Co. (1961).

JOHNSON, R. J. 'Choice in Classification: On the Subjectivity of Objective Methods', *Annals of the Association of American Geographers,* Vol. 58 (1968), pp. 575–90.

KATES, R. W. 'Perceptual Regions and Regional Perception in Flood Plain Management', *Papers, Regional Science Association,* Vol. 11 (1963), pp. 217–27.

KATES, R. W. and WOHWILL, J. (Eds.) 'Man's response to the Physical Environment', *Journal of Social Issues,* Vol. 22, No. 4 (October, 1966).

KIRK, W. 'Historical Geography and the Concept of the Behavioural Environment', *Indian Geographical Journal,* Silver Jubilee Edition (1951).

LANE, D. S. 'Variations in Revolutionary Bolshevik Ideology: A Regional Statistical Analysis', *East Lakes Geographer,* Vol. 3 (1968), pp. 21–28.

LEWIS, G. M. 'Regional Ideas and Reality in the Cis Rocky Mountain West'. *Transactions of the Institute of British Geographers,* Vol. 38 (1966), pp. 135–50.

LUKERMANN, F. 'The Concept of Location in Classical Geography', *Annals of the Association of American Geographers,* Vol. 51 (2) (1961), pp. 194–210.

LUKERMANN, F. 'The Calcul des Probabilitiés and the école française de géographie', *Canadian Geographer,* Vol. 9 (1965), pp. 128–38.

MERRENS, H. ROY. 'The Physical Environment of Early America: Images and Image Makers in Colonial South Carolina', *Geographical Review,* Vol. 59 (1969), pp. 530–56.

PETERSON, G. L. 'A Model of Preference: Quantitative analysis of the Perception and Visual Appearance of Residential Neighbourhoods', *Journal of Regional Science*, Vol. 7 (1967), pp. 19-31.

ROBERTS, F. S. and SUPPES, P. 'Some Problems in the Geometry of Visual Perception', *Synthese*, Vol. 17 (1967), pp. 173–201.

SEWELL, W., DERRICK, R., KATES, R. W. and PHILLIPS, L. R. 'Human Response to Weather and Climate: Geography Contributions', *Geographical Review,* Vol. 58 (April 1968), pp. 262–80.

SMITH, D. D. 'Modal Attitude Clusters: A Supplement for Study of National Character', *Social Forces,* Vol. 44 (June 1966), pp. 526–33.

SOMMER, R. 'Studies in Personal Space', *Sociometry*, Vol. 22 (1959), pp. 247–60.

SOMMER, R. 'The Ecology of Privacy', *Library Quarterly*, Vol. 36 (1966), p. 234.

SOMMER, R. 'Man's Proximate Environment', *Journal of Social Issues*, Vol. 22 (October 1966), pp. 59–70.

TUAN, YI-FU. 'Discrepancies between Environmental Attitude and Behaviour', *Canadian Geographer*, Vol. 12 (1968), pp. 176–91.

WARNETZ, W. 'Global Science and the tyranny of space', *Papers – Regional Science Association,* Vol. 19 (1967), pp. 7–19.

WATKINS, J. W. N. 'Ideal types and historical explanation', *British Journal of the Philosophy of Science,* Vol. 3 (1952), pp. 22–43.

WATSON, J. WREFORD. 'The role of Illusion in North American Geography', *Canadian Geographer*, Vol. 13 (1969), pp. 10–27.

WEBBER, M. 'Culture, Territoriality and the Elastic Mile', *Papers – Regional Science Association*, Vol. 13 (1964), pp. 59–69.

WILLIS, R. H. 'Finnish Images of the Northern Lands and Peoples', *Acta Sociologia,* Vol. 7, No. 2 (1964), pp. 73–88.

WILSON, E. B. 'Review of Scientific Explanation', *Journal of the American Statistical Association,* Vol. 50 (1955), pp. 1354–7.

WOLD, H. 'Causality and Econometrics', *Econometrica,* Vol. 22 (1954), pp. 162–77.

WRIGHT, J. K. *Human Nature in Geography,* Cambridge (Mass.): Harvard University Press (1966).

1 Theory, science and geography[1]

Wayne K. D. Davies
Department of Geography, University of Wales, Swansea, U.K.

Throughout its history geography has been characterized by an unceasing methodological debate upon its scope and content, a debate that has occasionally scorched the pages of its varied journals.[2] Today an apparently new perspective has been opened up under the impact of the so-called quantitative revolution.[3] Statistical methods have been introduced to attain a desired level of objectivity, and a search for laws and theories has proceeded apace. All are devoted to the fundamental conception of geography as a science. Indeed, many of the exponents of this new perspective adopt an almost Lutheran stand. 'The basic approach to geography is to assume that geography is a strict science, and then proceed to examine the substantive results of such a conclusion'.[4] Such attitudes of conviction tend to give a false impression of the fact that geography has always been considered as a science – Humboldt[5], Ritter[6] and Davis,[7] and even Vidal de la Blache[8] have all evoked the necessity of considering its scientific nature and searching for laws, whilst Hartshorne has enshrined this general conception. 'Geography accepts the universal scientific standards of precise, logical reasoning based on specifically defined, if not standardized concepts. It seeks to organize its field so that scholarly procedures of investigation and presentation may make possible, not an accumulation of unrelated fragments of individual evidence, but rather the organic growth of repeatedly checked and constantly reproductive research'.[9] Such a statement of faith contrasts strikingly with the descriptive and subjective approaches that generally characterize the present. As Ackerman observes 'the end product of geographic research still has been the contemplation of the unique. Small wonder that the subject was open to characterization as an art. The only ready way of integrating unlike entities has appeared to be through an intuitive process, and geography appears to be concerned with unlikes at a critical step'.[10] He proceeds to point out that the intuitive 'leaps' of areal differentiation that were made especially by the teachers of the early years of this century were certainly admirable, but 'at the same time they confuse our view of the frontier of fundamental research in the discipline'.[11]

Reconciliation or explanation of this apparent contradiction between the state of the discipline, and its scientific nature, can only be made by the use of the unfashionable historical method. An appreciation must be made of the changing nature of 'science' and especially scientific method. Hanson[12] has recently demonstrated how facts are very much affected by the sort of attitude that the observer possesses, so that the views of past geographers need to be interpreted in the light of current knowledge about their attitudes, convictions and prejudices.

It is significant that Humbolt and Ritter both died in the year that Darwin published his *Origin of Species*. Their organization of material was certainly based upon the prevalent philosophical views of their age, for they sought to interpret the underlying unity of nature. This unity was conceived in a teleological sense by Ritter, but was more aesthetic in the case of Humboldt.[13] Similarly their method of analysis was in line with current thinking, for Ritter rejected the *a priori* theories of the rationalists, believing that 'one must not proceed from opinion or hypothesis to observation, but from observation to observation'.[14] This essentially inductive approach had long been characteristic of scientific thinking, certainly going back to Bacon[15] and it achieved a new vigour in the educational principles of Pestalozzi.[16] He indicated the necessity of acquiring facts, their integration and finally the discernment of the general system. In geography the integration was spatial, but the laws were certainly not forthcoming from this sort of approach. Indeed it is significant that it was not until 1865 that Bacon's inductive process was attacked with any success by Liebig, the first to do so from the standpoint of modern science.[17] Yet while this inductive approach was considered the ideal, it is without doubt that the major scientific advances were not conceived in such a manner. Galileo, Newton and Darwin all possessed in considerable degree a theoretical insight by which their observations were correlated, their 'flash of intuition' was certainly not inductively derived. But while Ritter and Humbolt observed, their work was devoted to describing a 'natural unity' of the world, a romantic conception that has survived to the present day.

The impact of Darwin's evolutionary thesis was not confined to biology, for its implications and methodological conceptions permeated science. The mechanism he postulated as an explanation of the variation of organisms was that of the survival of the fittest, a survival that depended upon adaptation to the environment. This cause-effect doctrine led, in the determinist school of geography, to the human phenomena in space being considered as determined by their physical background. Causal connections were substituted in cases where the interrelation was only casual. Yet despite its over elaboration, and the often untenable positions into which it floundered, it is significant that the determinists used the logical methodology that lies at the heart of science today. Its conclusions were expressed as laws that were deductively connected into theories, while the importance of the generalization was stressed. Thus although W. M. Davis can be criticized for maintaining that the relationships he described were seen from 'the modern principle of evolution – the adaptation of all the earth's inhabitants to earth',[18] he went on to express views upon the relation of the general to the particular that would not be contradicted today. 'I believe it is possible to discover and establish general principles in geography (as in physics and geometry) and to teach individual items chiefly as illustrations of the principles under which they fall . . . they may, perhaps, be . . . aided in perceiving the proper relation of the specific to the general in their own subject'.[19]

Rejection of the determinist thesis by the possibilist school unfortunately led to the abandonment of much of this scientific methodology – the structure or logic was discarded because of the inadequacy of the premises. The historical perspective certainly corrected the corruption of relations that the determinist school had exemplified in their more rigid works. However, many of the ills that have been diagnosed by certain young geographical doctors of philosophy, Bunge[20] and Berry,[21] certainly stem from this rejection. Lip service was paid to law development and scientific description, and the stress upon man's free will inevitably led to the contemplation of the unique case,[22] and to the enshrinement of regional description as the true goal of geography.[23] To apply orderly logical analysis and experimentation to the sort of description that stressed the 'personality' of areas was impossible. The intuitive process that integrated the unique crystallized as an art.

A rapprochement with modern science has taken place at an accelerating pace within the last ten years. It has been realized that social phenomena can be regarded as law-giving, if the laws are regarded as generalizations, not chains of command.[24] But it must be stressed that the critics of the established field have not attempted to reject all that has gone before. The attempt has been made to reformulate the material upon a more systematic basis and in a scientific fashion.

Although it has been shown that the conception of geography as a science has meant different things at different times, it seems necessary at this stage to answer the question 'What is a science'? The Oxford Dictionary defines science as 'systematic and formulated knowledge; pursuit of this or principles regulating such pursuit',[25] an interpretation that is wide enough to include the classificatory lawfulness of Aristotle[26] and the scientific method of Bacon.[27] This does, however, lend support to the view that the aim of science has probably been the same at all time. 'For science in its totality the ultimate goal is the creation of a monistic system in which . . . the world's enormous multiplicity is reduced to something like unity'[28] or as Whitehead has put it 'to see what is general in what is particular'.[29] However the method by which this aim is arrived at has changed over the centuries. Bacon's inductive approach was designed to prepare the mind for the ultimate truth by cleansing it of anticipation and prejudices.[30] Modern scientific methodology, however, differs. 'Science never pursues the illusory aims of making its answers final . . .',[31] indeed it is essentially a way of thinking that creates a body of empirically supported propositions that are related within a particular theoretical structure. This creation has occurred as a result of experimentation and observation that are fruitful for further investigation.[32]

Davis has observed that 'geography will become more and more a scientific study in proportion to the use that is made of the fully developed

scientific method'.[33] Looking at the results of the fifty years since this statement was made, and within the context of the modern, not a Baconian conception of a science, then it must be accepted that geography is still ill
qualified to consider itself a science. The mere accumulation of facts, even if
obtained objectively,[34] does not make a science in the modern sense. It is the
way of approaching these facts that distinguishes a science. 'Out of the uninterrupted sense experiences, science cannot be distilled, no matter how in-
dustriously we gather and sort them'.[35] This is the essential difference between
alchemy purged of its magic and chemistry.

Essentially, therefore, it is by the standard of current organizational
techniques used by fellow workers that the science of geography must be
judged. We cannot afford to agree with Hartshorne that 'the essential
characteristics of geography have been determined for us by its past development . . . (and) . . . can only be considered as binding or restrictive on our
freedom of research only if one fails to observe the full expanse and depth of
the field which that tradition provides'.[36] Such a view is pre-scientific in that
it projects attitudes to the golden past, not to the future. To adhere to such a
historically determined methodology is to erect a guild system upon our
thought. Indeed one of the essential characteristics of modern science is that
it is cumulative and progressive, ever re-examining its concepts in the light of
evidence that has been discovered. An analogy is useful to demonstrate this
point. What would be the current position of physics if its concern had been
limited to gross matter? The heart of physics is now seen to be the science of
events, not particles, and the general principle of indeterminancy has had a
profound influence on all scientific thinking.

Inevitably some of the findings of the other sciences are influencing
geographical thought so that it is not surprising that the Hettner-Hartshorne
conception of methodology[37] is being challenged. Bunge, following on from
Schaefer,[38] has observed that a single methodology cannot incorporate the
general and the unique, at least not in the sense that Hartshorne seems to
contemplate, and still remain a science.[39] This dichotomy is not the only one
that has arisen in geographical thinking and Hartshorne has clearly demonstrated how ill-founded many of them are.[40] Yet this general-unique difference
remains, and it is pertinent, in view of its basic significance, to examine its
origin and nature.

Geography studies the spatial variations of phenomena,[41] and with a
romantic conception of the unity of nature then, these variations *were* perceived on the basis of their integration in regions. A region is, however,
essentially an intellectual concept,[42] though the possibilist monographs
seemed to give them an identity of their own, so that they seemed far more
than an illuminatory device. Yet once one accepts the region as being the
prime unit of study, i.e. the place where the integration of spatial variation
occurs, then one is accepting the unique. It is bound to be unique because of
the multitude of possible combinations that could occur amongst all the
available phenomena. To consider the region as the culminating core of
geography is to elaborate the doctrine of the unique. Sets of discoveries

formulated upon such as basis cannot therefore be connected together. The only possible way of connecting these together is to hold some romantic view of the underlying unity of nature as in the mid nineteenth century or to propose some deterministic philosophy. In the terms of the latter then it is the physical basis in any area that has determined the human manifestation. From this base the spatial variation is explained. Yet it is the spatial variation in phenomena and their connections that is ultimately the prime cause of these regional differences, not vice versa. It is not place that gives these differences, but the space that holds them. An analogy with the concepts of atomic structure will demonstrate this point.

By the early nineteenth century chemists realized that the different elements that they could recognize were constructed of atoms with certain properties – hence the atomic weights table. Subsequent research has shown that each atom is merely a special case of a number of basic particles, the distinctive atom being the result of a special connection of these particles. So it is the organization of the structure, not the early conception of the building block, that is important and basic to understanding and further work. Different planes of understanding have therefore been elaborated: material, element, atom, particle etc., but it is the organization of the structure at any level that gives it distinctiveness. The properties of any structure can be described at each level, but one is only accounting for the differences, not illustrating how they are derived. So by conceiving of the phenomenon in terms of its structure and connections, one can proceed to a different level of understanding. Although the analogy can be pushed too far, regions in geography could be conceived as attempts at different planes of understanding, that is if we ignore any direct connection with atomic structure. Hence the connections and structure of the regions are all important. To formulate a region, or look at a material, and then to simply describe it will not get us far, for this is an elaboration of the unique. Science can only deal with populations; only then do its laws hold good and the scientific method of orderly analysis can be applied. Inevitably the process of analysis breaks down when the object of study is changed from a population of electrons to a single one. Yet this is what geography is attempting in its enshrinement of the region – it is building up the unique and then looking for laws – little wonder they have not been forthcoming.

In essence it must be accepted, as Bunge puts it, that 'a single methodology cannot embrace both the unique and the general'.[43] This has involved some re-thinking about the concept of a region, but once it is conceived as a generic, not a unique fact, then the difficulty is resolved. So Schaefer is able to observe that 'regional geography is like the laboratory in which the theoretical physicists' generalizations must stand the test of use and truth'.[44]

The paradoxical situation that began this discussion has now been resolved. Acceptance of geography as a science can only be made if it places itself within the current methodological conception of a science, and uses its techniques, instead of harking back to the science of other eras. But the two schools of thought are not as opposed as one might expect at first glance, and the new generation of geographers does not reject all that has gone before, quite the reverse.[45] It is possible to achieve useful work and results even under a different philosophy. What is radical about the new approach is the desire to make the discipline more productive, and it is necessary to demonstrate how this is being achieved.

Objections are sometimes made that descriptions are unscientific. However all science began as description. Facts relevant to each particular perspective were chosen and they were then organized. The advances came when the discoveries were fitted together into theoretical structures and tested. The early geographers also began by describing the facts relevant to their discipline, and their organization showed well conceived spatial notions and stimulating conceptions as to the interrelations between objects. In essence this organization is a theory, or at least a hypothesis: the connection of facts is carried out by means of an intuitive theory.[46] However these intuitive connections were rarely formulated in a way that made the testing of the propositions possible. For any discipline to advance scientifically, the theories, once suggested, should be removed from the intuitive level and be made specific.[47] The theory should then be formulated as precisely as possible and tested. This is what students of geography have rarely attempted except in limited fields. Concern with the region per se as the organizational framework has led to the compilation of inventories, a series of facts about a particular zone. This traditional approach, formalized in the geology, soils, climate, settlement etc. progression has had a stultifying effect upon geography, because, as we have seen, it leads to the contemplation of the unique. Inventories have been the organizational medium, not theories. Yet there is a way out of this impasse once the geographic variations are conceived in space rather than in place. Then description is not made equivalent to inventory, but is directed towards theory formulation and the translation of the intuitive into the specific. Goal-directed or problem-orientated programmes can then be conceived,[48] and with successive building upon different works the end process must be creative since it is not only conceived for itself. Thus the work of an individual can be built upon in a way that is impossible in art, since it is not scientifically creative. T. S. Eliot need not necessarily have had a Shakespeare before him in order to produce his work, but Newton certainly needed Galileo.

While the work of a scientist should be conceived as putting forward and testing theories, the initial stage does not necessarily call for logical analysis. Einstein has observed that 'the search for those highly universal . . . laws from which a picture of the world can be obtained cannot come by pure logic. There is no logical path leading to those . . . laws. They can only be reached by intuition . . .'[49] Koestler has recently revealed the nature of this intuitive act, calling it the bisociative act,[50] and although he attempts to apply it to both science and arts there is an essential difference; in science the act leads to testable predictions. Admittedly mechanical methods and set formulas can eliminate mistakes, but eventual advance is by an intuitive flash. The importance of this for geography should be apparent: these intuitive flashes are only of use if the material has been organized in a testable way, at least if a claim to be a science is made. Problem orientation at, and between places, not inventory, is the way out of the geographical dilemma, otherwise inventory leads to the enshrinement of place, the unique.

This sort of argument leads into a necessary discussion of the need for quantitative methods. They are not only significant in making geography more objective, but also are important in that the theories formulated in any description can be made explicit.

The use of mathematical techniques is not necessarily a revolutionary approach:[51] it is merely a development of one that has a long history in geography. It is by the logical connection of maps and words that geographers have, in the past, linked the facts with their broad generalizations. Yet statistics act in a similar way: they represent another logical connection, but now the validity of the argument can be tested.[52] So the use of statistical techniques can fulfil three functions: the summary of data (descriptive statistics), the testing of statements that are made, and also in the formulation of models where their use as a logical process is more apparent.[53] Enough has been written upon the usefulness of these techniques to make any greater elaboration at this stage out of place, but it is certainly necessary to consider their spread as only one more example of the diffusion of scientific techniques into the modern world.

It seems to be one of the fundamental results of science that this diffusion of new techniques brings about a changing attitude about objects, for they can be manipulated in different ways. This is certainly true today, for the work of the theoretical physicists has revealed that events at the atomic level cannot be observed with certainty. A new principle, that of uncertainty, has been introduced.[54] The laws of probability, of chance, have replaced the strict determinism of Newton's laws, and quantum mechanics have given the most complete picture to date of the understanding of the world. The relation of this principle to geography has already been sketched by Emrys Jones[55] who, following Kaufmann,[56] pointed out that there are, perhaps, two kinds of laws, the cause: effect laws of classical physics, adequate at a macroscopic level and microscopic laws or quantum laws. 'If the actions of a large number of human beings follow a pattern, however variable the human motivation, then generalizations or broad principles can be drawn up. But however broad the generalization, it might fail in strict application to any single phenomenon. Any pattern which emerges does so as the statistical mean of the behaviour of a mass of human individuals, and any generalization which the human geographer might find useful must be based on this behaviour'.[57] This idea will incorporate the concept of human free will since 'the course of human activity cannot . . . be adequately predicted'.[58] The analogy with the physicist is apt, for electrons are so small that individual behaviour cannot be observed without upsetting them. Thus, the uncertainty principle, if applied to the world of human action, enables us to appreciate that there is a difference between action in mass, and action at an individual plane. The latter is not determined, and the former represents the answer to 'how' and not 'why'.

In other words the generalization is 'a provisional hypothesis to account for observed phenomena'.[59] The result of this is that probability can be applied to human geography. The approach 'replaces the concept of an inevitable effect by that of a probable trend'.[60] Several workers have, by their simulation techniques, already adopted this sort of approach in their geographical writings,[61] and it is apparent that many of the models that are being used for testing data are incorporated in a stochastic framework.

Geography is, therefore, certainly being pushed down the path of modern science, at least in some quarters. The physicist or chemist who is engaged in research can go to the crux of the matter, to the heart of an organized structure for a set of scientific doctrines. In consequence a collection of generally accepted problem situations are already in existence. The hope is that geography can develop in such a way. In case the necessity of a theoretical framework is overlooked, it must be emphasized that it is only when experimentation is organized on such a basis that advances are made. As Popper observes 'experiment is planned action in which every step is governed by theory'.[62] Organization is therefore the key to progress, and this involves methodological questions that have been obscured in the glorification of 'the region'. Yet the type of organization chosen does affect the results. There are many kinds of theory, so that the results must be considered upon the basis of the structure used. A recent study has outlined some of the difficulties with respect to geomorphology.[63] But these are not the only problems: science as a whole is suffering from the specialization of its branches. Fortunately the so-called 'General Systems Theory'[64] has been formulated in order to re-integrate the diverse aspects. Mutual relations can then be perceived, with the consequent possibility of cross fertilization. Berry[65] has recently applied the technique to show the variety of possible approaches to regional analysis, and as a by-product has shown the inadequacy of present approaches.

What is significant about these new frontiers is not so much the results that have been achieved to date, but the attitude of mind that is engendered. The search for organizational concepts and theories[66] is developing apace, fresh perspectives are being opened up upon old attitudes, problems and dilemmas. All this need not be restrictive for its ultimate aim is clarification. It is not before time that geography developed a productive methodology with an organized point of view, theoretical structure and a problem orientation. At present its methodology is largely pre-quantum. Yet the subject is not alone in its struggles. For instance, sociology is experiencing all the same internal conflicts in its advance towards the status of a science. In geography, as in sociology,[67] and like all young sciences, a great deal of energy will be devoted towards descriptive study but the value of this is only fragmentary until they can be unified with other descriptive findings. What is needed at all levels of generality is the formulation and testing of significant hypotheses. The alternative is to sink back into a morass of regional description at a subjective level. So if geography is to be a discipline in the true sense of the word, then no longer can we share the view of Wooldridge and East that 'the scope of geography is distressingly wide and its aim is far from clear'.[68] That the scope is wide should be cause of jubilation not distress, whilst its aim should be, and is being clarified.

References

1   Reprinted from *Tijdschrift voor Economische en Sociale Geografie*, Vol. 57 (July–August 1966), Rotterdam, Netherlands.

2   For instance the reply by Hartshorne to Schaefer's article:
HARTSHORNE, R. 'Exceptionalism in Geography Re-examined', *Annals of the Association of American Geographers*, Vol. 45 (1955), pp. 204–44.
SCHAEFER, F. 'Exceptionalism in Geography; a Methodological Examination', *Annals of the Association of American Geographers*, Vol. 43 (1953), pp. 226–49.

3   BURTON, I. 'The Quantitative Revolution and Theoretical Geography', *The Canadian Geographer*, Vol. 7 (1963), pp. 151–62.

4   BUNGE, W. 'Theoretical Geography', *Lund Studies in Geography*, Series C, No. 1, Lund (Sweden): Gleerup (1962), p. x, introduction.

5   HARTSHORNE, R. 'The Nature of Geography', *A.A.G. Reprint* (1958), see pp. 69 and 77 for Humboldt's views on geography and the sciences.

6   Ibid, p. 54 for Ritter's views.

7   DAVIS, W. M. 'Possibly . . . geography is the most complex of all sciences', p. 38 and note discussion on pp. 41–2 in: *Geographical Essays*, edited by JOHNSON, D. W. New York: Dover Publications (1954).

8   VIDAL DE LA BLACHE, P. He possessed 'a conception of the earth as a whole, whose parts are unco-ordinated, where phenomena follow a definite sequence and obey general laws to which particular cases are related', in *Principles of Human Geography* (translated by M. T. Bingham), London (1926), p. 7.

9   HARTSHORNE, R. 'The Nature of Geography', *A.A.G. Reprint* (1958), p. 464.

10   ACKERMAN, E. A. 'Geography as a Fundamental Research Discipline', University of Chicago, *Department of Geography Research Paper*, No. 53 (1958), p. 16.

11   Ibid., p. 16.

12   HANSON, N. R. *Patterns of Discovery*, Cambridge University Press (1958).

13*a*. HARTSHORNE, R. op. cit., p. 65.
  b. SINNHUBER, K. 'Carl Ritter', *Scottish Geographical Magazine*, Vol. 75 (1959), pp. 153–63.

14   Quoted in HARTSHORNE, R. op. cit., p. 54.

15   BACON, FRANCIS (1561–1626), see Chapter 7, pp. 526–30 in: RUSSELL, BERTRAND, *History of Western Philosophy*, London: Allen and Unwin, new edition (1961).

16   SINNHUBER K. op. cit., p. 156.

17   LIEBIG, *Induktion und Deduktion*, quoted in: POPPER, K. *The Logic of Scientific Discovery*, London: Hutchinson (1959), pp. 25–32.

18   DAVIS, W. M. op. cit., p. 32.

19   Ibid., pp. 30–1.

20   BUNGE, W. op. cit.

21   BERRY, B. J. L. 'Approaches to Regional Analysis: A Synthesis', *Annals of the Association of American Geographers*, Vol. 54 (1964), pp. 2–11.

22   The importance of the often quoted formulas of the possibilists in this respect cannot be overstressed. Thus, 'there are no necessities, but everywhere possibilities; and man as the master of these possibilities is the judge of their use', in: FEBVRE, L. *A Geographical Introduction to History* (1925), p. 236.

23a.   DICKINSON, R. E. and HOWARTH, O. J. R. 'The culminating point and, as many would claim, the essential aim of modern geography is thus the latest phase in the development of the subject', in *The Making of Geography*, Oxford (1933), p. 233.

  b.   Note also the view of GILBERT, E. W. 'It is through the region that new life has been given to the dead bones of geography. In the view of the writer geography is the art of describing the personalities of regions', in: 'Geography and Regionalism', Chapter 15 of *Geography in the Twentieth Century*, edited by TAYLOR, G., London: Methuen 1957, p. 346.

24   JONES, E. 'Cause and Effect in Human Geography', *Annals of the Association of American Geographers*, Vol. 46, No. 4 (1956), pp. 369–77.

25   *The Concise Oxford Dictionary*, fourth edition (revised by E. McIntosh), Oxford University Press (1950).

26   See Chapters on Aristotle, Book 1, Chapters 19–23, in: RUSSELL, B. op. cit., pp. 173–217.

27   See Chapter on Bacon, Book 3, Chapter 7, in: RUSSELL, B. op. cit., pp. 526–30.

28   HUXLEY, A. *Literature and Science,* Chatto and Windus, 1963.

29   WHITEHEAD, A. N. 1911, quoted in: RUSSELL, B. op. cit.

30   BACON see Book 3, Chapter 7, in: RUSSELL, B. op. cit., pp. 526–30.

31   POPPER, K. op. cit., p. 281.

32   Note NAGEL, F. *The Structure of Science,* London: Routledge and Kegan Paul (1961), pp. 1–14.

33   DAVIS, W. M. op. cit., p. 59.

34   The distinction between objective and subjective follows the definition used by Kant. 'A justification is objective, if in principle it can be tested and understood by anybody'. 'Subjective is applied to feelings of conviction'. Quoted in: POPPER, K. op. cit., p. 44.
        In view of the discussion which follows it must be emphasized that a 'conviction' can be converted into an 'objective justification' only by the use of adequate testing procedures. The intuitive thought can then become a specific and accepted fact.

35   POPPER, K. op. cit., p. 280.

36   HARTSHORNE, R. 'Perspective on the Nature of Geography', *A.A.G.,* London, Murray (1960), p. 183.

37   For instance, SCHAEFER, F. op. cit., p. 226 and BERRY, B. J. L. op. cit., p. 2.

38   SCHAEFER, F. op. cit.

39   BUNGE, W. op. cit., p. 9.

40   HARTSHORNE, R. op. cit., p. 79.

41   'Geography is concerned with the arrangement of things and with the associations of things that distinguish one area from another. It is concerned with the connections and movements between areas.' JAMES, P. E. 'American Geography at Mid Century', *29th Yearbook of the National Council for the Social Studies*, Washington (1959), p. 10.
        'A more explicit conception of this view is given by BERRY op. cit., pp. 2–5. He observes that 'the integrating concepts and processes of the geographer relate to spatial arrangements and distribution to spatial integration, to spatial interaction and organization and to spatial process.' This spatial view is, however, limited 'to the world wide eco-system of which man is the dominant part. Whilst this perspective of viewing a particular system is sufficient to differentiate geography from the physical and biological sciences, Berry observes that the social sciences also study the man-made environment. However, other distributional and organizational theories are more central to the social scientists than this spatial one, so it is this spatial perspective that is important. Berry concludes by observing that 'these environments are not studied in their totality by geographers, only in their spatial aspects.'

Unless geography intends to claim the whole of science this ought to be accepted. Schaefer (op. cit., p. 227) has also drawn attention to this by emphasizing that geography and history have claimed too much: they are not *the* integrating sciences.

42 'A region is not an objective fact: rather it is an intellectual concept', JAMES, P. E. op. cit., p. 17.

43 BUNGE, W. op. cit., p. 12.

44 SCHAEFER, F. op. cit., p. 230.

45 BUNGE, W. op. cit., p. 6.

46 STEBBING, S. L. *A Modern Elementary Logic*, London: Methuen University Paperbacks (1961), p. 170.

47 Most anthropologists also emphasize this attitude by showing the need for a clear conceptual framework and adequate theoretical basis, e.g. FORSTER, G. M. *Traditional Cultures*, New York: Harper and Row (1962), pp. 195–217.

48*a*. FORSTER, G. M. ibid., p. 201.

   b. BERRY, B. J. L. 'Further Comments Concerning "Geographic" and "Economic" Economic Geography', *Professional Geographer,* Vol. 11 (1959), No. 1, Part 1, p. 12.

49 A. Einstein quoted in: POPPER, K. op. cit., p. 32.

50 KOESTLER, A. *The Act of Creation*, London: Hutchinson (1964).

51 BURTON, I. op. cit.

52 BUNGE, W. op. cit. p. 37.

53 BURTON, I. op. cit., p. 1.

54 HEISENBERG, W. *Nuclear Physics*, Philosophical Library (1953).

55 JONES, E. op. cit., p. 370.

56 KAUFMANN, F. A. *Methodology of the Social Sciences,* Oxford (1944).

57 JONES, E. op. cit., p. 373.

58 Ibid., p. 373.

59 Ibid., p. 374.

60 BRONOWSKI, J. *The Common Sense of Science,* New York, Randon House (1959).

61 One of the early works in this field was: HÄGERSTRAND, T. 'The Propagation of Innovation Waves', *Lund Studies in Geography*, Series B, No. 4 (1952), University of Lund, Sweden.

62 POPPER, K. op. cit., p. 280.

63 CHORLEY, R. J., 'Geomorphology and General System Theory', *Geological Survey Professional Paper* 500-B, U.S. Govt. Printing Office, Washington (1962).

64 BERTALANFFY, L. VON 'General Systems Theory: A New Approach to the Unity of Science', *Human Biology*, Vol. 23 (1951), pp. 303–61.

65 BERRY, B. J. L. 'Approaches to Regional Geography: A Synthesis', *Annals of the Association of American Geographers,* Vol. 54 (1964), pp. 2–10.

66 For example: CHORLEY, R. J. 'Geography and Analogue Theory', *Annals of the Association of American Geographers,* Vol. 54 (1964), pp. 127–37.

67 MADGE, J. *The Origins of Scientific Sociology*, London: Tavistock Publication (1963).

68 WOOLDRIDGE, S. W. and EAST, W. G. *The Spirit and Purpose of Geography,* London: Hutchinson's University Press (1952), p. 13.

2   Space and process[1]

J. M. Blaut

Clark University, Worcester, Mass., U.S.A.

'Space,' said Whittlesey, is 'the basic organizing concept of the geographer.'[2] But space is a treacherous philosophical word, and every empirical and metaphysical notion of pure, timeless space has been discarded by scientific philosophy. Where, then, does geography stand?

Root concepts. The Greek Atomists reduced everything to permanent bodies drifting and colliding in empty space, the Void. Thus arose the notion of absolute space as a distinct, physical, and eminently real or empirical entity in itself.[3] This doctrine became enshrined in the cosmology of Newton, with its 'absolute space in its own nature . . . always similar and immovable' and its 'absolute, true and mathematical time.'[4] Newton believed in 'a space composed of points, and a time composed of instants, which had an existence independent of the bodies and events that occupied them.'[5] This existence was physical, not abstract or mental: Newtonian space propagated light and conveyed forces, and was 'the sensorium of God'. But it was timeless and immaterial.

Opposing the absolutist doctrine was that of relative space, the idea that space by itself is inconceivable: 'What is empty is nothing. What is nothing cannot be.'[6] Apparently even older than the belief in absolute space, the relativist doctrine seems to derive from pragmatic concepts of measurement. The Sumerians, for example, viewed 'areal extension . . . from the anthropocentric aspect of the labor involved.'[7] The relative concept of space as merely a relation between events or an aspect of events, and thus bound to time and process, was later elaborated by Leibniz, whose philosophy of space 'forms the basis of the [modern] theory of relativity.'[8]

Philosophers down through Kant and the German Idealists sought to weave metaphysical theories round the idea of space. Some, including Aristotle, tried to mediate between absolute and relative space by denying empirical reality to space, and instead granting it metaphysical reality as pure form, conceptually, if not physically, distinct from time and matter.[9] Others, believing wholeheartedly in absolute space, tried to establish the belief as necessarily, aprioristically true. Among these was Immanuel Kant.

Spatial extension, according to Kant, is not a property of the things of the external world, but is assigned to incoming sense-data by the transcendental outer sense, the intuition (*Anschauung*) of space. Next, the inner sense, the *Anschauung* of time, separately assigns them temporality. Finally, they are led into the 'understanding,' where the Kantian categories attach to them the 'concepts' – causality, individuality, substance, and so forth. These processes are transcendental rather than psychological, and space is thus a fusion of synthetic (empirical, non-logical) perception and an *a priori* 'form of the understanding'. Implicit in this view are the complete separation of space and time and causality; the synthetic, non-logical nature of Euclidean geometry; the metaphysical, rather than physical, basis of absolute space; and the divisibility of sciences into those of the outer sense, presenting Nature in space, the inner sense, presenting Soul or Mind in time, and the faculty of pure reason, telling us what a thing really is.[10]

The post-Kantian German Idealists were heavily influenced by Kant, but sought to soften his doctrine of the synthetic *a priori*, and thus of absolute space and time. Schelling, in particular, rejected Kant's opposition of outer to inner, Nature to Soul, and partly reunited space and time. Schelling's Romantic panpsychism extended Mind to the whole universe, and Mind implied time. Continuity, life, and progress were central to his philosophy, and space became scale, or breadth, or striving outward, or unity in diversity. But Romantic Idealism was more poetic than convincing, and could not survive counterattack by the Neo-Kantians (e.g., Windelband) and by Newtonian science.[11]

The nineteenth century was therefore dominated by what Whitehead calls 'the intellectual "spatialization" of things',[12] science holding with the absolute space of Newton and philosophy with the even more absolute space of Kant. During this same period, however, non-Euclidean geometries were throwing into question the synthetic *a priori*; advances in logic, epistemology, and psychology were showing Kant's transcendental theory of knowledge to be bad psychology rather than good metaphysics; and Newtonian space was succumbing to new physical theories. Long before Einstein, in fact, the relativistic revolution had been set in motion. By the beginning of the present century, absolute space had been discarded throughout most of science and philosophy. Today, says Poincaré, 'whoever speaks of absolute space employs a word devoid of meaning.'[13]

Relative space is inseparably fused to relative time, the two forming what is called the space-time manifold, or simply *process*. Nothing in the physical world is purely spatial or temporal; everything is process. The time dimension may be neglected, but it is always implied. Pure space cannot even survive as an empirical abstraction: it is not, after all, a quality like color (are there non-spatial qualities?), or a class (are there non-spatial classes?), or even a concept (can a concept be devoid of meaning?). Pure space is in fact relegated to pure mathematics, and *every empirical concept of space must be reducible by a chain of definitions to a concept of process*. Thus, for example, a word like 'area' either signifies process – a spatio-temporal segment of process or the operation of measuring it – or it signifies some entirely non-empirical, mathematical construct. Even the concept of simultaneity (as between two events on a map) loses its apparent meaning as a timeless slice of space. Simultaneity implies space-time overlap, the co-ordination of two or more space-times to a single frame of reference – e.g. by sending a message from one event to a second, receiving a reply, and recording the second as existing during a specified interval in the space-time of the first. This relation is obviously not timeless.[14]

Geographic concepts.   Methodological issues, Hartshorne reminds us, often boil down to 'different interpretations placed on ambiguous words'.[15] The word 'space' is nothing if not ambiguous, and some very fundamental disputes turn on its meaning. An attempt at clarification thus seems in order.

Consider the following spatial expressions. Space and other geometric terms have been appropriated by geography to designate segments of the earth at its surface. Thus we, as geographers, refer to 'spaces' and 'areas' and *Erdräume*, and argue that geography is the 'study of areas' or is 'concerned with areas.' Sometimes, also, 'area' is used to describe a physical surface, usually the surface of the land. We speak further of physical 'distances', and of 'places', 'positions,' and 'points' on the earth's surface. Space-word expressions like these have been current at least since Humboldt and Ritter. Ritter spoke of '*irdisch erfüllten Räume*' and Humboldt of the '*Raum-erfüllenden*',[16] and ever since geographers have employed phrases assigning properties (usually logical predicates) to space-words, thus: the 'content of area'; 'things in space'; 'phenomena of area.' Other expressions labelled 'spatial' indicate some connection between discrete phenomena, but a connection which is somehow spatial thus: 'space relations'; 'distance relations'; 'spatial association'; Michotte's 'principle of extension'.[17] Still other expressions refer to spatial qualities as properties: the spatial 'aspect' of something; the 'location' of X; Marthe's '*räum-lich . . . erscheinende Qualitäten an den Objekten*' and De Geer's 'abstract spatial qualities';[18] and numerous ideas of the spatial 'form' or 'morphology' of something. An additional group of expressions ties space to one or another notion of cause, or force, or explanation, thus: the 'factor' of distance; locational 'element'; spatial 'variable'; the 'significance' of space; and, of course, the old '*Raum-beeinflussung*', '*Raum als Schicksal*', and the like, of the geopoliticians.[19] Then there are spatial expressions indicating scale or size, externality or environment, uniqueness, and any number of other things. The list is neither complete nor systematic, because these are not genuine, precisely defined concepts; instead, they are merely expressions employing space-words, and they can be interpreted almost as one chooses. If given absolutistic interpretations, they become meaningless or metaphysical pseudo-concepts.[20] No sensible debate can be generated by a pseudo-concept of space, and certainly no such notion can serve as geography's basic organizing concept.

Absolutistic interpretations lead in two directions: toward belief in the existence of a Newtonian space, timeless, objectless, but physical; and toward belief in Kant's or some other metaphysical space. The first interpretation is particularly alluring to geographers, because it seems to provide us with space *as subject matter*, allowing us to imagine that a sentence of the form 'Geography is a science of space' has the same kind of meaning as verbally equivalent sentences about other sciences – botany as the science of plants, geology as the science of rocks, and so forth. Few geographers have carried such interpretations to the extreme of setting up a category of phenomena made up only of pure space, but many have talked about timeless, immaterial space as though it were real and substantial enough to be viewed and studied empirically. All that seems to be required is a belief that withdrawing the temporal dimension from a section of reality, along with all objects, some how leaves a 'something spatial' behind for the geographer to study. One source of this belief is the structure of common-sense language: when we speak of something as *in* space, or of the content *of* space, it is a simple enough matter to assume that space – rather like the Cheshire Cat's grin – remains after the contents have been removed.[21] Another source is the Newtonian idea of space as empty-but-with-mysterious-properties, a legacy from nineteenth-century science. Still another source is our unfortunate habit of expressing ideas about simultaneity in terms of slices and cross-sections. If, as it were, we slice through time at one timeless instant, we do *not* get a purely spatial cross-section; we get nothing at all.[22] Another source is the spurious spatiality of maps. The map-thing, the ink-on-paper sign-vehicle itself, is of course relatively unchanging, and beguiles us into imagining that the map-meaning, the signification of the map, is something other than process. Further confusion is added by the fact that maps portray simultaneity directly, pictorially, whereas time-depth is represented only (in most cases) by inference.

Newtonian notions of absolute space are far from uncommon in geographical literature. The reader will, I think, find ideas of the following sorts: space as a real, perceptible quality; as a phenomenon sepaɪable from time and matter; as a causal force; and as a concept of distance which is neither one of measurement nor process-connection nor scale. Such ideas are asserted by Marthe, De Geer, Granö, Lösch, many geopoliticians, Huntington and other environmentalists (whose characteristic error was that of mistaking proximity for influence), and Warntz, to name a few exponents of absolutism.[23] Lukermann's analysis of the Newtonian assumptions in Warntz's writings is particularly instructive.[24]

The arguments against Newtonian absolute space were summarized previously, and only the methodological consequences need be criticized here. Any concept of space which refers to the empirical world, i.e. is not purely analytic (mathematical), must either be a concept of spatio-temporal process or one of measurement. Thus, *Erdraum* is at the same time *Erd-Zeitraum*; and the geographer's 'geomer', 'compage', or 'total integration' is indistinguishable from the historian's 'period', except in what things he looks for – i.e., in subject matter. The content of area *is* the area. Sequent-occupance stages, where they exist, and all successive present-day geographies fit neatly together, as Hartshorne has emphasized, into a continuous process.[25] Space relations, if expressing only distances-apart, are simply reports of direct or indirect measurement; if meant as an indication of influence *across* distance,

they translate into the spatio-temporal flow of events *through* distance –
what Isard calls, in one context, 'the friction of physical distance'. Spatial
factors and spatial variables may be impinging, external (environmental)
factors or variables, or they may simply be impressively extensive processes,
but they remain, nevertheless, ordinary causal (or conditioning) variables.

With only moderate injustice to Kant, we can label as Kantian any
metaphysical notion of absolute space. Kant's own view of space leads directly
into his conception of geography, his argument running about as follows:
knowledge about the spatial location of objects is quite distinct from knowledge about their true nature and the natural laws governing them. The latter
sorts of knowledge are eternal and universal, are truly scientific, and arise
from the 'understanding,' which, as we have seen, supplies the required *a
priori* 'concepts'. Thus, space and time coordinates are separate and rather
secondary attributes of objects, and spatial and temporal arrangement of
objects is not a matter for science, but for natural history – the 'systematic
presentation [of natural things] at different times and in different places',[26]
which, says Kant, is merely a propaedeutic or introduction to true scientific
knowledge. According to this schema, geography cannot be concerned with
identification, classification, generalization, or explanation; these are functions of the 'understanding'. Rather, geography deals with spatial arrangement: 'geographical description of nature shows where [things] are to be
found on earth'.[27] Since it deals only with things which *nebeneinander liegen*,
which are simultaneous, geography is not properly comparative. Since it is
tied to the outer sense, it is artifically segregated from social science, as
Ratzel has noted.[28] If we deny validity to Kant's curious idea of space, and
even more curious idea of object, the entire schema collapses; the spatial
aspect and the true nature of things simply cannot be separated. That Kant's
theory, even so, has aided the growth of geographic thought probably reflects
the unconscious substitution of relativistic spatial concepts for Kantian ones
by later methodologists (and sometimes by Kant himself!): a measurement
concept transforms the Kantian schema into a fair definition of cartography;
perhaps also of the most descriptive regional geography. A concept of process
transaction yields at least the rudiments of modern explanatory geography.

Structure and process, form and matter, being and becoming – each
of these ancient metaphysical oppositions is woven into the contrast between
spatial morphology and the content of space, between, on the one hand, a
kind of permanent and rigid spatial framework, distinct from the material
world yet providing it with shape and pattern, purely geometric yet somehow
registering on maps, and, on the other hand, the flux of changing process.
This contrast between spatial structure and process is essentially Kantian,
but also derives in part from Goethe's rather Platonic conception of morphology as persistence, permanence-amidst-change, of form (e.g., the arche-
typal plant), and in part from the Kant- and Goethe-influenced spatial ideas
of Windelband and other post-Kantians.[29] The notion of space involved
here is absolutistic in a metaphysical rather than a physical sense, because
space is not viewed as an existent, physically separable from process, but as

an aspect of process which is apprehended in a different and mysterious way, and by a different kind of empirical science. Those who believe that geography is concerned with spatial morphology generally accord it the strange Kantian status of a mathematical (analytic) discipline which is nevertheless empirical (synthetic). Thus, W. Cramer spoke of geography as 'exclusively concerned with form', as an 'entirely mathematical discipline'. Krebs adopted a similar position, to which Sauer at one time seems to have been attracted.[30] Of many other arguments along the same lines, the most recent is that of Schaefer, who attempted the impossible task of supporting a Goethean position with Positivist arguments. Apparently misinterpreting Bergmann's valid distinction between process laws and cross-section laws (laws about simultaneous processes, but nonetheless processes), Schaefer claimed that geography is essentially morphological, that its laws contain no reference to time and change, that it is concerned with 'the spatial arrangement of phenomena ... and not so much ... the phenomena themselves', with 'spatial relations ... and no others'.[31]

All of these views assume that structure and process are two different things, which they are not; structures of the real world are simply slow processes of long duration,[32] the more slowly changing elements in any spatio-temporal segment, e.g., landforms as contrasted with the men traversing them. If by structure or morphology is meant *model*, the contrast to concrete process is obvious and trivial; it is merely that between a sign and its meaning, between the map and the territory, and no empirical science using a model, a map, or any other sign-system ignores the empirical things these signify. The things geographers call structural may be the things we customarily map, but anything can be mapped.

Some methodological implications. Deprived of the glamor of metaphysics and the false imagery of Newtonian atoms-in-the-Void, space becomes a most unprepossessing word. It labels certain purely mathematical notions, certain operations of measurement, and certain aspects of or relations among processes. Mathematical spaces and distance-measurements are the property of all science. If, therefore, the word space is to designate any peculiarly geographic concepts, these must be concepts of process. But can any such qualify as our basic organizing concept?

Three spatial concepts seem particularly important for geography. One is a straight-forward concept of relative scale or magnitude of process, impeccably relativistic if rather Schellingian in flavor. Following Hartshorne, we can state that the minimal geographic problem has as its domain, its universe of discourse, an integration consisting of two objects, unit-processes, which are causally (processually) connected.[33] Each object by itself is likely to be the domain of another science. Therefore, the scale or magnitude of a geographic problem is *necessarily* greater than that of any problem concerning the objects themselves. The same applies with greater force for more complex integrations, up to the limiting case of geomer, compage, or total integration. Many of our spatial concepts seem merely to designate either the integration itself or its peculiarly geographic scale, thus: 'area' (an integration); 'spatial setting' (an integration excluding only the object-of-reference)'; 'spatial variable' (a causal factor operating at the scale of the integration, not the included object); 'spatial circulation' (movement of an object against the reference-frame of the including integration); 'spatial association' (association among the objects in an integration, not among the parts of a single object); and the like. Since scale is implied in the very concept of integration, this idea of space cannot be our basic organizing concept; and, if geography is a science of space in this sense, it is a science of certain types of space-time integrations.

A second concept of space, closely allied to the first, designates various notions of between-ness, of physically intervening variables. The problem of explaining or predicting a spatial location minimally involves: the object of reference; all other objects influencing the object of reference and forming, with it, an integration (as in a nodal region); and certain real processes intervening between and connecting the object of reference with others in the integration. These last might be called the ground against which the objects – the more clearly discriminated unit-processes – play; they commonly appear on maps or location diagrams as naked lines connecting the objects, or they simply disappear into the blank spaces. Space as intervening ground-processes (e.g., those intervening between farm and market) is obviously an important concept for, among other things, it introduces the friction of distance, but it is not really a basic organizing concept for anyone but location theorists. However, neglect of this concept has had at least two unfortunate effects. First, it has generated much spatial mysticism, since the space between objects must be filled either with ground-processes, or with strange Forces, Influences, or with the Void. Second, it has led, in the writer's opinion, to needless disputation over the place of simple, single-element, distribution studies in geography. Hettner maintained that 'the Where of something is a property of the object', not of its surroundings, and that a simple distribution is therefore a problem for the specialist in that type of object.[34] But space is a property (relation) of the integration as a whole, i.e., distributed objects, and ground, and other kinds of objects together. Thus, explaining or predicting a simple distribution involves analysis not only of relations among the distributed objects, but also, as Ackerman has noted, relations with their environment (Ackerman's 'element of space'), and is eminently geographic.[35] Of course, wherever interaction among the distributed objects themselves is the only important locational factor, other kinds of objects having no significance, and the ground generating no severe friction, as is the case with most non-material culture elements and some population distributions, the problem is not geographic, because only the one kind of object is directly involved.

The third critical spatial concept is the cross-sectional relation of simultaneity. Simultaneity among objects *within* an integration is built into the definition of integration, and requires no comment (it is about as surprising as the fact that a horse has four simultaneous legs), except for one caution: since an integration is spatio-temporal, only the structural elements need be simultaneous. But simultaneity *between* integrations involves some special issues. Since Kant, geographers have been accustomed to thinking of areal comparisons, comparisons of place with place, and analysis of areal differences and similarities, as essentially limited to simultaneously occurring events, to space in the sense of cross-sections through time. But if geography is to become a generalizing science (and no geographer, to the writer's knowledge, has ever forbidden us this goal), it must take its comparisons where it finds them, comparing here – now with there – then as well as there – now. If one wishes to classify farming systems, cities, or floods and to discover the laws that generate and differentiate each class, one accomplishes less than half the task by investigating only those examples co-existing at a given time. The former simultaneity concept is merely an implicate of the Hartshornian concept of integration (if two objects affect each other, they *must* be simultaneous); the latter, that of cross-sectional comparisons, is painfully confining. Therefore, we are still without a spatial notion which qualifies as a suitable basic organizing concept. I can think of no other candidates.

References

Part of this paper was read at the 1957 annual meeting, Association of American Geographers.

1   Reprinted from *The Professional Geographer*, Vol. 13, No. 4 (July, 1961).
2   WHITTLESEY, D. 'The Regional Concept and the Regional Method,' in James, P. and Jones, C. F., (Eds.), *American Geography: Inventory and Prospect*, Syracuse (1957), p. 28.
3   For discussion of Greek ideas about space, see JAMMER, MAX *Concepts of Space: A History of Theories of Space in Physics*, Cambridge, Mass. (1954); Burnet, John, *Early Greek Philosophy*, 4th ed., New York (1930); Russell, Bertrand, *A History of Western Philosophy*, New York (1945).
4   NEWTON, ISAAC *Principia*. For discussions of Newtonian space, see Jammer, op. cit.; Burtt, E. A., *Metaphysical Foundations of Modern Science*, 2nd ed., New York (1932): Reichenbach, Hans, *The Philosophy of Space and Time*, transl., New York (1958) and *The Rise of Scientific Philosophy*, Berkeley (1951).
5   RUSSELL, op. cit., p. 40.
6   MELISSUS, quoted in BURNET, op. cit., p. 323.
7   JAMMER, op. cit., p. 7.
8   REICHENBACH, op. cit., *The Philosophy* . . . p. 20.
9   See ARISTOTLE's *Physics*, Book 4.
10 I have simplified Kant's views without, I believe, misrepresenting them. For discussions of Kantian space, see Kant's *Critique of Pure Reason and Metaphysical Foundations of Natural Science;* Russell, op. cit., Reichenbach, op. cit., *The Philosophy* . . .; Garnett, G. B., *The Kantian Philosophy of Space*, New York (1939).
11 See WATSON, JOHN *Schelling's Transcendental Idealism*, Chicago (1882); Gode, A. *Natural Science in German Romanticism*, New York (1941); Mead, G. H., *Movements of Thought in the Nineteenth Century*, Chicago (1936).
12 WHITEHEAD, A. N. *Science and the Modern World*, New York (1925), pp. 74, 93–124, 212. See also RUSSELL, BERTRAND *The Analysis of Matter*, London

(1927), p. 14; Dingle, H., 'The Scientific Outlook in 1851 and 1951', *British Journal of Philosophy and Science*, Vol. 2 (1951), pp. 85–104.

13 POINCARÉ, HENRI *Science and Method*, transl., New York (1921), p. 95.

14 See WHITEHEAD, A. N. *The Principle of Natural Knowledge*, Cambridge (1919), Secs. 2 and 11; Russell, Bertrand, *Philosophy*, New York (1927), Ch. 10.

15 HARTSHORNE, RICHARD *The Nature of Geography*, 2nd ed., Lancaster, Pa., (1946), p. 149.

16 RITTER, CARL *Erdkunde* (1817), quoted in Hartshorne, op. cit., p. 57; Humboldt, A.V. *Kosmos*, Vol. 1, Stuttgart (1845), p. 59.

17 MICHOTTE, PAUL 'L'Orientation Nouvelle en Géographie', *Bulletin de la Société Royale Belge Géographique*, Vol. 45 (1921), pp. 15–17.

18 MARTHE, F. 'Begriff, Ziel, und Methode der Geographie,' *Zeit. Gesell. Erdkunde Berlin*, Vol. 12 (1877), p. 433; De Geer, Sten, 'On the Definition, Method and Classification of Geography', *Geografiska Annaler*, Vol. 5 (1923), pp. 3–7.

19 WUST, W. 'Raum und Welt-Anschauung', in Haushofer, K. (Ed.), *Raumüberwindende Mächte*, Leipzig (1934), p. 161; Grabowsky, A. *Raum als Schicksal*, Berlin (1933).

20 See KRAFT, VIKTOR *The Vienna Circle*, transl., New York (1953), p. 33.

21 See AYER, A. J. *Language, Truth, and Logic*, 2nd ed., London (1946), p. 42; Carnap, R., *The Logical Syntax of Language*, London (1937), Secs. 72–8.

22 See WHITEHEAD, A. N. *The Concept of Nature*, Cambridge (1920); Poincaré, op. cit., p. 95; Hartshorne, Richard, *Perspective on the Nature of Geography*, Chicago (1959), p. 82.

23 MARTHE, op. cit., pp. 422–67; De Geer, op. cit., pp. 1–37; Granö, J. G., 'Reine Geographie', *Acta Geographica*, Vol. 2, (1929), pp. 1–34, 174 ff, and 'Geographische Ganzheiten,' *Petermann's Mitt.*, Vol. 81 (1935), p. 297; Lösch, August, *The Economics of Location*, transl., New Haven (1954), pp. 3, 508; Huntington, Ellsworth and Cushing, S. W., *Principles of Human Geography*, 3rd ed., New York (1924), Ch. 1 (location along with landforms, etc., is a factor of the environment); Warntz, William, *Toward a Geography of Price*, Philadelphia (1959), pp. 13, 26, 41, 98, 101–4, and (with Stewart, J. Q.) 'Macrogeography and Social Science', *Geographical Review*, Vol. 48 (1958), pp. 168, 175–7, 180–1.

24 LUKERMANN, F. 'Geography: De Facto or De Jure,' 23 pp., mimeo (1960) (?). This excellent paper analyzes the work of Warntz, Schaefer, and a few other recent writers in terms of their ideas about absolute space, holism, and the like.

25 HARTSHORNE, op. cit., *Perspective* ... Ch. 8.

26 KANT, op. cit., *Metaphysical Foundations* ... p. 136, in Bax, E. B., transl., London (1891).

27 KANT *Physische Geographie*, Rink (ed.), quoted in Schaefer, F. K., 'Exceptionalism in Geography,' *Annals of the Association American Geographers*, Vol. 43 (1953), p, 232.

28 RATZEL, FRIEDRICH *Anthropo-Geographie*, Vol. 1, Stuttgart (1882), p. 10.

29 See STEINER, RUDOLF *Goethe's Conception of the World*, transl., London (1928), Windelband, W., 'Logic', in Jones, Henry, (Ed.), *Encyclopedia of the Philosophical Sciences*, Vol. 1, London (1913), pp. 39–41, 43–6. 56–9.

30 CRAMER, W., reported in Wagner, Hermann, 'Bericht über die Methodik der Erdkunde,' *Geografisch Jahrbuch*, Vol. 9 (1882), pp. 678–9; Vol. 12 (1888), p. 422; Krebs, Norbert, 'Nature- und Kulturlandschaft,' *Zeit. Gesell. Erdkunde Berlin* (1923), p. 83; Sauer, Carl, 'The Morphology of Landscape', *University of California Publications in Geography*, Vol. 2 (1925), pp. 19–53.

31 SCHAEFER, op. cit., pp. 227–8, 231, 242–9.

32 BERTALANFFY, LUDWIG V. *Problems of Life*, London (1952), p. 134.

33   HARTSHORNE, op. cit., *Perspective* ... Ch. 9. I assume here that the term 'object' is a reductive level below 'integration' (see Oppenheim, P. and Putnam, H., 'Unity of Science as a Working Hypothesis', *Minnesota Studies in Philosophy and Science,* Vol. 2 (1958), pp. 3–36) and that both are 'systems' as defined by Bertalanffy, L. V. in 'Outline of General System Theory,' *British Journal of Philosophy and Science,* Vol. 1 (1950–51), pp. 134–65.

34   HETTNER, ALFRED 'Das Wesen und die Methoden der Geographie', *Geog. Zeit.,* Vol. 11 (1905), p. 557.

35   ACKERMAN, EDWARD 'Geographic Training, Wartime Research, and Immediate Professional Objectives', *Annals of the Association American Geographers,* Vol. 35 (1945), p. 134.

3   Darwin's impact on geography[1]

D. R. Stoddart[2]

University of Cambridge, U.K.

ABSTRACT   Four themes in Darwin's writings are significant in the development of geographical thought. 1. The idea of development through time strongly influenced the progress of geomorphology, pedology, ecology, and to some extent the social sciences; 2. Darwin's stress on the intimate relationships between organic life and habitat gave impetus to organismic interpretations of regions and states, which persisted in geography long after the decline of biological Vitalism; 3. the themes of selection and struggle were deterministically applied in both human and political geography; 4. a fourth theme in Darwin's writings, the random nature of original variations, was ignored by geographers until recently, partly because of Darwin's own equivocal position on this issue. Finally, Darwin's work so changed the nineteenth century world view that the development of geography as a science itself became possible.

At a time when many sciences are re-examining the impact of biological thinking, and particularly of Charles Darwin's writings, on their methods and theoretical foundations,[3] geographers have been strangely silent, and the Darwin centenary in geographical circles passed almost unremarked.[4] It is, in fact, strange that Darwin's name is not prominent in either of Hartshorne's volumes on geographic methodology, where only passing reference is made to the impact of the life sciences on geography.[5] Whereas the centenaries of the deaths of Humboldt and Ritter on 6 May and 28 September, 1859, were commemorated by geographers, the first publication of *On the Origin of Species* on 24 November in the same year remained unnoticed.

Much of the geographical work of the past hundred years, however, has either explicitly or implicitly taken its inspiration from biology, and in particular from Darwin. Many of the original Darwinians, such as Hooker, Wallace, Huxley, Bates, and Darwin himself, had been actively concerned with geographical exploration, and it was largely facts of geographical distribution in a spatial setting which provided Darwin with the germ of his theory. This paper seeks to trace the broad lines of the biological impact on geography since 1859, to assess in what respects this impact was Darwinian and, equally important, what essential features of Darwin's thought were ignored by geographers.

It is important to recall that Darwin's theory was not simply one of 'evolution,' a word which did not appear in *The Origin* until the fifth edition in 1869, but concerned a mechanism whereby random variations in plants and animals could be selectively preserved, and by inheritance lead to changes at the species level. In geography, however, Darwinism was interpreted primarily as evolution, in the sense of a 'continuous process of change in a temporal perspective long enough to produce a series of transformations.'[6] It was in this sense that many natural and social scientists welcomed evolution from about 1860 onwards. Darwin, however, was primarily concerned with the mechanism of the change or, as *The Origin* was subtitled, 'the preservation of favoured races in the struggle for life.' This element of struggle was applied in a deterministic way, particularly in human geography, at about the same period of time. The crux of Darwin's theory, the randomness of the initial variations,[7] passed almost unnoticed. In both physical and human geography, supposedly Darwinian ideas were applied in an eighteenth rather than a nineteenth century fashion, and geographers were still applying essentially Newtonian views of causation well into the twentieth century. Why the central theme of Darwin's work was thus neglected is, therefore, a fundamental problem in the history of ideas. Finally, the Darwinian revolution gave fresh impetus to concepts of biological origin which date back to Ritter and before, and the subsequent development of ecology led to new insights in some branches of geographical thinking.

In this paper four themes are taken to be especially significant contributions to geographical thought from biology and, particularly, from Darwin.[8] They are:

1   the idea of change through time;
2   the idea of organization;
3   the idea of struggle and selection;
4   the randomness or chance character of variations in nature.

Each of these four themes is examined in turn from the geographical point of view, to determine in what sense such views were biological or Darwinian in origin, how geography reacted to them, and what geographical insights they stimulated.

Time and evolution

The first part of the nineteenth century, culminating in Lyell's *Principles of Geology*,[9] saw the breakdown of the medieval view of the age of the earth, just as the Copernican revolution had revised ideas on its position in the universe four centuries earlier. The expansion of physical geography towards the end of the century drew on a double inspiration: that of the early geologists from Hutton to Lyell, and that of evolutionary biology, which was itself dependent on the earlier breakdown of restrictive cosmological ideas.

The strongest and most explicit impact of evolution was in the study of landforms, a field in which Darwin had worked during the *Beagle* years, when he formed his theory of the transformation of fringing reefs into barrier reefs and then into atolls by the slow operation of subsidence of their foundation through time.[10] The initial deduction and subsequent development of this theory, as Gruber[11] observed, closely resembles the later development of Darwin's biological ideas, and it could serve as the archetype for the 'cyclic' ideas later developed in geomorphology. Huxley[12] himself published on the new subject of 'physiography' in the 1870's, but it was Davis who took evolution as his inspiration in the idea of the geographical cycle. Earlier workers, faced with the bewildering complexity of landforms, had sought to reduce them to order by nominal classification, much as had Linnaeus in taxonomy, but the failure to supply any unifying principle to such study reduced it to cataloguing.[13] In his first paper on the development of landforms, however, Davis referred to a 'cycle of life,' and used such terms as birth, youth, adolescence, maturity, old age, second childhood, infantile features, and struggle to emphasize the analogy of an organism undergoing a sequence of changes in form through time.[14] The power of evolutionary thinking to bring diverse facts into new meaningful relationships fascinated Davis: writing of the cycle in 1900, he stated that[15]

> in a word it lengthens our own life, so that we may, in imagination, picture the life of a geographical area as clearly as we now witness the life of a quick growing plant, and thus as readily conceive and as little confuse the orderly development of the many parts of a landform, its divides, cliffs, slopes, and water courses, as we now distinguish the cotyledons, stems, buds, leaves, flowers, and fruit of a rapidly maturing annual that produces all these forms in appropriate order and position in the brief course of a single summer. The time is ripe for the introduction of these ideas. The spirit of evolution that has been breathed by the students of the generation now mature all through their growing years, and its application in all lines of study is demanded.

Throughout his working life, Davis emphasized this theme of orderliness and development through time, which he termed evolution, but it is perhaps significant that he took his illustrations not from the species or the population, but from the individual. So successful was Davis in promoting this view that in his hands geomorphology became more the study of the origin of landforms than of landforms themselves,[16] and was thus readily channeled into the restricted field of denudation chronology.

Darwin was, of course, deeply influenced by Lyell's *Principles*,[17] and the two distinct components of Lyell's uniformitarianism, gradualism, and actualism, are implicit in *The Origin*. It has been argued that a strict uniformitarianism had no place for progression or transmutation of species,[18] and Lyell himself emphatically rejected their mutability in the early editions of the *Principles*.[19] But in the sense of excluding catastrophic explanations, Huxley was certainly correct in his view that[20]

> consistent uniformitarianism postulates evolution as much in the organic as in the inorganic world,

and that[21]

> the *Origin of Species* is the logical sequence of the *Principles of Geology*.

Uniformitarianism in geology, and subsequently in biology, involved, as Hutton clearly saw, the need for time in excess of that allowed by theology. In a famous passage on the Alps, Hutton described the continuous mantle of waste from the mountaintops to the sea:[22]

> throughout the whole of this long course, we may see some part of the mountain moving some part of the way. What more can we require? Nothing but time.

Once the reality of small but cumulative variations was established in biology, a similar conclusion followed. Time became one of Darwin's chief requirements, to the extent that he refused to accept Lord Kelvin's apparent demonstration of the youth of the earth based on estimated rates of cooling and the second law of thermodynamics.[23] And it was Kelvin, not Darwin, who was later shown to be wrong. When Davis in 1899, therefore, wrote his paper on the cycle of erosion, with the trinity of factors – structure, process, and time – it was time which he singled out as[24]

> the one of the most frequent application and of a most practical value

in landform study. The key to the cyclic view in geomorphology lies in fact in systematic, irreversible change of form through time, and from this derives the biological analogy of aging used by Davis, Johnson, and their school. Davisian geomorphology was deductive, time-oriented, and imbued with mechanistic notions of causation, deriving its uniformitarianism from Lyell and its theme of change through time at least partly from a simplified view of evolution.

Closely similar views were being proposed at about the same time in plant geography and particularly in ecology. Hooker, among the founders of the subject, was an explorer in his own right; later workers such as Shelford, Cowles, and Tansley were members of professional geographic bodies; and one man, Clements, occupied in plant ecology a position similar to that of Davis in geomorphology. In soil science also, similar naive views of evolution as change through time were emphasized by Marbut and his school, in introducing the ideas of Dokuchaiev and Sibirtsev into the English literature.[25] Both conceptually, and in the imagery employed, plant ecologists and pedologists followed Davis's biological analogy for change through time. Clements emphasized succession as the[26]

> universal process of formation development . . . the life-history of the climax formation.[27]

The conceptual similarity to geomorphology was seized on by Cowles, a botanist trained in physiography by Salisbury and Chamberlin, who brought Davisian geomorphology and Clementsian ecology together in 'physiographical ecology,' following field work in the Chicago area on the coincidence of plant formations and physiographic units.[28] Plant ecologists and geomorphologists both adopted terms such as infancy, youth, maturity, and old age to describe development through time.

As in geomorphology, the time-framework has proved too restrictive for later investigations.[29] Whittaker, in his thorough review of the climax concept, perceptively summarizes Clements's contribution to plant ecology, but he could well have been speaking of Davis's role in the development of geomorphology or Marbut's in pedology:[30]

> It was the great contribution of Clements to have formulated a system, a philosophy of vegetation, which has been a dominating influence on American ecology as a framework for ecological thought and investigation. . . . Some negative aspects of Clements' system are . . . the superficial verbalism, the tendency to fit evidence by one means or another into the philosophical structure, the thread of non-empiricism which runs through his thought and work. . . . The Clementsian system had a certain symmetry about it, it was a fine design if its premises were granted; and for its erection Clements may rank as one of the truly creative minds of the field.

In the social sciences the development of a time-perspective awaited that of a historical tradition, especially the emergence of a concept of prehistory in the 1830's. That evolutionary ideas were in the air in 1859 is shown by the appearance of Sir Henry Maine's *Ancient Law* as early as 1861. E. B. Tylor's *Early History of Mankind* (1865) and *Origin of Civilization* (1870), and the Duke of Argyll's *Primeval Civilization* (1869) set a fashion in the developmental interpretation of prehistory and ethnology which dominated the subject until Malinowski's functional reinterpretation in the 1920's. In social anthropology also, MacLennan's *Primitive Marriage* (1865), Frazer's *Golden Bough* (1890), Westermarck on religion, and Lewis Morgan, Durkheim, and Lévy-Bruhl on social structures, established an evolutionary position over a period of decades which dominated thinking in these subjects until the reaction in the twentieth century.[31] A few workers, some of them influential in geography, maintained a developmental framework, ranging from the 'unilinear' school of White to the 'multilinear' evolution of Steward.[32] Childe[33] influenced historical interpretation of technological development, particularly among English geographers; the botanist Geddes's work on cities influenced early ideas on urban geography;[34] Taylor applied developmental principles to the study of race and culture history;[35] and Beaver and others attempted to introduce cyclic ideas into the interpretation of economic landscapes[36] though with little success.

Change through time has been a dominant theme in much human geography, particularly in the work of the Berkeley School on the settlement of the American southwest and other areas. Here Sauer's influence has been dominant, and it is interesting that he himself studied at Chicago under the physiographer Salisbury and the plant ecologist Cowles.[37] Another pupil of Cowles, who also studied under Davis and published in both geomorphology and vegetation geography, was the historical geographer Ogilvie, who carried their emphasis on time into regional studies.[38] The influence of plant ecology and the historical viewpoint was also clear, both in concept and language, in Whittlesey's idea of sequent occupance in the development of landscapes.[39]

The history of these narrowly evolutionary views in geography, however, resembled that in social anthropology: the early and enthusiastic application of time-frameworks to the data, and then a retreat from a developmental to a functional approach, or to a much modified and refined evolutionary interpretation. Primarily, however, geographers interpreted the biological revolution in terms of change through time: what for Darwin was a process became for Davis and others a history. This was powerfully reinforced not only when geology burst through theological restrictions on time, but also when man himself was found to have a history going back into antiquity. For a time 'evolution' implied little more than the idea of change, development, and 'progress,' and Darwin was in spite of himself seen as its author.

Organization and ecology

Darwin's second major contribution to geography was the idea of the inter-relationships and connections between all living things and their environment, developed in Haeckel's new science of *ecology*.[40] In the third chapter of *The Origin*, Darwin had been impressed by the 'beautiful' and 'exquisite' adaptation and interrelationships of organic forms in nature, and the theme of ecology is implicit if unstated in many of his writings:[41]

how infinitely complex and close-fitting

he wrote in *The Origin*[42]

are the mutual relations of all organic beings to each other and to the physical conditions of life.

This was the theme of ecology and, while Clements in America was forcing vegetation into a time-framework, European workers in general were more concerned with community structures and functions, culminating in Tansley's idea of the ecosystem. Perhaps Darwin's most significant contribution to ecological thinking, however, was to include man in the living world of nature. This had been becoming inevitable, with Boucher de Perthes' work on the Somme gravels after 1837, but it was in 1859 that the importance of finds of ancient man was formally recognized, in Prestwich's paper to the Royal Society.[43] Darwin deliberately left the implications of *The Origin* for the history of man unspoken in his first edition, but he was disappointed by Lyell's reluctance to draw the obvious conclusions in *The Antiquity of Man* four years later.[44] The theological difficulties were considerable for, although it was possible to reinterpret the biblical account of time, fundamental Christian beliefs such as the fall of man and original sin hinged on specific details of special creation: if the creation proved a myth, what of the theology? In spite of Darwin's own reticence, his theory soon took the popular title of 'the ape theory,' and controversy in the 1860's centred around the problem of man's ancestry rather than of variation and selection.[45] Huxley's *Man's Place in Nature* in 1863, for example, dealt not with man's ecological status, but with his relationship with the apes,[46] man had emphatically become a subject for scientific speculation. Darwin himself, in the *Expression of the Emotions in Man and Animals* (1868) and in *The Descent of Man* (1871), went further by treating modern man on the same level as other living things.

Haeckel used the term 'ecology' in 1869, and from about 1910 'human ecology' was used for the study of man and environment, not in a deterministic sense, but for man's place in the 'web of life' or the 'economy of nature.' Park's statement of the scope of human ecology[47] deals with the web of life, the balance of nature, concepts of competition, dominance and succession, biological economics and symbiosis: all concepts taken from plant and animal ecology. For Park, human ecology investigated the processes involved in biotic balance, in which man interacts with nature through culture and technology. McKenzie[48] expressed similar ideas with a more economic bias. The simplicity of human ecology as a methodological framework when stated in purely biological terms was echoed by Barrows in his presidential address to the Association of American Geographers in 1923:[49]

> . . . geography is the science of human ecology. . . . Geography will aim to make clear the relationships existing between natural environments and the distribution and activities of man. Geographers will, I think, be wise to view this problem in general from the standpoint of man's adjustment to environment, rather than from that of environmental influence. . . . The center of geography is the study of human ecology in specific areas. This notion holds out to regional geography a distinctive field, an organizing concept throughout, and the opportunity to develop a unique group of underlying principles.

Barrows' address, perhaps hastening the expulsion of geomorphology from geography in the United States, aroused considerable animosity and little positive support among geographers, and the sociologists themselves gradually moved away from Park's position. Bews, himself a botanist, followed Park closely,[50] but beginning with Alihan, sociologists turned to community as their field of study.[51] With some exceptions, such as White and Renner's textbooks,[52] the field delimited by Barrows and Park was abandoned by both geographers and sociologists,[53] though in America the Berkeley school adopted an ecological approach in the study of settlement in the American southwest.

Ecology has become, however, increasingly empirical in method, and in doing so it has run counter to, and ultimately has superseded, the synthetic geographical tradition of explanation by analogy, which attempted to understand the complexity and interrelationships of phenomena by reference to the even greater complexity of living organisms.[54] This is a theme which may be traced to classical writings and medieval scholasticism, and is in no sense Darwinian in origin but after Darwin such treatment lost its more extreme metaphysical implications and became more directly biological in expression. The organism analogy is explicit in both classical plant ecology and pedology, particularly in Clements's writings. For him, the plant community is

> a complex organism, or superorganism, with characteristic development and structure[55]. . . . As an organism the formation arises, grows, matures, and dies. . . . Each climax formation is able to reproduce itself.[56]

Clements believed that[57]

> this concept is the 'open sesame' to a whole new vista of scientific thought, a veritable *magna carta* for future progress.

Similarly, Shaler, Whitney, and others interpreted the functional interrelationships in soils as a phase in the 'higher estate of organic existence.'[58] Later workers, however, failed to demonstrate the discrete existence of organic unity in either soil or vegetation formations,[59] and the organic analogy in physical geography never carried the influence which it did in other branches of the subject.[60]

In geography as a whole the organism analogy operated on three distinct levels: those of the earth, its regions, and its states; and on each level its use long predates Darwinian evolutionary theory. Organic theories of the state were revived by such philosophers as Hobbes, and were thoroughly worked out by Ahrens in 1850.[61] Much of this earlier work was teleological, as in Ritter's conception of terrestrial unity and in the preface to Hobbes' *Leviathan*, but in the nineteenth century the details of the analogy were being pursued more closely. Bluntschli[62] attributed to states the properties of human organisms, even the details of sex and personality, and a fundamental precept of Comte's positivism was that sociology could only be understood in biological terms.[63] The impact of Darwinian thinking, and the writings of Spencer in England and of Worms in France, helped popularize these narrower organic analogies in the social sciences, and they retained vitality in geography long after they had been abandoned in other branches of human studies.

The idea of the organic unity of the earth can best be traced to Ritter.[64]

> The earth is one; . . . all its parts are in ceaseless action and reaction on each other. . . . The earth is therefore . . . a unit, an organism of itself: it has its own law of development, its own cosmical life; it can be studied in no one of its parts.

To him the earth was a[65]

> living work from the hand of a living God, (with) a close and vital connection, like that between body and soul, between nature and history.

For both Humboldt and Ritter, unity, harmony, and interdependence of parts constituted the organic analogy. Half a century later Vidal de la Blache reached similar conclusions and acknowledged his debt to Ritter, both at the earth and at the regional level.[66]

In regional geography the idea of organic unity served as a unifying theme in an increasingly particularistic discipline. Herbertson in 1905 used the term 'macro-organism' for the 'complex entity' of physical and organic elements of the earth's surface:[67]

> the soil itself the flesh, the vegetation its epidermal covering with its animal parasites, and the water the circulating life-blood automatically stirred daily and seasonally by the great solar heat. . . . If we regard the Earth as an individual, and these geographical regions, districts, localities, as representing organs, tissues, and cells, we perhaps get nearest to a useful comparison.

Subsequently Herbertson wrote that natural regions are[68]

> definite associations of inorganic and living matter with definite structures and functions, with as real a form and possessing as regular and orderly changes as those of a plant or an animal.

Like plants and animals, regions can be hierarchically ranked into species, genera, orders, and classes.[69] Similar thinking pervaded English regional methodology in the first half of the century. Unstead spoke of the evolution of regions, in the sense of increasing complexity, and of their pathology, in the sense of conditions harmful to man. He admitted that regions, unlike organisms, cannot be said to die, but he compared continuity of existence in a region with that of the 'germ-plasm of organisms through the successive generations,'[70] thus endowing regions at one time with the properties of individuals and at other times of populations. Similar organismic views were expressed by Bluntschli, Stevens, and most extremely by Swinnerton.[71]

In political geography the use of the organic analogy usually is associated with Ratzel, whose whole work is colored by Darwinian evolutionary thinking.[72] The first chapter of his *Politische Geographie* is entitled 'Der Staat als bodenständiger Organismus,' and the mystical conception of the indivisibility of people and land when organized into a state goes far beyond the formalistic comparison of lines of communication and arteries, seats of government and the brain, and so on, which Spencer outlined.[73] The organic quality of states depends on organization and interdependence of parts; it then assumes properties of growth and competition, and in so doing goes beyond the organic analogies of the earth and the geographical region.

The fundamental criterion used by geographers for applying the organic analogy at all levels has been the possession of properties of organization of constituent components into a functionally related, mutually interdependent complex which in spite of continuous flow of matter and energy is in apparent equilibrium, and which possesses properties as a whole which are more than the sum of the parts.[74] In this one may distinguish the influence of Vitalism in biology and the holistic philosophy of Smuts and Whitehead.[75] The Vitalist approach has the appearance of profound insight, but in essence it calls up undemonstrable, and hence unprovable causes such as the *entelechy* of Driesch or the *élan vital* of Bergson; to explain phenomena which are otherwise too complex to understand. Such procedures in biology preclude the rational formulation and testing of hypotheses, for they lie outside hypothesis; they pose no questions, and hence obtain no answers.[76] This was perceived by Vallaux in geography many years ago, and with recent advances in the study of molecular biology Vitalism now has no place in science.[77] The major objection to the organic approach in geography, however, is methodological, for it is a synthetic notion which gives no assistance in actual investigation, and it is, furthermore, an essentially idiographic concept in an increasingly nomothetic science.[78] The concept is thus reduced to a metaphor of dubious value, hinging on gross formal and functional comparisons between living matter and complexly interrelated facts in areas, and as such has dropped out of geographic work since 1939, except in occasional mention of Herbertson and Vidal de la Blache.[79]

Selection and struggle

Although the limitations of organic analogies require little demonstration, the problem of the effect of environment on man leads into the difficult fields of environmental influence, selection, and adaptation.[80] Most pre-Darwinian writers on the effects of environment were content to look for cause–effect relationships, without enquiring too closely into process, and this theme was taken up by Ratzel in the first volume of *Anthropogeographie*, and later by his students Miss Semple and Demolins. To the French school, imbued with Vidal's notions of harmony and interrelationship, this was too rigid a framework for analysis, but in America Davis attempted to carry simplistic ideas of causality into the definition of geography itself.[81]

> Any statement is of geographical quality if it contains a reasonable relation between some inorganic element of the earth on which we live, acting as a control, and some element of the existence, or growth, or behavior or distribution of the earth's organic inhabitants, serving as a response.

This suggestion gained little support among geographers, who realized that no science can take as its field of study a specific relationship rather than a body of data, for if it did the statement of its aims would presuppose the existence of the relationship itself.[82]

If causal relationship provided an unsound methodological principle, however, the problem of environmental influence remained.[83] Fleure, who came deeply under the influence of Darwinism in 1892–1895, stressed the need for the physiological study of environmental effects on man, and in his typology of human regions (regions of difficulty, of effort, of increment) came close to applying Darwinian ideas of natural selection through environmental influence to human groups.[84] The study of physiological effects, however, has become a specialized branch of biology outside geographical competence, and geographers have generally restricted themselves to the inference of causation from covariance on a coarser scale. Huntington particularly took up the problem of natural selection, environmental influences, and human population on a world scale, and Taylor explored the same theme in a series of studies of race, peoples, states, and towns, emphasizing their development through time under the influence of environmental factors.[85] The questions which these determinists raised, however, were posed in so gross a manner that they could only invite the grossest answers; most geographers realized this, and neither Taylor nor Huntington gained full academic acceptance. The questions which they asked could not be meaningfully answered in geographical terms, and the whole determinist–possibilist controversy, 'unreal and futile' as Hartshorne termed it, moved on to a philosophical rather than an empirical level.[86]

Darwin's theory was, of course, one of natural selection rather than of evolution, and basic to his thesis was the idea, taken from Malthus, that populations tended to expand at a geometric rate and thus outstrip resources.[87]

> Thus, from the war of nature, from famine and death, the most exalted object of which we are capable of conceiving, namely, the production of the higher animals, directly follows, (wrote Darwin in the last paragraph of *The Origin*): There is grandeur in this view of life.

Such views in turn influenced social thinking, particularly in America, where Spencer's idea of the survival of the fittest and Darwin's of the struggle for life were used to justify laissez-faire in politics and economics, particularly in the 'Social Darwinism' of Sumner.[88] Hofstadter has shown how the geologists Ward and Shaler denied the value of unrestricted competition in social life, and how they and the Russian geographer Kropotkin stressed cooperation and mutual aid in social development.[89] The idea of social selection was often somewhat crudely phrased, especially in geographical writing. Thus Turner's frontier hypothesis[90] and especially Roosevelt's book on *The Winning of the West* both took the naive view that frontier conditions selected all that was pioneering and democratic in a society, which then itself took on the pioneer spirit. It is interesting that, except in the idea of competition, and its implications, Darwinism had little effect in classical equilibrium economics, and both the implications of random variation and of development through time are relatively recent innovations.[91]

It was in political geography, however, that ideas of struggle and selection on a national level were most significant. In 1896 Ratzel developed his seven laws of the growth of states, from which derived the powerful concept of *Lebensraum*:[92]

> Just as the struggle for existence in the plant and animal world always centres about a matter of space, so the conflicts of nations are in great part only struggles for territory.

Although there is undoubtedly a danger that selective quotation of this sort may do violence to Ratzel's essentially scholarly position, as both Broek and Wanklyn argued,[93] it is clear that the organic analogy for Ratzel not only provided a simple and powerful model in analytical political geography, but also an apparently scientific justification, in Darwinian selection, for political behavior. It is interesting that Semple, in her exposition of Ratzel's own views, decided to omit the cruder Spencerian analogies as already outdated even in sociology,[94] but in spite of her disclaimers her writings are permeated by such thinking.

Ratzel's views served as a source for the *Geopolitik* developed in Europe between the wars. States for Kjellen were biological manifestations,[95] endowed not only with morality but also with 'organic lusts.' Herbert Spencer's writings are directly echoed in Kjellen's *Staten som Lifsform*.[96] The political usage made of the organic view of the state, and the ideas of struggle and *Lebensraum*, brought the subject into intellectual disgrace in the 1930's, as Troll[97] has outlined, and modern political geography is at pains to dissociate itself from any kind of organic analogy.

Randomness and chance

This review of biological ideas in geography has demonstrated that 'Darwinism' or 'evolution' was almost always interpreted by geographers either in the sense of change through time or of social struggle and selection. In both cases the application has been largely deterministic: it has in fact been said that simple geographical determinism, in its picture of causality was one of the last fields of operation of the Newtonian world-view in the twentieth century. Any discussion of the biological impact on geographical thinking must hinge on this central question: why was Darwinism, a theory for the selection of randomly occurring variants, interpreted in a deterministic and not a probabilistic sense?[98] Why was chance omitted in geography? The question is of more than historical interest, for a century after *The Origin*, geographers are beginning to recognize the importance of stochastic processes in geographic change.[99]

The problem is more remarkable because the study of random processes in the nineteenth century was by no means limited to Darwinian biology: indeed, Merz has written that[100]

> the study of this blind chance in theory and practice is one of the greatest scientific performances of the nineteenth century.

In the natural sciences Laplace laid the foundations of probability theory at the beginning of the century, and the theme was subsequently taken up by Adolphe Quetelet in the social sciences and used by Buckle in a work with which Darwin was certainly familiar.[101] The new kinetic theory of gases developed by Herapath, Clausius, and Clerk Maxwell was appearing at the same time as *The Origin*.[102] Boltzmann extended statistical conceptions in mechanics; and in biology itself Darwin's work stimulated a long series of statistical studies, from Galton and Pearson to Fisher and Haldane. Why, then in such an intellectual atmosphere, was the geographical interpretation so deterministic?[103]

Part of the answer lies in Darwin himself. Darwin's theory made a clear distinction between the way in which evolution was effected, and the course of evolution itself: geography seized on the latter and ignored the former. Darwin began with the idea of the selection of 'chance' variations, which are 'no doubt' governed by laws.[104] These laws Darwin failed to discover, and in time he came to emphasize chance less and less, and by the last edition of *The Origin* he was thinking of directional variation in a Lamarckian sense. Nowhere does he use the word 'random,' and in the fourth chapter of *The Origin* he states that the use of the word 'chance' is 'wholly incorrect'.[105] Although he undoubtedly believed that unfavourable variations could be as numerous as favorable ones, this became less clear with each successive edition. Darwin's difficulty was this: whereas his theory explained adaptation in nature by variation and natural selection, he could not, before the discovery of Mendel's work on genetics, offer any explanation of the basic variation, but the very facts of adaptation, which provided his strongest evidence, and which natural selection explained, had long been accounted for by the Church in terms of Design.[106] Early nineteenth century theology in England was a curious mixture of revelation and natural theology, exemplified in the Bridgewater Treatises and in William Paley. Paley wrote, for example, in 1802, that[107]

> There cannot be a design without a designer; contrivance without a contriver; order without choice; arrangement, without any thing capable of arranging; subserviency and relation to a purpose, without that which could intend a purpose; means suitable to an end, and executing their office in accomplishing that end, without the end ever having been contemplated, or the means accomodated to it. Arrangement, disposition of parts, subserviency of means to an end, relation of instruments to an use, imply the presence of intelligence and mind.

Lacking a mechanism for variation, and shaken by the theoretical objections in Jenkin's *North British Review* article in 1867,[108] Darwin changed his ground. Although maintaining privately that[109]

> the old argument of design . . . fails (and that) there is no more design . . . than in the course which the wind blows,

he still had doubts: the thought of the eye made him cold all over, the sight of feathers in a peacock's tail made him sick.[110] To Asa Gray he wrote in distress in 1860 on the problem of evil and the question of design:[111]

> I am inclined to look at everything as resulting from designed laws, with the details, whether good or bad, left to the working of what we may call chance.

But the effort to reconcile the unreconcilable was a failure:[112]

> A dog might as well speculate on the mind of Newton. . . . The more I think the more bewildered I become.

Darwin therefore abandoned the fundamental issue of random variation on which both the natural theologians and the exponents of revealed religion could unite, and concentrated on descent and on selection. If descent could be demonstrated, then the argument from design would appear much less plausible than that from evolution,[113] whereas selection could be demonstrated, for example in pigeons, on a purely empirical level. Darwin thus outflanked his opponents and deflected them from his most serious weakness, but at the same time he laid himself open to the charge of plagiarism and lack of originality. After all, *evolution* was not new, only Darwin's mechanism was, yet before Mendel Darwin could only defend the former, not the latter.[114] Darwinism in the sense of development or evolution through time was seized on in geography as a unifying principle to subsume vast quantities of otherwise discrete and apparently unrelated data: the clarity and order which this interpretation revealed had a remarkable effect on the progress of the sciences. But called Darwinism or not, it omitted Darwin's central theme. Mendel's work, and particularly the statistical treatment of heredity by Sir Ronald Fisher in *The Genetical Theory of Natural Selection* gave Darwinists the weapons they needed; but Fisher's book appeared seventy years after Darwin's and by that time the 'evolutionary' impact in geography and other sciences had been made.

Conclusion

Biological influences in geography during the past century, therefore, although often claiming descent from evolution or from Darwin, have been interpreted in ways which at times subtly and at times blatantly diverge from Darwin's actual philosophy. The major themes of change through time, of selection and struggle, and of the interrelatedness of things (the organic analogy, and later ecology), are all present in Darwin's writings, specifically in the eleventh, fourth, and third chapters of *The Origin of Species*, but the unique contribution of Darwin's theory, that of random variation, was, for religious and scientific reasons, neglected in geographical circles. It is interesting that methods which incorporate randomness are now being increasingly used by geographers.

The discussion of these four themes demonstrates that geographical thinking in the past hundred years has cut across biological thinking, incorporating some ideas into the corpus of thought derived by Hartshorne and Hettner from Kant and Humboldt, but neglecting others. Even in their most extreme statement, however, these themes never came to dominate geographical thinking, which, by concentrating on the interdependence of phenomena on the earth's surface, evolved a rationale of its own. In this, however, Darwin's influence can still be distinguished, in the impact which he made on the nineteenth century world view. Darwin established a sphere of scientific enquiry free from *a priori* theological ideas, and freed natural science from the arguments of natural theology. With the publication of *Essays and Reviews* in 1860,[115] theology itself began to turn away from science and to acknowledge that in this field the Bible was no authority. Darwin, by empirical argument and inductive method, thus dismissed teleology as a live issue in scientific explanation,[116] and though similar arguments persisted in Vitalist biology they were gradually reduced by the expansion of knowledge. Darwin, furthermore, sealed the acceptance of uniformitarianism and law in science, and completed the dismissal of Providential interference and catastrophism in scientific writing. And finally, and in this he was alone, Darwin established man's place in nature, both in Huxley's sense, and in Haeckel's, and in so doing made man a fit object for scientific study. Modern geography is inconceivable without these general advances, but their elaboration belongs to the study of intellectual history, not to that of geographical thought.[117]

References

1   Reprinted from *Annals of the Association of American Geographers,* Vol. 3 (December 1966), Washington, D.C.
2   I thank R. J. Chorley and P. Haggett, Cambridge University, for their comments on this paper.
3   LOEWENBERG, B. J. 'Darwin, Darwinism and History,' *General Systems,* Vol. 3, Ann Arbor: Society for General Systems Research (1958), pp. 7–17. For an excellent recent review, see Loewenberg, B. J., 'Darwin and Darwin Studies, 1959–63,' in Crombie, A. C. and Hoskin, M. A. (Eds.), *History of Science,* Vol. 4, Cambridge: Heffer (1965), pp. 15–54.
4   A meeting at the Royal Geographical Society for the Darwin Centenary did not consider Darwin's contribution to scientific thought: Darwin, Sir C. G., 'Darwin as a Traveller,' *Geographical Journal,* Vol. 126 (1960), pp. 129–36. See also f.n. 87. A paper entitled 'Ch. Darwin's Influence on the Progress of Science in Geography,' by Malicki, A. was announced for presentation at the XI International Congress of the History of Science, Warsaw, (1965), but no abstract of this paper was published (*Sommaires,* XIe Congrès International d'Histoire des Sciences, Cracow, 24–29 August (1965), 2 volumes, 594 pp.).
5   HARTSHORNE, R. *The Nature of Geography, a Critical Survey of Current Thought in the Light of the Past,* Lancaster, Pa.: Association of American Geographers (1939) and Hartshorne, R., *Perspective on the Nature of Geography* Chicago: Rand McNally and Co., for Association of American Geographers (1959).
6   SCOON, R. 'The Rise and Impact of Evolutionary Ideas,' in Persons, S. (Ed.), *Evolutionary Thought in America,* New Haven: Yale University Press (1950), pp. 4–42; reference on p. 5.
7   BUTLER, SAMUEL neatly phrased the issue: 'To me it seems that the 'Origin of Variation,' whatever it is, is the only true 'Origin of Species,' in *Life and Habit,* London: Trübner and Co. (1878), reference on p. 263.
8   No attempt is made to cover more general issues, such as the influence of Darwin's work on classification and taxonomy, with the resulting emphasis in

geography on 'genetic classification,' or such fundamentally biological fields as zoogeography, on which both Darwin and Wallace worked. For commentary on these, see particularly Grigg, D. B., 'The Logic of Regional Systems,' *Annals of the Association of American Geographers,* Vol. 55 (1965), pp. 465–91, and Darlington, P. J. Jr., 'Darwin and Zoogeography,' *Proceedings,* American Philosophical Society, Vol. 103 (1959), pp. 307–19.

9   LYELL, SIR CHARLES *Principles of Geology: Being an Attempt to Explain the Former Changes of the Earth's Surface by Reference to Causes now in Operation,* London: John Murray, 3 volumes (1830–33). See also Gillispie, C. C., *Genesis and Geology: a Study in the Relations of Scientific Thought, Natural Theology, and Social Opinion in Great Britain, 1790–1850,* Cambridge: Harvard University Press (1951).

10   DARWIN, C. R. 'Coral Islands,' Introduction, map, and remarks by Stoddart, D. R., *Atoll Research Bulletin,* No. 88 (1962), pp. 1–20, and Darwin, C. R., *The Structure and Distribution of Coral Reefs,* London: Smith, Elder and Co. (1842).

11   GRUBER, H. E. and GRUBER, V. 'The Eye of Reason: Darwin's Development during the *Beagle* Voyage,' *Isis,* Vol. 53 (1962), pp. 186–200.

12   HUXLEY, T. H. *Physiography, an Introduction to the Study of Nature,* London: Macmillan and Co. (1877).

13   See, for example, the writings of RECLUS, ELISÉE *The Earth: a Descriptive History of the Phenomena of the Life of the Globe,* edited by Keane, A. H., London: Virtue, J. S. and Co. Ltd. (1886).

14   DAVIS, W. M. 'Geographic Classification, Illustrated by a Study of Plains, Plateaus and their Derivatives,' *Proceedings, American Association for the Advancement of Science* (1884), pp. 428–32.

15   DAVIS, W. M. 'The Physical Geography of the Lands,' *Popular Science Monthly,* Vol. 57 (1900), pp. 157–70; reprinted in Davis, W. M., *Geographical Essays,* Johnson, D. W. (Ed.), Boston: Ginn and Company (1909) (hereafter cited as *Essays*), pp. 70–86; reference on pp. 85–6. See also Davis, W. M., 'The Relations of the Earth Sciences in view of their Progress in the Nineteenth Century,' *Journal of Geology,* Vol. 12 (1904), pp. 669–87, especially p. 675; 'The Physical Factor in General Geography,' *The Educational Bi-monthly,* Vol. 1 (1906), pp. 112–22; 'The Geographical Cycle,' *Geographical Journal,* Vol. 14 (1899), pp. 481–504, and *Essays,* pp. 249–78, especially pp. 249 and 254; 'Peneplains and the Geographical Cycle,' *Bulletin,* Geographical Society of America, Vol. 23 (1922), pp. 589–98, especially pp. 594–95.

16   SAUER, C. O. 'The Morphology of Landscape,' *University of California Publications in Geography,* Vol. 2 (1925), pp. 19–54; see p. 32.

17   As early as 1835 Darwin wrote to Fox, W. D. that 'I am become a zealous disciple of Mr. Lyell's views. . . . I am tempted to carry parts to a greater extent even than he does.' Darwin, C. R. to Fox, W. D., Lima, July 1835, in Darwin, Francis (Ed.), *Life and Letters of Charles Darwin, including an Autobiographical Chapter,* London: John Murray, 3 volumes (1887) (hereafter cited as *LLD*), reference in Vol. 1, p. 263.

18   CANNON, W. F. 'The Uniformitarian-Catastrophist Debate,' *Isis,* Vol. 51 (1960), pp. 38–55; Hooykaas, R., *Natural Law and Divine Miracle: The Principle of Uniformity in Geology, Biology and Theology,* Leiden: Brill, E. J. (1963), pp. 93–101.

19   See GAVIN DE BEER'S, comment: 'Lyell used the principle of uniformitarianism to prove that evolution was impossible because evolution involved progressionism and progressionism involved catastrophism and catastrophism must be rejected. Darwin used uniformitarianism to show that simple, existing causes produced and directed evolution, and that there was no link between catastrophism and progressionism. . . . The supreme paradox was, therefore,

that Darwin used Lyell's methods to show that Lyell's views on biology were wrong.' de Beer, G., *Charles Darwin: Evolution by Natural Selection,* London: Thomas Nelson and Sons Limited (1963), reference on p. 104.

20   HUXLEY, T. H. 'On the Reception of the "Origin of Species," ' *LLD,* Vol. 2 (1887), footnote 17, pp. 179–204; reference on p. 190.

21   HUXLEY, T. H. 'The Coming of Age of "The Origin of Species," ' in Huxley, T. H., *Science and Culture, and other Essays,* London: Macmillan and Co. (1882), pp. 310–24; reference on p. 315.

22   HUTTON, J. *Theory of the Earth, with Proofs and Illustrations,* London and Edinburgh: printed for Messrs. Cadell, Junior, and Davies, London, and William Creech, Edinburgh, 2 volumes (1795), reference in Vol. 2, p. 329.

23   THOMSON, W., BARON KELVIN OF LARGS 'On the Secular Cooling of the Earth,' in Thomson, W. and Tait, P. G. (Eds.), *Treatise on Natural Philosophy,* Oxford: at the Clarendon Press (1867), Vol. 1 (all published), pp. 711–27; and Thomson, W., Baron Kelvin of Largs, 'The "Doctrine of Uniformity" in Geology briefly Refuted,' *Popular Lectures and Addresses,* Vol. 2, *Geology and General Physics,* London: Macmillan and Co. (1894), pp. 6–9. Darwin admitted to being 'greatly troubled at the short duration of the world according to Thomson, Sir W.' Darwin, C. R. to Croll, James, 31 January 1869, in Darwin, F. and Seward, A. C. (Editors), *More Letters of Charles Darwin: a Record of his Work in a Series of Hitherto Unpublished Letters,* London: John Murray, 2 volumes (1903), reference in Vol. 2, p. 163. For an historical treatment of the problem of time, see Haber, F. C., *The Age of the World: Moses to Darwin,* Baltimore: Johns Hopkins Press (1959), especially pp. 265–90.

24   DAVIS, W. M. 'The Geographical Cycle,' *Geographical Journal,* Vol. 14 (1899), pp. 481–504, and *Essays,* footnote 15, pp. 249–78, reference on p. 249.

25   For Dokuchaiev's views, see GLINKA, K. D. *Treatise on Soil Science,* fourth edition, Jerusalem: Israel Program for Scientific Translations, for the National Science Foundation, Washington, D.C. (1963), p. 188; and, on the American school, Marbut, C. F., 'Soils of the Great Plains', *Annals of the Association of American Geographers,* Vol. 13 (1923), pp. 41–6.

26   CLEMENTS, F. E. *Plant Succession, an Analysis of the Development of Vegetation* Washington: Carnegie Institution (1916), reprinted in Clements, F. E., *Plant Succession and Indicators: a Definitive Edition of Plant Succession and Plant Indicators,* New York: Hafner Publishing Company (1963), reference on p. 3.

27   Ibid., pp. 6, 168, and 239.

28   COWLES, H. C. 'The Causes of Vegetative Cycles,' *Botanical Gazette,* Vol. 51 (1911), p. 161; 'The Causes of Vegetative Cycles,' *Annals of the Association of American Geographers,* Vol. 1 (1911), pp. 3–20, reference on p. 3. For Cowles's substantive work, see Cowles, H. C., 'The Physiographic Ecology of Chicago and Vicinity: a Study of the Origin, Development and Classification of Plant Societies,' *Botanical Gazette,* Vol. 31 (1901), p. 73. Cowles's teacher, Salisbury, was not himself an advocate of the cycle of erosion concept, but he used it as a teaching device.

29   WHITTAKER, R. H. 'A Consideration of Climax Theory: the Climax as a Population and Pattern,' *Ecological Monographs,* Vol. 23 (1953), pp. 41–78; Nikiforoff, C. C., 'Reappraisal of the Soil,' *Science,* Vol. 129 (1959), pp. 186–96.

30   WHITTAKER, R. H. op. cit., p. 26.

31   GOLDMAN, I. 'Evolution and Anthropology,' *Victorian Studies,* Vol. 3 (1959), pp. 55–75; Mac Rae, D. G., 'Darwinism and the Social Sciences,' in Barnett, S. A. (Ed.), *A Century of Darwin,* London: William Heinemann Limited (1958), pp. 296–312; Leach, E. R., 'Biology and Social Anthropology: the

Current Status of the Biological Analogy,' *Cambridge Review,* Vol. 85 (1964), pp. 248–51; Northrop, F. S. C., 'Evolution in its Relation to the Philosophy of Nature and the Philosophy of Culture,' in Persons S. (Ed.) op. cit., pp. 44–83; Gerard, R. W., Kluckholn, C. and Rapoport, A., 'Biological and Cultural Evolution: some Analogies and Explorations,' *Behavioral Science,* Vol. 1 (1956), pp. 6–34. For popular views, see Hays, H. R., *From Ape to Angel: An Informal History of Social Anthropology,* New York: Alfred A. Knopf (1958), and Daniel, G. E., *The Idea of Prehistory,* London: Watts and Co. (1962).

32   WHITE, L. 'Evolutionary Stages, Progress and the Evolution of Cultures,' *Southwestern Journal of Anthropology,* Vol. 3 (1947) pp. 165–92; Steward, J. H., *Theory of Culture Change: the Methodology of Multilinear Evolution,* Urbana: University of Illinois Press (1955).

33   CHILDE, V. G., *Man Makes Himself,* London: Watts and Co. (1936); Childe, V. G., *Social Evolution,* London: Watts and Co. (1951).

34   GEDDES, P. *Cities in Evolution: An Introduction to the Town Planning Movement and to the Study of Civics,* London: Williams and Norgate (1915); Taylor, T. G., *Urban Geography: A Study of Site, Evolution, Pattern and Classification in Villages, Towns and Cities,* London: Methuen and Co. (first edition 1949, second edition 1951), see pp. 7–9, 421–3.

35   TAYLOR, T. G. *Environment and Race: A Study of the Evolution, Migration, Settlement and Status of the Races of Man,* London: Humphrey Milford, Oxford University Press (1927); Taylor, T. G., *Environment and Nation: Geographical Factors in the Cultural and Political History of Europe,* Toronto: University of Toronto Press (1936); Taylor, T. G., *Environment, Race and Migration. Fundamentals of Human Distribution: with Special Sections on Racial Classification; and Settlement in Canada and Australia,* Toronto: University of Toronto Press (1937).

36   BEAVER, S. H. 'Technology and Geography,' *Advancement of Science,* Vol. 18 (1961), pp. 315–27; Bobek, H., 'Die Hauptstufen der Gesellschafts- und Wirtschaftsentfaltung in geographischer Sicht,' *Die Erde,* Vol. 90 (1959), pp. 259–98; Carol, H., 'Stages of Technology and their Impact upon the Physical Environment: A Basic Problem in Cultural Geography,' *Canadian Geographer,* Vol. 8 (1964), pp. 1–9. The time dimension is also emphasized in a different way by the Swedish school of human geography, for example by Hägerstrand, T., 'The Propagation of Innovation Waves,' *Lund Studies in Geography, Series B, Human Geography,* Vol. 4 (1952), pp. 3–19, and Morrill, R. L., 'Simulation of Central Place Patterns over Time,' *Lund Studies in Geography, Series B, Human Geography,* Vol. 24 (1962), pp. 109–20.

37   LEIGHLY, J. 'Introduction,' in Leighly, J. (Ed.), *Land and Life: A Selection from the Writings of Carl Ortwin Sauer,* Berkeley and Los Angeles: University of California Press (1963), pp. 1–8.

38   OGILVIE, A. G. 'The Time-Element in Geography,' *Transactions of the Institute of British Geographers,* No. 18 (1952), pp. 1–16; see p. 6. Similar views were expressed by Darby, H. C., 'On the Relations of Geography and History,' *Transactions of the Institute of British Geographers,* No. 19 (1953), pp. 1–11, but Gilbert, E. W. resisted the introduction of 'scientific' or 'evolutionary' ideas into historical geography, in 'What is Historical Geography?' *Scottish Geographical Magazine,* Vol. 148 (1932), pp. 129–36.

39   WHITTLESEY, D. 'Sequent Occupance,' *Annals of the Association of American Geographers,* Vol. 19 (1929), pp. 162–5.

40   HAECKEL, E. 'Entwicklunsgang und Aufgaben der Zoologie,' *Jenaische Zeitschrift,* Vol. 5 (1869), p. 353.

41   STAUFFER, R. C. 'Ecology in the Long Manuscript Version of Darwin's *Origin of Species* and Linnaeus' *Economy of Nature,' Proceedings, American Philo-*

sophical Society, Vol. 104 (1960), pp. 235–41, and Vorzimer, P., 'Darwin's Ecology and Its Influence upon his Theory,' *Isis*, Vol. 56 (1965), pp. 148–55.

42   DARWIN, C. R. *On the Origin of Species by Means of Natural Selection; or, The Preservation of favoured Races in the Struggle for Life*, London: John Murray (1859). Page references are given here to the reprint of the sixth edition, London: Geoffrey Cumberlege, Oxford University Press (1951); reference on p. 81.

43   PRESTWICH, J. 'On the Occurrence of Flint-Implements, Associated with the Remains of Extinct Mammalia, in Undisturbed Beds of a late Geological Period,' *Proceedings of the Royal Society*, Vol. 10 (1860), pp. 50–9, read 26 May 1859. Published in full with amended title in *Philosophical Transactions of the Royal Society*, Vol. 150 (1861), pp. 277–317.

44   LYELL, C. *The Geological Evidences of the Antiquity of Man, with Remarks on Theories of the Origin of Species by Variation*, London: John Murray (1863). See Darwin, C. R. to Hooker, J. D., 24 February 1863, in *LLD*, Vol. 3 (1887), f.n. 17, p. 9.

45   ELLEGÅRD, A. 'Darwin and the General Reader: The Reception of Darwin's Theory of Evolution in the British Periodical Press, 1859–1872,' *Acta Universitatis Gothoburgenesis: Göteborgs Universitets Arsskrift*, Vol. 64 (1958), pp. 1–394.

46   HUXLEY, T. H. *Evidence as to Man's Place in Nature*, London: Williams and Norgate (1863).

47   PARK, R. E. 'Human Ecology,' *American Journal of Sociology*, Vol. 42 (1936), pp. 1–15.

48   MCKENZIE, R. D. 'The Ecological Approach to the Study of the Human Community,' *American Journal of Sociology*, Vol. 30 (1924), pp. 287–301; McKenzie, R. D., 'The Scope of Human Ecology,' *Publications of the American Sociological Society*, Vol. 20 (1926), pp. 141–54.

49   BARROWS, H. H. 'Geography as Human Ecology,' *Annals of the Association of American Geographers*, Vol. 13 (1923), pp. 1–14.

50   BEWS, J. W. *Human Ecology*, London: Humphrey Milford, Oxford University Press (1935).

51   ALIHAN, M. *Social Ecology: A Critical Analysis*, Morningside Heights, New York: Columbia University Press (1938); Hawley, A. H., *Human Ecology: A Theory of Community Structure*, New York: The Ronald Press Company (1950).

52   WHITE, C. L. and RENNER, G. T. *Geography, an Introduction to Human Ecology*, New York: D. Appleton-Century Company, Inc. (1936), and *Human Geography, an Ecological Study of Society*, New York: Appleton-Century-Crofts, Inc. (1948).

53   SCHNORE, L. F. 'Geography and Human Ecology,' *Economic Geography*, Vol. 37 (1961), pp. 207–17.

54   For commentary on the geographic relevance of the concepts of community and ecosystem, see MORGAN, W. B. and MOSS, R. P. 'Geography and Ecology: The Concept of the Community and its Relationship to Environment,' *Annals of the Association of American Geographers*, Vol. 55 (1965), pp. 339–50, and Stoddart, D. R., 'Geography and the Ecological Approach: The Ecosystem as a Geographic Principle and Method,' *Geography*, Vol. 50 (1965), pp. 242–51.

55   CLEMENTS, F. E. and SHELFORD, V. E. *Bio-ecology*, New York: John Wiley and Sons, Inc. (1939), p. 24.

56   CLEMENTS, F. E., op. cit., footnote 25, p. 3.

57   CLEMENTS, F. E. and SHELFORD, V. E., op.cit., footnote 54, p. 24.

58   SHALER, N. S. 'The Origin and Nature of Soils,' *12th Annual Report, U.S. Geological Survey* (1890–91), Part 1, pp. 213–345.

59   See, for example, NIKIFOROFF, C. C. 'Reappraisal of the Soil,' *Science*, Vol.
     129 (1959), pp. 186–96, and Kira, T., Ogawa, H. and Yoda, K., 'Some
     Unsolved Problems in Tropical Forest Ecology,' *Proceedings, Ninth Pacific
     Science Congress*, Vol. 4 (1962), pp. 124–34.

60   But see the more extreme writings of, for example, STRICKLAND, C. *Deltaic
     Formation, with Special Reference to the Hydrographic Processes of the
     Ganges and Brahmaputra*, Calcutta and London: Longmans, Green and Co.,
     Ltd (1940), especially Chapter 2, 'The Sea in Pregnancy.'

61   AHRENS, H. *Die Organische Staatslehres auf philosophisch-anthropologischer
     Grundlage*, Vienna: G. Gerold & Sohn (1850).

62   BLUNTSCHLI, J. *Allgemeine Statslehre: Fünfte umgearbeitete Auflage des ersten
     Bandes des Allgemeinen Statsrechts (Lehre vom Modernen Stat, Erster Theil)*,
     Stuttgart: Verlag der J. G. Cotta'schen Buchhandlung (1875).

63   COMTE, A. *Cours de Philosophie positive*, Paris: Bachelier, Librairie pour les
     Mathématiques, Vol. 1, and Bachelier, Imprimeur-Libraire pour les Sciences,
     Vols. 2–6, 6 volumes (1830–1842), and commentary by COKER, F. W., 'Organis-
     mic Theories of the State: Nineteenth Century Interpretations of the State as
     Organism, or as Person,' *Columbia University Studies in History and Economics*
     Vol. 38 (1910). Also Spencer, H., *The Principles of Sociology* (London:
     Williams and Norgate, 3 volumes, 1876–96), especially Vol. 2, and Worms, R.,
     Organisme et Société, Paris: V. Giard et E. Brière, Bibliothèque sociologique
     internationale (1896); also published as a Thèse, Faculté des Lettres, Paris
     (1895).

64   RITTER, C. *Comparative Geography*, translated by Gage, W. L., Edinburgh
     and London: William Blackwood and Sons (1865), pp. 64, 65.

65   RITTER, C. *The Comparative Geography of Palestine and the Sinaitic Peninsula*,
     translated and adapted to the use of Biblical Students by Gage, W. L., Edin-
     burgh: T. and T. Clark, 4 volumes (1866), reference in Vol. 2, p. 4.

66   VIDAL DE LA BLACHE, P. 'Le Principe de la Géographie Générale,' *Annales de
     Géographie*, Vol. 5 (1896), pp. 129–42; 'Des Caractères distinctifs de la
     Géographie,' *Annals de Géographie*, Vol. 22 (1913), pp. 289–99.

67   HERBERTSON, A. J. 'The Higher Units: a Geographical Essay,' *Scientia*, Vol. 14
     (1913), pp. 203–12, reference on p. 205. This paper is reprinted in *Geography*,
     Vol. 50 (1965), pp. 332–42.

68   HERBERTSON, A. J. 'Natural Regions,' *Geographical Teacher*, Vol. 7 (1913–14),
     pp. 158–63, reference on pp. 158–9.

69   Ibid., p. 161.

70   UNSTEAD, J. F. 'Geographical Regions illustrated by reference to the Iberian
     Peninsula,' *Scottish Geographical Magazine*, Vol. 42 (1926), pp. 159–70,
     reference on p. 168.

71   BLUNTSCHLI, H. 'Die Amazonasniederung als harmonischer Organismus,'
     Geographische Zeitschrift, Vol. 27 (1921), pp. 49–67; Stevens, A. 'The
     Natural Geographical Region,' *Scottish Geographical Magazine*, Vol. 55
     (1939), pp. 305–17, see pp. 308–10; Swinnerton, H. H. 'The Biological
     Approach to the Study of the Cultural Landscape,' *Geography*, Vol. 23
     (1938), pp. 83–9.

72   On the nineteenth century background to Ratzel's thought, see particularly
     STEINMETZLER, J. 'Die Anthropogeographie Friedrich Ratzels und ihre
     ideengeschichtlichen Wurzeln,' *Bonner Geographischer Abhandlungen*, Bd. 19
     (1956), pp 1–151.

73   RATZEL, F. *Politische Geographie*, Munich: R. Oldenbourg (1897), and
     Spencer, H., cit.

74   FLEURE, H. J. *An Introduction to Geography*, London: Benn (1929), p. 13.

75   DRIESCH, H. *The Science and Philosophy of the Organism: the Gifford Lectures
     delivered before the University of Aberdeen in the year 1907* (Vol. 1) ... *and*

in the year 1908 (Vol. 2), London: Adam and Charles Black (1908); Smuts, J. C., *Holism and Evolution,* London: Macmillan and Co. Ltd. (1926); Whitehead, A. N., *Science and the Modern World: Lowell Lectures 1925,* Cambridge: at the University Press (1926).

76   BECK, W. S. *Modern Science and the Nature of Life,* London: Macmillan and Co. (1958; page references to the 1961 edition, Harmondsworth: Penguin Books). There is some similarity between Vitalist beliefs and Vidal's idea of the 'personnalité géographique' of regions. See Vidal de la Blache, P., *Tableau de la Géographie de la France,* Tome 1, première partie, of Lavisse, E., *Histoire de France illustrée depuis les Origines jusqu'à la Revolution,* Paris: Hachette et Cie (1911), especially première partie, 'Personnalité géographique de la France,' pp. 5–54.

77   VALLAUX, C. 'La Surface Terrestre assimilée à un Organisme,' Chapter 2, pp. 28–57, in *Les Sciences Géographiques,* Paris: Librairie Felix Alcan, nouvelle edition (1929), see p. 49; Ernst Caspari, 'On the Conceptual Basis of the Biological Sciences,' in Colodny, R. G. (Ed.), *Frontiers of Science and Philosophy,* London: George Allen and Unwin Ltd (1964), pp. 131–45; and Beck, op. cit.

78   SIDDALL, W. R. *Idiographic and Nomothetic Geography: The Application of Some Ideas in the Philosophy of History and Science to Geographic Methodology* (University of Washington, Ph.D. Thesis) (1959).

79   Crowe attacked the organism analogy in 1938: CROWE, P. R. 'On Progress in Geography,' *Scottish Geographical Magazine,* Vol. 54 (1938), pp. 1–19, see pp. 10–11. See also the discussions by Hartshorne, R., 'Is the Geographic Area an Organism?' in *The Nature of Geography,* op. cit., f.n. 5, pp. 256–60, and by Stoddart, D. R., 'Organism and Ecosystem as Geographical Models,' in Chorley, R. J. and Haggett, P. (Eds.), *Models in Geography,* London: Methuen and Co. (1967).

80   TATHAM, G. 'Environmentalism and Possibilism,' in Taylor T. G. (Ed.), *Geography in the Twentieth Century: A Study of Growth, Fields, Techniques, Aims and Trends,* London: Methuen, (1951), pp. 128–62; Febvre, L. with Bataillon, L., *A Geographical Introduction to History,* London: Kegan Paul, Trench, Trubner and Co. Ltd., (1925), being the translation of *La Terre et l'Évolution Humaine: Introduction géographique à l'Histoire,* Paris: La Renaissance du Livre (1922); Ratzel, F., *Anthropo-geographie oder Grundzüge der Anwendung der Erdkunde auf die Geschichte,* Stuttgart: J. Engelhorn (1882).

81   DAVIS, W. M. 'An Inductive Study of the Content of Geography,' *Bulletin of the American Geographical Society,* Vol. 38 (1906), pp. 67–84, and *Essays,* footnote 15, pp. 3–22, reference on p. 8. Also Davis, W. M., 'Systematic Geography,' *Proceedings of the American Philosophical Society* Vol. 41 (1902), pp. 235–59.

82   SAUER, C. O. op. cit., pp. 51–2.

83   BRIGHAM, A. P. 'Problems of Geographic Influence,' *Annals of the Association of American Geographers,* Vol. 5 (1915), pp. 3–25; Dryer, C. R., 'Genetic Geography: The Development of the Geographic Sense and Concept,' *Annals of the Association of American Geographers,* Vol. 10 (1920), pp. 3–16.

84   FLEURE, H. J. 'Geography and the Scientific Movement,' *Geography,* Vol. 22 (1937), pp. 178–88; Fleure, H. J., 'The Later Development in Herbertson's Thought: A Study in the Application of Darwin's Ideas,' *Geography,* Vol. 37 (1952), pp. 97–103; Fleure, H. J., 'Human regions,' *Scottish Geographical Magazine,* Vol. 35 (1919), pp. 94–105, revised from 'Régions humaines,' *Annales de Géographie,* Vol. 26 (1917), pp. 161–74.

85   HUNTINGTON, E. *Mainsprings of Civilization,* New York: John Wiley and Sons (1945); Huntington, E., 'Geography and Natural Selection,' *Annals of the*

Association of American Geographers, Vol. 14 (1924), pp. 1–16; Taylor,T. G., *Environment and Race: A Study of the Evolution, Migration, Settlement and Status of the Races of Man,* London: Humphrey Milford, Oxford University Press (1927); Taylor, T. G., *Environment and Nation: Geographical Factors in the Cultural and Political History of Europe,* Toronto: University of Toronto Press (1936); Taylor, T. G., *Urban Geography: A Study of Site, Evolution, Pattern and Classification in Villages, Towns and Cities,* London: Methuen and Co. Ltd. (1949).

86   HARTSHORNE, R. *Perspective on the Nature of Geography,* op. cit., f.n. 5, p. 57; Martin, A. F., 'The Necessity for Determinism, a Metaphysical Problem Confronting Geographers,' *Transactions of the Institute of British Geographers,* No. 17 (1951), pp. 1–11; for a review of the whole group of issues around environmental influence and determinism, see Lewthwaite, G. R., 'Environmentalism and Determinism: A Search for Clarification,' *Annals of the Association of American Geographers,* Vol. 56 (1966), pp. 1–23.

87   DARWIN, C. R. op. cit., footnote 42, p. 560.

88   Herbst has attempted to trace the dichotomy in American geography between physical and human studies to the early influence of Social Darwinism on, for example, Davis's definition of the nature of the subject. See HERBS, J., 'Social Darwinism and the History of American Geography,' *Proceedings of the American Philosophical Society,* Vol. 105 (1961), pp. 538–44.

89   HOFSTADTER, R. *Social Darwinism in American Thought,* Philadelphia: University of Pennsylvania Press (1944); revised edition, Boston: The Beacon Press (1955); the decline of the more extreme Social Darwinism in America closely paralleled the eclipse of Spencer's evolutionary philosophy by the pragmatism of William James and John Dewey in the later years of the nineteenth century. See especially Wiener, P. S., *Evolution and the Founders of Pragmatism,* Cambridge: Harvard University Press (1949).

90   TURNER, F. J. *The Frontier in American History,* New York: Henry Holt and Company (1920).

91   VEBLEN, T. 'Why is Economics not an Evolutionary Science?' *Quarterly Journal of Economics,* Vol. 13 (1898), pp. 373–97; Spengler, J. J., 'Evolutionism in American Economics, 1800–1946,' in Persons, S., op. cit., f.n. 5, pp. 202–66; Alchian, A., 'Uncertainty, Evolution, and Economic Theory,' *Journal of Political Economy* (1950), pp. 211–21; and, on development economics, Rostow, W. W., *The Stages of Economic Growth: a non-communist Manifesto,* Cambridge: at the University Press (1960).

92   The laws are set forth in RATZEL, F. 'Die Gesetze des räumlichen Wachstums der Staaten: ein Beitrag zur wissenschaftlichen politischen Geographie,' *Petermanns Mitteilungen,* Vol. 42 (1896), pp. 97–107, and also 'The Territorial Growth of States,' *Scottish Geographical Magazine,* Vol. 12 (1896), pp.351–61; the concept of *Lebensraum* is enunciated in Ratzel, F., op. cit., footnote 80, p. 458.

93   BROEK, J. O. M. 'Friedrich Ratzel in Retrospect,' mimeographed, abstract in *Annals of the Association of American Geographers,* Vol. 44 (1954), p. 207; Wanklyn, H., *Friedrich Ratzel: a Biographical Memoir and Bibliography,* Cambridge: at the University Press (1961).

94   SEMPLE, E. C. *Influences of Geographic Environment: on the Basis of Ratzel's System of Anthropogeography,* New York: Henry Holt and Company (1911), p. v.

95   KJELLEN, R. *Staten som Lifsform,* Vol. 3, No. 3 of *Politiska Handböcker,* Stockholm: H. Geber (1916), and the German editions, *Der Staat als Lebensform,* uebersetzt von Margarethe Landfeldt, Leipzig: S. Herzel (1917) and *Der Staat als Lebensform,* uebertragung von J. Sandmeier, Berlin-Grunewald: Vowinckel K. (1924), 4 Auflage. Also Weigert, H. W., *Generals and Geo-*

graphers: The Twilight of Geopolitics New York: Oxford University Press, (1942), pp. 106–7.

96   KJELLEN, RUDOLPH, op. cit., footnote 94, quoted by Fifield, R. H. and Pearcy, G. E., *Geopolitics in Principle and Practice*, Boston: Ginn and Company (1944), p. 11.

97   TROLL, C. 'Geographic Science in Germany during the Period 1933–45: a Critique and Justification,' *Annals of the Association of American Geographers*, Vol. 39 (1949), pp. 99–137.

98   FISHER, R. A. goes so far as to state that 'Darwin's chief contribution, not only to biology but to the whole of natural science, was to have brought to light a process by which contingencies *a priori* improbable, are given, in the process of time, an increasing probability, until it is their non-occurrence rather than their occurrence which becomes highly improbable.' See 'Retrospect of the Criticisms of the Theory of Natural Selection,' in Huxley, J., Hardy, A. C. and Ford, E. B. (Eds.), *Evolution as a Process,* London: George Allen and Unwin Ltd (1954), pp. 84–98, reference on p. 91.

99   For an early statement on probability in geography, see BRUNHES, J. 'Du Caractère propre et du Caractère complexe des Faits de Géographie Humaine,' *Annals de Géographie,* Vol. 33 (1913), pp. 1–40, and translation in 'The Specific Characteristics and Complex Character of the Subject-matter of Human Geography,' *Scottish Geographical Magazine,* Vol. 29 (1913), pp. 304–22, 358–74: 'Every truth concerning the relations between natural surroundings and human activities can never be anything but approximate; to represent it as something more exact than that is to falsify it' (pp. 362–3). 'All biological relations, all oecological truths, are, and can be, nothing more than statistical truths' (p. 364). For recent substantive work, see, for example, Hägerstrand, T., op. cit., and Morrill, R. L., op. cit.

100   MERZ, J. T. *A History of European Thought in the Nineteenth Century,* Volume 2, Edinburgh: William Blackwood and Sons Ltd (1928), p. 624. For a historical review, see Nagel, E., 'Principles of the Theory of Probability,' *International Encyclopedia of Unified Science,* Vol. 1, No. 6, Chicago: University of Chicago Press (1939).

101   DE LAPLACE, P. S. *Théorie analytique des Probabilités,* Paris: Mme. Courcier, Ve., Imprimeur-Librairie pour les Mathématiques (1812); de Laplace, P. S., *Essai philosophique sur les Probabilités,* Paris: Bachelier, Successeur de Mme. Courcier, Ve., Librairie pour les Mathématiques (1814): Quetelet, L. A. J., *Sur l'Homme et le Développement de ses Facultés: ou essai de Physique sociale,* Bruxelles: L. Hauman et Compe., 2 volumes, (1836); Herschel, J., 'Quetelet on Probabilities,' *Edinburgh Review,* Vol. 42 (1850), pp. 1–57; Buckle, H. T., *History of Civilisation in England,* Volume 1, London: J. W. Parker and Son (1857), and Darwin's comments in *LLD,* Vol. 2 (1888), footnote 17, pp. 110 and 386. One may of course argue that these earlier workers used statistical analysis as a tool to overcome error, and incompleteness in our perception of the world, rather than recognizing that the real world is itself subject to chance. See the comments by Hesse, M. B. on Gillispie, C. C., 'Intellectual factors in the Background of Analysis by Probabilities,' in Crombie, A. C. (Ed.), *Scientific Change, Symposium on the History of Science, University of Oxford, 9–15 July 1961,* London: Heinemann (1963), pp. 430–53 and comments pp. 471–6.

102   CLERK MAXWELL, J. 'Illustrations of the Dynamical Theory of Gases, Part I,' *London, Edinburgh and Dublin Philosophical Magazine and Journal of Science,* Series 4, Vol. 19 (1860), pp. 19–32, read 21 September, 1859.

103   LUKERMANN, F. in an interesting recent discussion, has drawn attention to the dependence of the French school of human geography on the work of French statisticians and natural scientists in the nineteenth century, from Laplace to

Cournot and later Henri Poincaré, and the intellectual milieu in which they worked. The French possibilists thus form an exception to the generalizations in this paragraph. See Lukermann, F., 'The "Calcul des Probabilités" and the École Française de Géographie,' *Canadian Geographer,* Vol. 9 (1965), pp. 128–37.

104 DARWIN, C. R. to Hooker, J. D., 23 December 1856, *LLD,* Vol. 2 (1888), footnote 17, p. 87.

105 DARWIN, op. cit., footnote 41, p. 138. Huxley, of course, interpreted even 'chance' variations deterministically, op. cit. (1887), footnote 20, pp. 199–201.

106 See on this theme ELLERGÅRD, A., op. cit., footnote 44.

107 PALEY, W. *Natural Theology, or, Evidences of the Existence and Attributes of the Deity collected from the Appearances of Nature,* London: printed for R. Faulder (1802), p. 12.

108 See the anonymous article by JENKIN, FLEEMING 'The Origin of Species,' *North British Review,* Vol. 92 (1867), pp. 277–318, and discussion by Vorzimer, P., 'Charles Darwin and Blending Inheritance,' *Isis,* Vol. 54 (1963), pp. 371–90.

109 DARWIN, C. R. *The Autobiography of Charles Darwin 1809–1882,* edited by Barlow, Nora, London: Collins (1958), p. 87.

110 DARWIN, C. R. to Gray, Asa, April 1860, *LLD,* Vol. 2 (1888), footnote 17, p. 296. Perhaps he recalled Sturmius's remark, quoted by Paley, 'that the examination of the eye was a cure for atheism,' in Paley, W., op. cit., p. 35.

111 DARWIN to Gray, Asa, *LLD,* Vol. 2 (1888), footnote 16, p. 312, 22 May 1860.

112 Ibid.

113 ELLERGÅRD, A. op. cit.

114 FLEMING, D. 'The Centenary of the *Origin of Species,*' *Journal of the History of Ideas,* Vol. 20 (1959), pp. 437–46; On the general theme of Darwin's theological difficulties, see Mandelbaum, M., 'Darwin's Religious Views,' *Journal of the History of Ideas,* Vol. 20 (1958), p. 363–78, and Fleming, D., 'Charles Darwin, the Anaesthetic Man,' *Victorian Studies,* Vol. 4 (1961), pp. 219–36. For an alternative interpretation, deriving ideas of randomness from natural theology, see Cannon, W. F., 'The Bases of Darwin's Achievement: a Revaluation,' *Victorian Studies,* Vol. 5 (1961), pp. 109–34.

115 TEMPLE, F. and others, *Essays and Reviews,* London: John W. Parker and Son (1860).

116 The role of teleology in Darwin's thought is notoriously difficult to assess, especially in the later editions of *The Origin* as Darwin shifted his ground over mechanism. These, together with the much-quoted last paragraph, have led to the argument that Darwinian evolution was in fact of a teleological nature. The situation is complicated by the curious reaction in some theological circles, which saw in this interpretation a way out of the crisis which the publication of *The Origin* had caused. A reviewer has drawn my attention to G. Himmelfarb's account of this in *Darwin and the Darwinian Revolution,* London: Chatto and Windus (1959), pp. 325–9. Asa Gray, among others, was acutely aware of the teleology issue, and his attempt to interpret *The Origin* teleologically led to growing estrangement from Darwin himself. Gray saw natural selection as purposive, 'and to most minds Purpose will imply Intelligence'; Gray, A., 'Relation of Insects to Flowers,' *Contemporary Review,* Vol. 41 (1882), p. 609. This quotation is taken from the elegant treatment of Gray's position in 'A theist in the Age of Darwin,' Chapter 18, pp. 355–83, in Dupree, A. H., *Asa Gray 1818–1888,* Cambridge: Belknap Press of Harvard University Press (1959).

117 In preparing this paper the following discussions have been valuable: Gillispie, C. C., *Genesis and Geology: A Study in the Relations of Scientific*

Thought, Natural Theology, and Social Opinion in Great Britain, 1790–1850, Cambridge: Harvard University Press (1951); Ellergård, A., *Darwin and the General Reader* (footnote 45); Eiseley, L., *Darwin's Century: Evolution and the Men who discovered it,* London: Victor Gollancz (1959); Himmelfarb, Gertrude, *Darwin and the Darwinian Revolution,* London: Chatto and Windus (1959); Greene, J. C., *Darwin and the Modern World View: the Rockwell Lectures, Rice University,* Baton Rouge: Louisiana State University Press (1961); Bronowski, Jacob, 'Introduction,' in Banton, M. (Ed.), *Darwinism and the Study of Society,* London: Tavistock Publications (1961); and the writings of Darwin himself, particularly *The Origin of Species* (1859), the *Life and Letters* (1888), and the *Autobiography* (1858).

Since this paper initially went to press, I have seen an early treatment of the geographical content of Darwin's own writings by Marinelli, Giovanni, 'Carlo Roberto Darwin e la Geografia,' *Atti dell' Istituto Veneto de Scienze, Lettere ed Arti,* Serie 5, Vol. 8 (1882), pp. 1279–1321. Marinelli treats Darwin's coral reef work at length, and in analyzing the geographical nature of Darwin's other writings, he concludes that their principle was essentially chorological. The paper is reprinted in *Rivista de Filosofia scientifica,* Vol. 2 (1882–1883), pp. 385–410, and in *Scritti minori di Giovanni Marinelli,* Vol. 1, *Metodo e Storia della Geografia,* Firenze: Tipografia de M. Ricci (1908), pp. 99–141. See also Ule, Willi, 'Darwin's Bedeutung in der Geographie,' *Deutsche Rundschau für Geographie und Statistik,* Vol. 31 (1909), pp. 433–43.

4   Geography, experience, and imagination: towards a geographical epistemology[1]

David Lowenthal
American Geographical Society, New York, U.S.A.

'The most fascinating *terræ incognitæ* of all are those that lie within the minds and hearts of men.' With these words, John K. Wright concluded his 1946 presidential address before the Association of American Geographers. This paper considers the nature of these *terræ incognitæ*, and the relation between the world outside and the pictures in our heads.[2]

The general and the geographical world view

Neither the world nor our pictures of it are identical with geography. Some aspects of geography are recondite, others abstruse, occult, or esoteric; conversely, there are many familiar features of things that geography scarcely considers. Beyond that of any other discipline, however, the subject matter of geography approximates the world of general discourse; the palpable present, the everyday life of man on earth, is seldom far from our professional concerns. 'There is no science whatever,' wrote a future president of Harvard, a century and a half ago, 'which comes so often into use in common life.'[3] This view of geography remains a commonplace of contemporary thought. More than physics or physiology, psychology or politics, geography observes and analyzes aspects of the milieu on the scale and in the categories that they are usually apprehended in everyday life. Whatever methodologists think geography ought to be, the temperament of its practitioners makes it catholic and many-sided. In their range of interests and capacities – concrete and abstract, academic and practical, analytic and synthetic, indoor and outdoor, historical and contemporary, physical and social – geographers reflect man generally. 'This treating of cabbages and kings, cathedrals and linguistics, trade in oil, or commerce in ideas,' as Peattie wrote, 'makes a congress of geographers more or less a Committee on the Universe.'[4]

Geographical curiosity is, to be sure, more narrowly focused than mankind's; it is also more conscious, orderly, objective, consistent, universal, and theoretical than are ordinary queries about the nature of things. Like geography, however, the wider universe of discourse centers on knowledge and ideas about man and milieu; anyone who inspects the world around him is in some measure a geographer.

As with specifically geographical concepts, the more comprehensive world of ideas that we share concerns the variable forms and contents of the earth's surface, past, present, and potential – 'a torrent of discourse about tables, people, molecules, light rays, retinas, air-waves, prime numbers, infinite classes, joy and sorrow, good and evil.'[5] It comprises truth and error, concrete facts and abstruse relationships, self-evident laws and tenuous hypotheses, data drawn from natural and social science, from history, from common sense, from intuition and mystical experience. Certain things appear to be grouped spatially, seriated temporally, or related causally: the hierarchy of urban places, the annual march of temperature, the location of industry. Other features of our shared universe seem unique, amorphous, or chaotic: the population of a country, the precise character of a region, the shape of a mountain.[6]

UNIVERSALLY ACCEPTED ASPECTS OF THE WORLD VIEW

However multifarious its makeup, there is general agreement about the character of the world and the way it is ordered. Explanations of particular phenomena differ from one person to another, but without basic concurrence as to the nature of things, there would be neither science nor common sense, agreement nor argument. The most extreme heretic cannot reject the essence of the prevailing view. 'Even the sharpest dissent still operates by partial submission to an existing consensus,' reasons Polanyi, 'for the revolutionary must speak in terms that people can understand.'[7]

Most public knowledge can in theory be verified. I know little about the geography of Sweden, but others are better informed; if I studied long and hard enough I could learn approximately what they know. I cannot read the characters in Chinese newspapers, but hardly doubt that they convey information to the Chinese; assuming that there is a world in common, other peoples' ways of symbolizing knowledge must be meaningful and learnable.

The universe of geographical discourse, in particular, is not confined to geographers; it is shared by billions of amateurs all over the globe. Some isolated primitives are still ignorant of the outside world: many more know little beyond their own countries and ways of life; but most of the earth's inhabitants possess at least rudiments of the shared world picture. Even peoples innocent of science are privy to elements of our geography, both innate and learned: the normal relations between figure and ground; the distinctive setting of objects on the face of the earth; the usual texture, weight, appearance, and physical state of land, air, and water; the regular transition from day to night; the partition of areas by individual, family, or group.

Beyond such universals, the geographical consensus tends to be additive, scientific, and cumulative. Schools teach increasing numbers that the world is a sphere with certain continents, oceans, countries, peoples, and ways of living and making a living; the size, shape, and general features of the earth are known by more and more people. The general horizon of geography has expanded rapidly. 'Until five centuries ago a primal or regional sense of space dominated human settlements everywhere'; today, most of us share the conception of a world common to all experients.[8]

THE GENERAL CONSENSUS NEVER COMPLETELY ACCEPTED

The whole of mankind may in time progress, as Whittlesey suggests, to 'the sense of space current at or near the most advanced frontier of thought.' But no one, however inclined to pioneer, visits that frontier often, or has surveyed more than a short traverse of it. 'Primitive man,' according to Boulding, 'lives in a world which has a spatial unknown, a dread frontier populated by the heated imagination. For modern man the world is a closed and completely explored surface. This is a radical change in spatial viewpoint.'[9] But the innovation is superficial; we are still parochial. 'Even in lands where geography is part of a compulsory school curriculum, and among people who possess considerable information about the earth,' Whittlesey points out, 'the world horizon is accepted in theory and rejected in practice.'[10]

The 'dread unknowns' are still with us. Indeed, 'the more the island of knowledge expands in the sea of ignorance, the larger its boundary to the unknown.'[11] Primitive world views were simple and consistent enough for every participant to share most of their substance. Within Western scientific society, no one really grasps more than a small fraction of the public, theoretically communicable world view. The amount of information an individual can acquire in an instant or in a lifetime is finite, and miniscule compared with what the milieu presents; many questions are too complex to describe, let alone solve, in a practicable length of time. The horizons of knowledge are expanding faster than any person can keep up with. The proliferation of new sciences extends our powers of sense and thought, but their rigorous techniques and technical languages hamper communication; the common field of knowledge becomes a diminishing fraction of the total store.[12]

On the other hand, we tend to assume things are common knowledge which may not be; what seems to me the general outlook might be mine alone. The most devoted adherents to a consensus often mistake their own beliefs for universal ones. For a large part of our world view, we take on faith much of what we are told by science. But we may have got it wrong; as Chisholm points out, 'We are all quite capable of believing falsely at any time that a given proposition is accepted by the scientists of our culture circle.'[13] In our impressions of the shared world view we all resemble the fond mother who watched her clumsy son parade, and concluded happily, 'Everyone was out of step but my Johnnie.'

THE WORLD VIEW NOT SHARED BY SOME

The most fundamental attributes of our shared view of the world are confined, moreover, to sane, hale, sentient adults. Idiots cannot suitably conceive space, time, or causality. Psychotics distinguish poorly between themselves and the outside world. Mystics, claustrophobics, and those haunted by fear of open space (agoraphobia) tend to project their own body spaces as extensions of the outside world; they are often unable to delimit themselves from the rest of nature. Schizophrenics often underestimate size and overestimate distance. After a brain injury, invalids fail to organize their environ-ments or may forget familiar locations and symbols. Impairments like aphasia, apraxia, and agnosia blind their victims to spatial relations and logical connections self-evident to most. Other hallucinatory sufferers may identify forms but regularly alter the number, size, and shape of objects (polyopia, dysmegalopsia, dysmorphopsia), see them always in motion (oscillopsia), or locate everything at the same indefinite distance (porr-hopsia).[14]

A fair measure of sensate function is also prerequisite to the general view of the common world. No object looks quite the way it feels; at first sight, those born blind not only fail to recognize visual shapes but see no forms at all, save for a spinning mass of colored light. They may have known objects by touch, but had nothing like the common conception of a space with objects in it. A purely visual world would also be an unreal abstraction; a concrete and stable sense of the milieu depends on synesthesia, sight combined with sound and touch.[15]

To see the world more or less as others see it, one must above all grow up; the very young, like the very ill, are unable to discern adequately what is themselves and what is not. An infant is not only the center of his universe, he *is* the universe. To the young child, everything in the world is alive, created by and for man, and endowed with will: the sun follows him, his parents built the mountains, trees exist because they were planted. As Piaget puts it, everything seems intentional; 'the child behaves as if nature were charged with purpose,' and therefore conscious. The clouds know what they are doing, because they have a goal. 'It is not because the child believes things to be alive that he regards them as obedient, but it is because he believes them to be obedient that he regards them as alive.' Asked what something is, the young child often says it is *for* something – 'a mountain is for climbing' – which implies that it has been *made* for that purpose.[16]

Unable to organize objects in space, to envisage places out of sight, or to generalize from perceptual experience, young children are especially poor geographers. To learn that there are other people, who perceive the world from different points of view, and that a stable, communicable view of things cannot be obtained from one perspective alone, takes many years. Animism and artificialism give way only gradually to mechanistic outlooks and expla-nations. 'No direct experience can prove to a mind inclined towards animism that the sun and the clouds are neither alive nor conscious'; the child must first realize that his parents are not all-powerful beings who made a universe centered on himself. Piaget traces the development in children of perceptual and conceptual objectivity, on which even the most primitive and parochial geographies depend.[17] Again in old age, however, progressive loss of hearing, deficiencies of vision, and other infirmities tend to isolate one from reality and to create literally a second geographical childhood.[18]

Different as they are from our own, the perceived milieus, say, of most children of the same age (or of many schizophrenics; or of some drug addicts) may closely resemble one another. But there is little communication or mutual understanding of a conceptual character among children. No matter how many features their pictures of the world may have in common, they lack any *shared* view of the nature of things.

MUTABILITY OF THE GENERAL CONSENSUS

The shared world view is also transient: it is neither the world our parents knew nor the one our children will know. Not only is the earth itself in constant flux, but every generation finds new facts and invents new concepts to deal with them. 'You cannot step twice into the same river,' Heraclitus observed, 'for fresh waters are ever flowing in upon you.' Nor does anyone look at the river again in the same way: 'The vision of the world geographers construct must be created anew each generation, not only because reality changes but also because human preoccupations vary.'[19]

Because we cherish the past as a collective guide to behavior, the general consensus alters very slowly. Scientists as well as laymen ignore evidence incompatible with their preconceptions. New theories which fail to fit established views are resisted, in the hope that they will prove false or irrelevant; old ones yield to convenience rather than to evidence. In Eiseley's phrase, 'a world view does not dissolve overnight. Rather, like . . . mountain ranges, it erodes through long centuries.'[20] The solvent need not be truth. For example, in the seventeenth century many scholars believed that the earth – the 'Mundane Egg' – was originally 'smooth, regular, and uniform; without Mountains, and without a Sea'; to chastise man for his sins, at or before the Deluge, God crumpled this fair landscape into continents and ocean deeps, with unsightly crags and chasms; modern man thus looked out on 'the Ruins of a broken World.' This version of earth history was overthrown, not by geological evidence, but principally by a more sanguine view of God and man, and by a new esthetic standard: to eighteenth century observers, mountains seemed majestic and sublime, rather than hideous and corrupt.[21]

ANTHROPOCENTRIC CHARACTER OF THE WORLD VIEW

Mankind's best conceivable world view is at most a partial picture of the world – a picture centered on man. We inevitably see the universe from a human point of view and communicate in terms shaped by the exigencies of human life. ' "Significance" in geography is measured, consciously or unconsciously,' says Hartshorne, 'in terms of significance to man'; but it is not in geography alone that man is the measure. 'Our choice of time scale for climatology,' according to Hare, 'is conditioned more by the length of our life span than by logic'; the physics of the grasshopper, Köhler points out, would be a different physics from ours.[22] 'All aspects of the environment,' as Cantril puts it, 'exist for us only in so far as they are related to our purposes. If you leave out human significance, you leave out all constancy, all repeatability, all form.'[23]

Purpose apart, physical and biological circumstances restrict human perception. Our native range of sensation is limited; other creatures experience other worlds than ours. The human visual world is richly differentiated, compared with that of most species, but others see better in the dark, perceive ultraviolet rays as colors, distinguish finer detail, or see near and distant scenes together in better focus. To many creatures the milieu is more audible and more fragrant than to us. For every sensation, moreover, the human perceptual world varies within strict limits; how bright the lightning looks, how loud the thunder sounds, how wet the rain feels at any given moment of a storm depends on fixed formulae, whose constants, at least, are unique to man.[24]

The instruments of science do permit partial knowledge of other milieus, real or hypothetical. Blood ordinarily appears a uniform, homogeneous red to the naked eye; seen through a microscope, it becomes yellow particles in a neutral fluid, while its atomic substructure is mostly empty space. But such insights do not show what it is actually like to see normally at a microscopic scale. 'The apparently standardized environment of flour in a bottle,' Anderson surmises, 'would not seem undifferentiated to any investigator who had once been a flour beetle and who knew at firsthand the complexities of flour-beetle existence.'[25] The perceptual powers and central nervous systems of many species are qualitatively, as well as quantitatively, different from man's. We can observe, but never experience, the role of surface tension and molecular forces in the lives of small invertebrates, the ability of the octopus to discriminate tactile impressions by taste, of the butterfly to sense forms through smell, or of the jellyfish to change its size and shape.

The tempo of all varieties of experience is also specific. Time yields humans on the average eighteen separate impressions, or instants, every second; images presented more rapidly seem to fuse into continuous motion. But there are slow-motion fish that perceive separate impressions up to thirty each second, and snails to which a stick that vibrates more than four times a second appears to be at rest.[26]

As with time, so with space; we perceive one of many possible structures, more hyperbolic than Euclidean.[27] The six cardinal directions are not equivalent for us: up and down, front and back, left and right have particular values because we happen to be a special kind of bilaterally symmetrical, terrestrial animal. 'It is one contingent fact about the world that we attach very great importance to things having their tops and bottoms in the right places; it is another contingent fact [about ourselves] that we attach more importance to their having their fronts and backs in the right places than their left and right sides.'[28] Up and down are everywhere good and evil: heaven and hell, the higher and lower instincts, the heights of sublimity and the depths of degradation, even the higher and the lower latitudes have ethical spatial connotations. And left and right are scarcely less differentiated.

Other species apperceive quite differently. Even the fact that physical space seems to us three-dimensional is partly contingent on our size, on the shape of our bodies (an asymmetrical torus), and, perhaps, on our semi-circular canals; the world of certain birds is effectively two-dimensional, and some creatures apprehend only one.[29]

Man's experienced world is, then, only one tree of the forest. The difference between this and the others is that man knows his tree is not the only one; and yet can imagine what the forest as a whole might be like. Technology and memory extend our images far beyond the bounds of direct sensation; consciousness of self, of time, of relationship, and of causality overcome the separateness of individual experiences.[30] Thanks to what has been likened to 'a consummate piece of combinatorial mathematics,'[31] we share the conception of a common world. Whatever the defects of the general consensus, the shared world view is essentially well-founded. 'We are quite willing to admit that there may be errors of detail in this knowledge,' as Russell wrote, referring to science, 'but we believe them to be discoverable and corrigible by the methods which have given rise to our beliefs, and we do not, as practical men, entertain for a moment the hypothesis that the whole edifice may be built on insecure foundations.'[32]

Personal geographies

Separate personal worlds of experience, learning, and imagination necessarily underlie any universe of discourse. The whole structure of the shared picture of the world is relevant to the life of every participant; and anyone who adheres to a consensus must personally have acquired some of its constituent elements. As Russell put it, 'If I believe that there is such a place as Semipalatinsk, I believe it because of things that have happened to *me*.'[33] One need not have been in Semipalatinsk; it is enough to have heard of it in some meaningful connection, or even to have imagined (rightly or wrongly) that it exists, on the basis of linguistic or other evidence. But if the place did not exist in some – and potentially in all – personal geographies, it could scarcely form part of a common world view.

INDIVIDUAL AND CONSENSUAL WORLDS COMPARED

The personal *terra cognita* is, however, in many ways unlike the shared realm of knowledge. It is far more localized and restricted in space and time: I know nothing about the microgeography of most of the earth's crust, much less than the sum of common knowledge about the world as a whole and larger parts, but a great deal about that tiny fraction of the globe I live in – not merely facts that might be inferred from general knowledge and verified by visitors, but aspects of things that no one, lacking my total experience, could ever grasp as I do. 'The entire earth,' as Wright says, is thus 'an immense patchwork of miniature *terræ incognitæ*'[34] – parts of private worlds not incorporated into the general image. Territorially, as otherwise, each personal environment is both more and less inclusive than the common realm.

COMPLEX NATURE OF PERSONAL MILIEUS

The private milieu is more complex and many aspects of it are less accessible to inquiry and exploration than is the world we all share. 'Like the earth of a hundred years ago,' writes Aldous Huxley, 'our mind still has its darkest Africas, its unmapped Borneos and Amazonian basins. . . . A man consists of . . . an Old World of personal consciousness and, beyond a dividing sea, a series of New Worlds – the not too distant Virginias and Carolinas of the personal subconscious . . . ; the Far West of the collective unconscious, with its flora of symbols, its tribes of aboriginal archetypes; and, across another, vaster ocean, at the antipodes of everyday consciousness, the world of Visionary Experience. . . . Some people never consciously discover their antipodes. Others make an occasional landing.'[35]

To be sure, the general world view likewise transcends objective reality. The hopes and fears of mankind often animate its commonsense perceptions. The supposed location and features of the Garden of Eden stimulated medieval mapmakers; many useful journeys of exploration have sought elusive El Dorados. Delusion and error are no less firmly held by groups than by individuals. Metaphysical assumptions, from original sin to the perfectibility of man, not only color but shape the shared picture of the world. But fantasy plays a more prominent role in any private milieu than in the general geography. Every aspect of the public image is conscious and communicable, whereas many of our private impressions are inchoate, diffuse, irrational, and can hardly be formulated even to ourselves.

The private milieu thus includes much more varied landscapes and concepts than the shared world, imaginary places and powers as well as aspects of reality with which each individual alone is familiar. Hell and the Garden of Eden may have vanished from most of our mental maps, but imagination, distortion, and ignorance still embroider our private landscapes. The most compelling artifacts are but pale reflections of the lapidary architecture of the mind, attempts to recreate on earth the visionary images ascribed by man to God; and every marvel unattained is a Paradise Lost.[36]

In each of our personal worlds, far more than in the shared consensus, characters of fable and fiction reside and move about, some in their own lands, others sharing familiar countries with real people and places. We are all Alices in our own Wonderlands, Gullivers in Lilliput and Brobdingnag. Ghosts, mermaids, men from Mars, and the smiles of Cheshire cats confront us at home and abroad. Utopians not only make mythic men, they rearrange the forces of nature: in some worlds water flows uphill, seasons vanish, time reverses, or one- and two-dimensional creatures converse and move about. Invented worlds may even harbor logical absurdities: scientists swallowed up in the fourth dimension, conjurors imprisoned in Klein bottles, five countries each bordering on all the others.[37] Non-terrestrial geometries, topographical monsters, and abstract models of every kind in turn lend insight to views of reality. If we could not imagine the impossible, both private and public worlds would be the poorer.

THE EXTENT TO WHICH PRIVATE WORLDS ARE CONGRUOUS WITH 'REALITY'

Though personal milieus in some respects fall short of and in others transcend the more objective consensual reality, yet they at least partly resemble it. What people perceive always pertains to the shared 'real' world; even the landscapes of dreams come from actual scenes recently viewed or recalled from memory, consciously or otherwise, however much they may be distorted or transformed. Sensing can take place without external perception (spots before the eyes; ringing in the ears), but 'so expressive a phrase as "the mind's eye" ' is current, Smythies points out, because there is 'something very like seeing about having sensory mental images'.[38]

Illusions do not long delude most of us; 'we see the world the way we see it because it pays us and has paid us to see it that way.'[39] To find our way about, avoid danger, earn a living, and achieve basic human contacts, we usually have to perceive what is there. As the Sprouts express it, 'the fact that the human species has survived (so far) suggests that there must be considerable correspondence between the milieu as people conceive it to be, and as it actually is.'[40] If the picture of the world in our heads were not fairly consistent with the world outside, we should be unable to survive in any environment other than a mental hospital. And if our private milieus were not recognizably similar to one another, we could never have constructed a common world view.

THE RANGE AND LIMITS OF PERSONAL KNOWLEDGE OF THE WORLD

However, a perfect fit between the outside world and our views of it is not possible; indeed, complete fidelity would endanger survival. Whether we stay put or move about, our environment is subject to sudden and often drastic change. In consequence, we must be able to see things not only as they are, but also as they might become. Our private milieus are therefore flexible, plastic, and somewhat amorphous. We are physiologically equipped for a wide range of environments, including some of those that we create. But evolution is slow; at any point in time, some of our sensate and conceptual apparatus is bound to be vestigial, better suited to previous than to present milieus.

As individuals, we learn most rapidly about the world not by paying close attention to a single variable, but by superficially scanning a great variety of things. 'Everyday perception tends to be selective, creative, fleeting, inexact, generalized, stereotyped' just because imprecise, partly erroneous impressions about the world in general often convey more than exact details about a small segment of it.[41] The observant are not necessarily most accurate: effective observation is never unwaveringly attentive. As Vernon emphasizes, 'changing perceptions are necessary to preserve mental alertness and normal powers of thought.' Awareness is not always conducive to survival. He who fails to see a tiger and hence does not attract its attention 'may escape the destruction which his more knowing fellow invites by the very effects of his knowledge.' So, Boulding concludes, 'under some circumstances, ignorance is bliss and knowledge leads to disaster.'[42]

Essential perception of the world, in short, embraces every way of looking at it: conscious and unconscious, blurred and distinct, objective and subjective, inadvertent and deliberate, literal and schematic.

Perception itself is never unalloyed: sensing, thinking, feeling, and believing are simultaneous, interdependent processes. A purely perceptual view of the world would be as lame and false as one based solely on logic, insight, or ideology. 'All fact,' as Goethe said, 'is in itself theory.' The most direct and simple experience of the world is a composite of perception, memory, logic, and faith. Looking down from a window, like Descartes, we say that we see men and women, when in fact we perceive no more than parts o hats and coats. The recognition of Mt. Monadnock, Chisholm demonstrates, is a conceptual as well as a visual act:[43]

> Suppose that you say to me, as we are riding through New Hampshire, 'I see that that is Mt. Monadnock behind the trees.' If I should ask, 'How do you know it's Monadnock?' you may reply by saying, 'I've been here many times before and I can *see* that it is.' . . . If I still have my doubts about what you claim to see . . . I may ask, 'What makes you *think* that's Monadnock that you see?' . . . An appropriate answer would be this: 'I can see that the mountain is shaped like a wave and that there is a little cabin near the top. There is no other mountain answering to that description within miles of here.' . . . What you now claim to see is, not that the mountain is Monadnock, but merely that it has a shape like a wave, and that there is a cabin near the top. And this new 'perceptual statement' is coupled with a statement of independent information ('Monadnock is shaped like a wave and there is a cabin near the top; no other mountain like that is within miles of here') – information acquired prior to the present perception.

And each succeeding perceptual statement can similarly be broken down into new perceptual claims and other additional information, until 'we reach a point where we find . . . no *perceptual* claim at all.'

Uniqueness of private milieus

Despite their congruence with each other and with the world as it is, private milieus do diverge markedly among people in different cultures, for individuals within a social group, and for the same person as child and as adult, at various times and places, and in sundry moods. 'The life of each individual,' concludes Delagado, 'constitutes an original and irreversible perceptive experience.'[44]

Each private world view is unique, to begin with, because each person inhabits a different milieu. 'The fact that no two human beings can occupy the same point at the same time and that the world is never precisely the same on successive occasions means,' as Kluckhohn and Mowrer put it, that 'the physical world is idiosyncratic for each individual.' Experience is not only unique; more significantly, it is also self-centered; I am part of your milieu, but not of my own, and never see myself as the world does. It is usually one's self to which the world attends; 'we will assume that an eye looks at us, or a gun points at us,' notes Gombrich, 'unless we have good evidence to the contrary.'[45]

Each private world view is also unique because everyone chooses from and reacts to the milieu in a different way. We elect to see certain aspects of the world and to avoid others. Moreover, because 'everything that we know about an object affects the way in which it appears to the eye,' no object is apt to seem quite the same to any two percipients.[46] Thus 'in some respects,' as Clark says, 'each man's appraisal of an identical situation is peculiarly his own.'[47]

CULTURAL DIFFERENCES IN ASPECTS OF WORLD VIEWS

Appraisals are, of course, profoundly affected by society and culture. Each social system organizes the world in accordance with its particular structure and requirements; each culture screens perception of the milieu in harmony with its particular style and techniques.[48]

Consider social and cultural differences in habits of location and techniques of orientation. Eskimo maps, Stefansson reports, often show accurately the number and shape of turns in routes and rivers, but neglect lineal distances, noting only how far one can travel in a day. The Saulteaux Indians do not think of circular motion, according to Hallowell; to go counterclockwise is to move, they say, from east to south to west to north, the birth order of the four winds in their mythology. To find their way about, some peoples utilize concrete and others abstract base points; still others edges in the landscape, or their own locations. The Chukchee of Siberia distinguish twenty-two compass directions, most of them tied to the position of the sun and varying with the seasons. The precise asymmetrical navigation nets of Micronesian voyagers made use of constellations and islands. Tikopians, never far from the ocean, and unable to conceive of a large land mass, use *inward* and *seaward* to help locate anything: 'there is a spot of mud on your seaward cheek.'[49] In the Tuamotus, compass directions refer to winds, but places on the atolls are located by reference to their direction from the principal settlement. Westerners are more spatially egocentric than Chinese or Balinese. The religious significance of cardinal directions controls orientation indoors and out on the North China plain, and the Balinese give all directions in terms of compass points. Where we would say 'go to the left,' 'towards me,' or 'away from the wall,' they say 'take the turn to the West,' 'pull the table southward,' or, in case of a wrong note on the piano, 'hit the key to the East of the one you are hitting.'[50] Disorientation is universally disagreeable; but inability to locate north quite incapacitates the Balinese. The English writer Stephen Potter was amazed to find that most Americans neither knew nor cared what watershed they were in, or which way rivers flowed – facts he maintained were second nature to Englishmen.[51]

Apperception of shape is also culturally conditioned. According to Herskovits, an electrical engineer working in Ghana complained that 'When a trench for a conduit must be dug, I run a line between the two points, and tell my workers to follow it. But at the end of the job, I invariably find that the trench has curves in it.' In their land 'circular forms predominate. . . . They do not live in . . . a carpentered world, so that to follow a straight line marked by a cord is as difficult for them' as drawing a perfect freehand circle is for most of us.[52] Zulus tested with the Ames trapezoidal window actually saw it as a trapezoid more often than Americans, who usually see it as a rectangle; habituated to man-made rectangular forms, we are apt unconsciously to assume that *any* four-sided object is a rectangle.[53]

Territoriality – the ownership, division, and evaluation of space – also differs from group to group. In American offices, workers stake out claims around the walls and readily move to accommodate new employees; but the Japanese gravitate toward the center of the room, and many Europeans are loathe to relinquish space once pre-empted. Eastern Mediterranean Arabs distinguish socially between right and left hand sides of outer offices, and value proximity to doors. In seeing and describing landscapes, Samoans emphasize the total impression, Moroccans the details. The Trukese sharply differentiate various parts of open spaces, but pay little attention to dividing lines or edges – a trait which makes land claims difficult to resolve.[54]

As with shapes, so with colors. Our most accustomed hues, such as blue and green, are not familiar in certain other cultures; whereas gradations scarcely perceptible to us may be part of their common experience. 'There is no such thing as a "natural" division of the spectrum,' Ray concludes. 'Each culture has taken the spectral continuum and has divided it into units on a quite arbitrary basis. . . . The effects of brightness, luminosity, and saturation are often confused with hue; and the resulting systems are emotional and subjective, not scientific.'[55] Among the Hanunóo of Mindoro, Conklin shows, the most basic color terms refer to degrees of wetness (saturation) and brightness; hue is of secondary interest.[56]

As the diverse views of color suggest, it is not merely observed pheno-mena that vary with culture, but whole categories of experience. A simple percept here may be a complex abstraction there. Groupings of supreme importance in one culture may have no relevance in another. The Aleuts had no generic name for their island chain, since they did not recognize its unity. The Aruntas organize the night sky into separate, overlapping constellations, some out of bright stars, others out of faint ones. To the Trukese, fresh and salt water are unrelated substances. The gauchos of the Argentine are said to have lumped the vegetable world into four named groups: cattle fodder, bedding straw, woody material, and all other plants – including roses, herbs, and cabbages.[57] There is no natural or best way to classify anything; all categories are useful rather than true, and the landscape architect rightly prefers a morphological to a genetic taxonomy. The patterns people see in nature also vary with economic, ethical, and esthetic values. Esthetically neutral to Americans, colors have moral connotations to Navahos; an Indian administrator's attempt to use colors as impartial voting symbols came to grief, since the Navahos viewed blue as good and red as bad.[58]

THE SIGNIFICANCE OF LINGUISTIC DIFFERENCES IN APPERCEPTION OF THE MILIEU

The very words we use incline us toward a particular view of the universe. In Whorf's now classic phrase, 'We dissect nature along lines laid down by our native languages. . . . We cut nature up, organize it into concepts and ascribe significances as we do, largely because we are parties to an agreement to organize it in this way – an agreement that holds throughout our speech community.'[59] To be sure, language also adjusts to the world view, just as environment molds vocabulary: within a single generation the craze for skiing has given us almost as many different words for *snow* as the Eskimos have.

Linguistic patterns do not irrevocably imprison the senses, but rather, Hoijer judges, 'direct perception and thinking into certain habitual channels.'[60] Things with names are easier to distinguish than those that lack them; the gauchos who used only four floristic terms no doubt saw more than four kinds of plants, but 'their perceptual world is impoverished by their linguistic one.'[61] Classifications into animate or inanimate, masculine, feminine, or neuter, and mass (sand, flour, grass, snow) or particular nouns (man, dog, thimble, leaf) variously affect the way different speech communities view things. We tend to think of waves, mountains, horizons, and martinis as though they were composed of discrete entities, but conceive surf, soil, scenery, and milk as aggregates, principally because the former terms are plurals, the latter indefinite nouns.[62]

The structural aspects of language influence ways of looking at the world more than do vocabularies. Seldom consciously employed, usually slow to change, syntax pervades basic modes of thought. In Shawnee, La Barre suggests, 'I let her have one on the noggin' is grammatically analogous to 'The damned thing slipped out of my hand.'[63] Lacking transitive verbs, Greenlanders tend to see things happen without specific cause; 'I kill him,' in their language, becomes 'he dies to me.' In European tongues, however, action accompanies perception, and the transitive verb animates every event with purpose and cause. The Hopis have subjectless verbs, but most Indo-European subjects have objects, which gives expression a dualistic, animistic stamp. In Piaget's illustration, to say 'the wind blows' 'perpetrates . . . the triple absurdity of suggesting that the wind can be independent of the action of blowing, that there can be a wind that does not blow, and that the wind exists apart from its outward manifestations.'[64] Important differences also occur within linguistic families. The French distinction between the imperfect tense (used for things and processes) and the perfect (used for man and his actions) contrasts the uniformity of nature with the uniqueness of man in a way that English does not ordinarily express.

That such distinctions can all be conveyed in English shows that language does not fetter thought; with sufficient care and effort, practically everything in any system of speech can be translated. Nevertheless, a concept that comes naturally and easily in one tongue may require awkward and tedious circumlocution in another. The difference between what is customary for some but difficult for others is apt to be crucial in terms of habits of thought and, perhaps, orders of events. European scientists, whose languages lump processes with substances as nouns, took much longer to account for vitamin deficiencies than for germ diseases, partly because 'I have a germ' was a more natural locution than 'I have a lack of vitamins.' In short, as Waismann says, 'by growing up in a certain language, by thinking in its semantic and syntactical grooves, we acquire a certain more or less uniform outlook on the world. ... Language shapes and fashions the frame in which experience is set, and different languages achieve this in different ways.'[65]

PERSONAL VARIATIONS IN ASPECTS OF THE WORLD VIEW

Private world views diverge from one another even within the limits set by logical necessity, human physiology, and group standards. In any society, individuals of similar cultural background, who speak the same language, still perceive and understand the world differently. 'You cannot see things until you know roughly what they are,' comments C. S. Lewis, whose hero on the planet Malacandra at first perceives 'nothing but colours – colours that refused to form themselves into things.'[66] But what you think you know depends both on what is familiar to you and on your proclivities. When the well-known is viewed from fresh perspectives, upside down or through distorting lenses, form and color are enhanced, as Helmholtz noted; the unexpected has a vivid, pictorial quality. On the other hand, prolonged observation may change red to apparent green, or shrink a figure in proportion to its surroundings.[67]

The purpose and circumstances of observation materially alter what is seen. The stage electrician cares how the lights look, not about the actual colors of the set; the oculist who tests my eyes is not interested in what the letters are, but in how they appear to me. Intent modifies the character of the world.[68]

Outside the laboratory, no two people are likely to see a color as the same unless they similarly identify the thing that is colored. Even then, preconceptions shape appearances, as Cornish points out: 'The exquisite colours which light and atmosphere impart to a snowy landscape are only half seen by many people owing to their opinion that "snow is really white."'[69]

Such stereotypes may outweigh other physiological facts. The United States Navy was advised to switch the color of survival gear and life jackets from yellow to fluorescent red, not so much to increase visibility as to buoy the confidence of the man lost at sea; dressed in red, he imagines, 'They can't fail to see me.'[70]

The way a landscape looks depends on all the attendant circumstances, for each sense is affected by the others. Velvet looks soft, ice sounds solid, red feels warm because experience has confirmed these impressions. The sight of gold and blond beech trees lit by sunlight made Cornish forget that he was cold; but he could not appreciate a 'frosty' blue landscape seen from a cold railway carriage. 'Quite often,' notes H. M. Tomlinson, 'our first impression of a place is also our last, and it depends solely upon the weather and the food.'[71]

Circumstance apart, each person is distinctively himself. 'The individual carries with him into every perceptual situation . . . his characteristic sensory abilities, intelligence, interests, and temperamental qualities,' according to Vernon; and his 'responses will be coloured and to some extent determined by these inherent individual qualities.'[72] Ability to estimate vertical and horizontal correctly, for example, varies with sex and personality as well as with maturity: strongminded men are better at telling which way is up than are women, neurotics, and children, whose kinesthetic sense reinforces visual perception less adequately.[73] The story of the Astronomer-Royal, Maskeleyne, who dismissed a faithful assistant for persistently recording the passage of stars more than half a second later than he did, is often told to illustrate the inevitability of perceptual divergence under the best of circumstances.[74] Each of us warps the world in his own way and endows landscapes with his particular mirages.

People at home in the same environments, for example, habitually select different modes of orientation. There is only one published 'New Yorker's Map of the United States,' but Trowbridge found a great variety of personal imaginary maps. Individual deviations of direction ranged from zero to 180 degrees off course; some were consistent, others more distorted at Times Square than at the Battery, or accurate about Albany but not about Chicago. Still others assumed that streets always point towards cardinal directions, or imagined all distant places as lying due east or west. A few know which direction they face the moment they emerge from subways and theatres, others are uncertain, still others are invariably mistaken. Lynch characterizes structural images of the environment as positional, disjointed, flexible, and rigid, depending on whether people orient themselves principally by distant landmarks, by memories of details in the landscape, by crossings, street turns, or directions, or by maps.[75]

SUBJECTIVE ELEMENTS IN PRIVATE GEOGRAPHIES

Another reason why private world views are irreducibly unique is that all information is inspired, edited, and distorted by feeling. Coins look larger to the children of the poor,[76] the feast smells more fragrant to the hungry, the mountains loom higher to the lost. 'Had our perceptions no connexion with our pleasures,' wrote Santayana, 'we should soon close our eyes on this world.'[77] We seldom differentiate among people, places, or things until we have a personal interest in them. One American town is much like another to me, unless I have a good motive for telling them apart. The most exhaustive study of photographs and ethnological evidence does not enable us to distinguish among individuals of another race with the ease, speed, and certainty generated by strong feeling. All Chinese may look the same to me, but not to the man – however foreign – with a Chinese wife. Only the flea circus owner can tell you which is which among his performing fleas.[78]

Stereotypes influence how we learn and what we know about every place in the world. My notions of Australia and Alaska are compounds of more or less objective, veridical data and of the way I happen to feel about deserts, icefields, primitive peoples, pioneers, amateur tennis, and American foreign policy. Similar evanescent images come readily to mind; to Englishmen in the 1930's, according to one writer, Kenya suggested 'gentleman farmers, the seedy aristocracy, gossip columns and Lord Castlerosse'; South Africa 'Rhodes and British Empire and an ugly building in South Parks Road and Trafalgar Square.'[79] Education and the passage of time revise but never wholly displace such sterotypes about foreign lands and people. The present consensus of teen-aged geography students in an English school is that 'South Africans break off from the Boer War to eat oranges, make fortunes from gold and diamonds, and oppress natives, under a government as merciless as the ever-present sun.'[80] Those who think of China as an abode of laundrymen, France as a place where people eat snails, and the Spanish as hotblooded are only a trifle more myopic than anyone else; it is easier to deplore such generalizations than to replace them with more adequate and convincing images.

Because all knowledge is necessarily subjective as well as objective, delineations of the world that are purely matter-of-fact ordinarily seem too arid and lifeless to assimilate; only color and feeling convey versimilitude. Besides unvarnished facts, we require fresh firsthand experience, individual opinions and prejudices. 'The important thing about truth is not that it should be naked, but what clothes suit it best.'[81] The memorable geographies are not compendious texts but interpretative studies embodying a strong personal slant. A master at capturing the essence of a place, Henry James did so by conveying 'less of its appearance than of its implications.'[82] In Blake's lines,[83]

> This Life's dim Windows of the Soul
> Distorts the Heavens from Pole to Pole
> And leads you to Believe a Lie
> When you see with, not thro', the Eye.

The ideal traveller, according to one critic, ought to be 'aware not only of the immediate visual aspect of the country he visits, its history and customs, its art and people, but also of his own relation to all these, their symbolic and mythic place in his own universal map.'[84] We mistrust science as the sole vehicle of truth because we conceive of the remote, the unknown, and the different in terms of what is near, well-known, and self-evident for us, and above all in terms of ourselves. What seems to us real and true depends 'on what we know about ourselves and not only on what we know about the external world. Indeed,' writes Hutten, 'the two kinds of knowledge are inextricably connected.'[85]

THE ROLE OF THE INDIVIDUAL PAST IN APPERCEPTION OF THE MILIEU

Personal as well as geographical knowledge is a form of sequent occupance. Like a landscape or a living being, each private world has had a career in time, a history of its own. Since personality is formed mainly in the earliest years, 'we are determined, simultaneously, both by what we were as children and by what we are experiencing now.' In Quine's words, 'We imbibe an archaic natural philosophy with our mother's milk. In the fullness of time, what with catching up on current literature and making some supplementary observations of our own, we become clearer on things. But . . . we do not break with the past, nor do we attain to standards of evidence and reality different in kind from the vague standards of children and laymen.'[86]

The earlier mode of thought continues throughout life. According to Portmann, we all remain to some extent pre-Copernican: 'The decisive early period in our contact with nature is strongly influenced by the Ptolemaic point of view, in which our inherited traits and responses find a congenial outlet. . . . Nor is the Ptolemaic world merely a phase to be outgrown, a kind of animal experience; it is an integral part of our total human quality.'[87]

As every personal history results in a particular private milieu, no one can ever duplicate the *terra cognita* of anyone else. An adult who learns a foreign word or custom does not start from *tabula rasa*, but tries to match concepts from his own language and culture – never with complete success. Among 'children, exposed serially to two cultures,' notes Mead, '. . . the premises of the earlier may persist as distortions of perception into later experience, so that years later errors in syntax or reasoning may be traced to the earlier and 'forgotten' cultural experience.'[88]

We are captives even of our adult histories. The image of the environment, as Boulding says, 'is built up as a result of all past experience of the possessor of the image. Part of the image is the history of the image itself.' I have touched on this in connection with color perception: 'The color in which we have most often seen a thing is imprinted ineffaceably on our memory and becomes a fixed attribute of the remembered image,' says Hering. 'We see through the glass of remembered colors and hence often differently than we should otherwise see them.'[89] The sitter's family invariably complain that the painter has made him look too old, because they view as a composite memory the face the painter confronts only today.

'Everyone sees the world as it was in the past, reflected in the retarding mirror of his memory.'[90]

Memory need not be conscious to influence images; as Hume pointed out, aspects of our past that we fail to recall also leave their imprint on mental maps. 'The unconscious inner world,' writes Money-Kyrle, 'is peopled by figures and objects from the past, as they are imagined often wrongly to have been.' Correct or not, recollections can virtually efface aspects of the actual contemporary landscape. Pratolini's *Il Quartiere* portrays inhabitants of a razed and empty section of Florence who instinctively continued to follow the lines of the former streets, instead of cutting diagonally across the square where buildings had stood.[91]

Memory likewise molds abstract ideas and hypotheses. Everything I know about America today is in part a memory of what I used to think about it. Having once conceived of the frontier as a cradle of democracy, it is quite another thing for me to learn that it was not, than it is for someone else to learn the 'true' fact without the old error. What we accept as true or real depends not only on what we think we know about the external world but on what we have previously believed.

Shared perspectives of whole cultures similarly incorporate the past. 'Meanings may reflect not the contemporary culture but a much older one. The landscape in general,' Lynch remarks, 'serves as a vast mnemonic system for the retention of group history and ideals.'[92]

Conclusion

Every image and idea about the world is compounded, then, of personal experience, learning, imagination, and memory. The places that we live in, those we visit and travel through, the worlds we read about and see in works of art, and the realms of imagination and fantasy each contribute to our images of nature and man. All types of experience, from those most closely linked with our everyday world to those which seem furthest removed, come together to make up our individual picture of reality.[93] The surface of the earth is shaped for each person by refraction through cultural and personal lenses of custom and fancy. We are all artists and landscape architects, creating order and organizing space, time, and causality in accordance with our apperceptions and predilections. The geography of the world is unified only by human logic and optics, by the light and color of artifice, by decorative arrangement and by ideas of the good, the true, and the beautiful. As agreement on such subjects is never perfect nor permanent, geographers too can expect only partial and evanescent concordance. As Raleigh wrote, 'It is not truth but opinion that can travel the world without a passport.'[94]

References

1   Reprinted from *Annals of the Association of American Geographers,* Vol. 51,
      No. 3 (September 1961), Washington, D.C.
          This is an expanded version of a paper read at the XIXth International
      Geographical Congress, Stockholm, August, 1960. For encouragement,
      advice, and criticism, I am grateful to George A. Cooper, Richard Harts-
      horne, William C. Lewis, William D. Pattison, Michael G. Smith, Philip L.
      Wagner, William Warntz, J. W. N. Watson, and John K. Wright, Richard F.
      Kuhns, Jr., has kindly read and commented on several drafts of the manu-
      script, and I am indebted to him for numerous suggestions and references.

2   WRIGHT, JOHN K. 'Terrae Incognitae: the Place of the Imagination in Geo-
      graphy,' *Annals of the Association of American Geographers,* Vol. 37 (1947), pp.
      1–15, on p. 15. The phrase 'The World Outside and the Pictures in Our
      Heads' is the name of the first chapter in Lippmann, Walter, *Public Opinion,*
      New York: Macmillan (1922). As my subtitle suggests, this is not a study of
      the meaning or methods of geography, but rather an essay in the theory of
      geographical knowledge. Hartshorne's methodological treatises analyze and
      develop logical principles of procedure for geography as a professional science,
      'a form of "knowing," ' as he writes, 'that is different from the ways in which
      we "know" by instinct, intuition, *a priori* deduction or revelation' (Hartshorne,
      Richard, *Perspective on the Nature of Geography,* Chicago: Rand McNally
      for *Association of American Geographers* (1959), p. 170). My epistemological
      inquiry, on the other hand, is concerned with *all* geographical thought,
      scientific and other: how it is acquired, transmitted, altered, and integrated
      into conceptual systems; and how the horizon of geography varies among
      individuals and groups. Specifically, it is a study in what Wright calls *geosophy:*
      'the nature and expression of geographical ideas both past and present . . .
      the geographical ideas, both true and false, of all manner of people – not only
      geographers, but farmers and fishermen, business executives and poets,
      novelists and painters, Bedouins and Hottentots' ('Terrae Incognitae,' p. 12).
      Because geographers are 'nowhere . . . more likely to be influenced by the
      subjective than in their discussions of what scientific geography ought
      to be' (ibid.), epistemology helps to explain why and how methodologies
      change.

3   SPARKS, JARED MS. in Sparks Collection (132, Misc. Papers, Vol. I, 1808–14),
      Harvard College Library; quoted in Brown, Ralph H., 'A Plea for Geography,
      1813 Style,' *Annals of the Association of American Geographers,* Vol. 41 (1951),
      p. 235. For similar nineteenth century views see my 'George Perkins Marsh on
      the Nature and Purpose of Geography,' *Geographical Journal,* Vol. 126 (1960),
      pp. 413–17.

4   PEATTIE, RODERICK *Geography in Human Destiny,* New York: George W.
      Stewart (1940), pp. 26–7. 'In the broadest sense,' Hartshorne, Richard notes,
      'All facts of the earth surface are geographical facts,' *The Nature of Geography:
      a Critical Survey of Current Thought in the Light of the Past,* Lancaster, Pa.:
      Association of American Geographers (1939), p. 372. On the interests and
      capacities of geographers, see Whitaker, J. Russell, 'The Way Lies Open,'
      Annals of the Association of American Geographers, Vol. 44 (1954), p. 242; and
      Meynier André, 'Réflexions sur la spécialisation chez les Géographes,'
      Norois, Vol. 7 (1960), pp. 5–12.
          Most of the physical and social sciences are, both in theory and in
      practice, more generalizing and formalistic than geography. The exceptions
      are disciplines which, like geography, are in some measure humanistic:
      notably anthropology and history. The subject-matter of anthropology is as

diversified as that of geography, and more closely mirrors the everyday concerns of man; but anthropological research still concentrates predominantly on that small and remote fraction of mankind – 'primitive' or nonliterate, traditional in culture, homogeneous in social organization – whose ways of life and world views are least like our own (Berndt, Ronald M., 'The Study of Man: an Appraisal of the Relationship between Social and Cultural Anthropology and Sociology,' *Oceania,* Vol. 31 (1960), pp. 85–99). More particularistic, more concerned with uniqueness of context than geography, history also comprehends more matters of common interest (especially the acts and feelings of individuals); but because the whole realm of history lies in the past, more historical data is secondary, derivative. Although 'geography cannot be strictly contemporary' (James, Preston E., 'Introduction: the Field of Geography,' in *American Geography: Inventory and Prospect,* Syracuse University Press, for Association of American Geographers (1954), p. 14), geography is usually *focused* on the present; direct observation of the world plays a major role in geography, a trifling one in history. In theory, at least, the remote in space is everywhere (on the face of the earth) personally accessible to us, the remote in time accessible only through memories and artifacts.

5   QUINE, W. V. 'The Scope and Language of Science,' *British Journal for the Philosophy of Science,* Vol. 8 (1957), pp. 1–17, on p. 1.

6   For various combinations of geographical facts and relationships, see Wright, John K., ' "Crossbreeding" Geographical Quantities,' *Geographical Review,* Vol. 45 (1955), pp. 52–65. For the varieties of data that comprise knowledge in general, see Carnap Rudolph, 'Formal and Factual Science,' in Feigl, Herbert and Brodbeck, May (Eds.), *Readings in the Philosophy of Science,* New York: Appleton-Century-Crofts (1953), pp. 123–8; Popper, Karl R. *The Logic of Scientific Discovery,* New York: Basic Books (1959), appendix x, pp. 420–41: Waismann, Friedrich, 'Analytic-Synthetic,' *Analysis,* Vol. 11 (1950–51), pp. 52–6; Watkins, J. W. N., 'Between Analytic and Empirical,' *Philosophy,* Vol. 32 (1957), pp. 112–31.

7   POLANYI, MICHAEL, *Personal Knowledge: Towards a Post-Critical Philosophy,* Chicago: University of Chicago Press (1958), pp. 208–9.

8   WHITTLESEY, DERWENT, 'The Horizon of Geography,' *Annals of the Association of American Geographers,* Vol. 35 (1945), pp. 1–36, on p. 14.

9   BOULDING, KENNETH E. *The Image,* Ann Arbor: University of Michigan Press (1956), p. 66.

10   WHITTLESEY, op. cit., pp. 2, 14.

11   RODBERG, L. S. and WEISSKOPF, V. F., 'Fall of Parity,' *Science,* Vol. 125 (1957), pp. 627–33; on p. 632.

12   POLANYI, *Personal Knowledge,* p. 216; Delgado, Rafael Rodriguez, 'A Possible Model for Ideas,' *Philosophy of Science,* Vol. 24 (1957), pp. 253–69, on p. 255. 'The organism has a definite capacity for information which is a minute fraction of the physical signals that reach the eyes, ears, and epidermis' (Cherry, Colin, *On Human Communication: a Review, a Survey, and a Criticism,* New York: Wiley (1957), p. 284). See also Miller, George A., 'The Magical Number Seven, Plus or Minus Two: Some Limits on Our Capacity for Processing Information,' *Psychological Review,* Vol. 63 (1956), pp. 81–97; Quastler Henry, 'Studies of Human Channel Capacity,' in Cherry, Colin, Ed., *Information Theory; Papers Read at the Third London Symposium, 1955,* New York, Academic Press (1956), pp. 361–71.

13   CHISHOLM, RODERICK M. *Perceiving: a Philosophical Study,* Ithaca: Cornell University Press (1957), p. 36. Personal surprise and disappointment are evidence to most of us that our private worlds are not, in fact, identical with the common world view (Money-Kyrle, R. E., 'The World of the Unconscious and the World of Commonsense,' *British Journal for the Philosophy of*

Science, Vol. 7 (1956), pp. 86–96, on p. 93). Birks, G. A., 'Towards a Science of Social Relations,' (ibid., Vol. 7 (1956), pp. 117–28, 206–21) shows what happens when private ideas about the world have to be adjusted to conform with the consensus.

14 For the effects of various types of illness and injury on perception and cognition of the milieu see FENICHEL, OTTO *The Psychoanalytic Theory of Neurosis,* New York: W. W. Norton (1945), p. 204; de la Garza, C. O. and Worchel, Philip, 'Time and Space Orientation in Schizophrenics,' *Journal of Abnormal and Social Psychology,* Vol. 52 (1956), pp. 191–4; Weckowicz, T. E. and Blewett, D. B., 'Size Constancy and Abstract Thinking in Schizophrenia,' *Journal of Mental Science,* Vol. 105 (1959), pp. 909–34; Eysenck, H. J., Granger, G. W., and Brengelmann, J. C., *Perceptual Processes and Mental Illness,* Institute of Psychiatry, Maudsley Monographs No. 2, London: Chapman and Hall (1957); Granger, G. W., 'Psychophysiology of Vision,' in *International Review of Neurobiology,* Vol. 1 (1959), New York: Academic Press, pp. 245–98; Paterson, Andrew and Zangwill, O. L., 'A Case of Topographic Disorientation Associated with a Unilateral Cerebral Lesion,' *Brain,* Vol. 68 (1945), pp. 188–212; Luria, A. R., 'Disorders of "Simultaneous Perception" in a Case of Bilateral Occipito-Parietal Brain Injury,' *Brain,* Vol. 82 (1959), pp. 437–49.

15 'From a perception of only 3 senses . . . none could deduce a fourth or fifth' (BLAKE, WILLIAM 'There Is No Natural Religion: First Series,' in *Selected Poetry and Prose of William Blake,* New York: Modern Library (1953), p. 99); the congenital deaf-mute does not know how music sounds even though he knows that tones exist. For the effects of sensory deprivation, see Deutsch, Felix, 'The Sense of Reality in Persons Born Blind,' *Journal of Psychology,* Vol. 10 (1940), pp. 121–40; von Fieandt, Kai, 'Toward a Unitary Theory of Perception,' *Psychological Review,* Vol. 65 (1958), pp. 315–20; Révész, Géza, *Psychology and Art of the Blind,* London: Longmans Green (1950); Young, J. Z., *Doubt and Certainty in Science: a Biologist's Reflection on the Brain,* Oxford: Clarendon Press (1951), pp. 61–6.

16 PIAGET, JEAN, *The Child's Conception of the World,* Paterson, N. J.: Littlefield and Adams (1960), pp. 248, 357.

17 *Child's Conception of the World,* pp. 384–5; *Construction of Reality in the Child,* New York: Basic Books (1954), pp. 367–9. Piaget and his associates have worked chiefly with schoolchildren in Geneva. How far their categories and explanations apply universally or vary with culture and milieu remains to be determined. Margaret Mead ('An Investigation of the Thought of Primitive Children, with Special Reference to Animism,' *Journal of the Anthropological Institute,* Vol. 62 (1932), pp. 173–90) found that Manus children rejected animistic explanations of natural phenomena. They were more matter-of-fact than Swiss children (and Manus adults) because their language was devoid of figures of speech, because they were punished when they failed to cope effectively with the environment, because their society possessed no machines too complex for children to understand, and because they were barred from animistic rites until past puberty. In Western society, on the other hand, 'the language is richly animistic, children are given no such stern schooling in physical adjustment to a comprehensible and easily manipulated physical environment, and the traditional animistic material which is decried by modern scientific thinking is still regarded as appropriate material for child training' (p. 189). (Indeed, books written for children show clearly that adults think children *ought* to be animists.) Elsewhere, however, child animism appears to be significant and tends to decline with age and maturity (Jahoda, Gustav, 'Child Animism: I. A. Critical Survey of Cross-Cultural Research,' *Journal of Social Psychology,* Vol. 47 (1958), pp. 197–212).

18   The decline of sensory perception leads the elderly to make false judgments about the environment, and often arouses feelings of isolation and apathy. See Weiss, Alfred D., 'Sensory Functions,' and Braun, Harry W., 'Perceptual Processes,' in Birren, J. E. (Ed.), *Handbook of Aging and the Individual: Psychological and Biological Aspects,* Chicago: University of Chicago Press (1959), pp. 503–42 and 543–61, respectively.

19   BÉLANGER, MARCEL 'J'ai choisi de devenir géographe,' *Revue Canadienne de Géographie,* Vol. 13 (1959), pp. 70–2, on p. 70. This version of Heraclitus is in Russell, Bertrand, *A History of Western Philosophy,* New York: Simon and Schuster (1945), p. 45; a somewhat different phrasing appears in Plato's 'Cratylus' (*The Dialogues of Plato,* Jowett, B., tr., 2 vols. New York: Random House (1937), Vol. 1, p. 191).

20   EISELEY, LOREN *The Firmament of Time,* New York: Atheneum (1960), p. 38. Scientists at the French Academy in the seventeenth century denied evidence for the fall of meteorites, obvious to most observers, because they opposed the prevalent superstition that meteorites came by supernatural means. For this and other instances of how 'the most stubborn facts will be set aside if there is no place for them in the established framework of science,' see Polanyi, *Personal Knowledge,* pp. 138–58.

21   NICOLSON, MARJORIE HOPE *Mountain Gloom and Mountain Glory: The Development of the Aesthetics of the Infinite,* Ithaca: Cornell University Press (1959); quotations from Thomas Burnet's *Sacred Theory of the Earth,* London (1684) on pp. 198, 206.

22   HARTSHORNE, *Perspective on the Nature of Geography,* p. 46; Hare, F. Kenneth, 'The Westerlies,' *Geographical Review,* Vol. 50 (1960), p. 367; Köhler, Wolfgang, *The Place of Value in a World of Facts* (1938 Ed.) New York: Meridian Books (1959). 'There is no ultimate source for the physicist's concepts,' adds Köhler, 'other than the phenomenal world' (p. 374).

23   CANTRIL, HADLEY 'Concerning the Nature of Perception,' *Proceedings of the American Philosophical Society,* Vol. 104 (1960), pp. 467–73, on p. 470. 'The environment with which we are concerned is not the one which is measured in microns, nor that which is measured in light years, but that which is measured in millimeters or meters . . . [It] is not that of particles, atoms, molecules, or anything smaller than crystals. Nor is it that of planets, stars, galaxies, or nebulae. The world of man . . . consists of matter in the solid, liquid, or gaseous state, organized as an array of surfaces or interfaces between matter in these different states', Gibson, James J., 'Perception as a Function of Stimulation,' in Koch, Sigmund, (Ed.), *Psychology: a Study of a Science, Study I. Conceptual and Physiological Foundations,* New York: McGraw-Hill (1959), pp. 456–501, on p. 469.

24   STEVENS, S. S. 'To Honor Fechner and Repeal His Law,' *Science,* Vol. 133 (1961), pp. 80–86. For human and animal sensory and perceptual capacities see Portmann, Adolf, 'The Seeing Eye,' *Landscape,* Vol. 9 (1959), pp. 14–18; Baumgardt, Ernest, 'La vision des insectes,' *La Nature,* Vol. 90 (1960), pp. 96–9; Griffin, Donald R., 'Sensory Physiology and the Orientation of Animals,' *American Scientist,* Vol. 41 (1953), pp. 208–44; Wells, M. J., 'What the Octopus Makes of It: Our World from Another Point of View,' *Advancement of Science,* Vol. 17 (1961), pp. 461–71; von Uexküll, J., *Umwelt und Innenwelt der Tiere,* Berlin: Julius Springer (1909); von Frisch, Karl, *Bees: their Vision, Chemical Senses and Language,* Ithaca: Cornell University Press (1950), pp. 8–12, 34–6; Griffin, Donald R., *Listening in the Dark: the Acoustic Orientation of Bats and Men,* New Haven: Yale University Press, (1958); von Frisch, K., 'Über den Farbsinn der Insekten,' and Viaud, G., 'La Vision chromatique chez les animaux (sauf les insectes),' in *Mechanisms of Colour Discrimination,* New York: Pergamon Press (1960), pp. 19–40 and

41–66, respectively; Mueller, Conrad G., 'Visual Sensitivity,' Pollack, Irwin, 'Hearing,' and Beidler, Lloyd M., 'Chemical Senses,' in *Annual Review of Psychology*, Vol. 12 (1961), pp. 311–34, 335–62, and 363–88, respectively.

25   ANDERSON, EDGAR 'Man as a Maker of New Plants and New Plant Communities,' in Thomas, William L. Jr., (Ed.), *Man's Role in Changing the Face of the Earth,* Chicago: University of Chicago Press (1956), pp. 763–77, on p. 776.

26   VON UEXKÜLL, J. *Theoretical Biology,* London: Kegan Paul (1926), pp. 66–8; von Bertalanffy, Ludwig, 'An Essay on the Relativity of Categories,' *Philosophy of Science,* Vol. 22 (1955), pp. 243–63, on p. 249.

27   Visual space is Euclidean only locally; for normal observers with binocular vision, space has a constant negative curvature corresponding with the hyperbolic geometry of Lobachevski. See Luneburg, Rudolph K., *Mathematical Analysis of Binocular Vision,* Princeton: Princeton University Press (1947), and Blank, Albert A., 'Axiomatics of Binocular Vision. The Foundations of Metric Geometry in Relation to Space Perception,' *Journal of the Optical Society of America,* Vol. 48 (1958), pp. 328–34. But under optimal conditions, Gibson maintains, perceptual space is Euclidean ('Perception as a Function of Stimulation,' pp. 479–80).

28   MAYO, BERNARD, 'The Incongruity of Counterparts,' *Philosophy of Science,* Vol. 25 (1958), pp. 109–15, on p. 115; Takala, Martti 'Asymmetries of Visual Space' *Annales Academiae Scientarium Fennicae,* Ser. B., Vol. 72, No. 2, Helsinki (1951). Because gravity, unlike bilateral symmetry, affects everything on earth, people adapt more rapidly to distorting spectacles that invert up and down than to those that reverse left and right. 'Hochberg, Julian E., 'Effects of the Gestalt Revolution: the Cornell Symposium on Perception,' *Psychological Review,* Vol. 64 (1957), pp. 74–6.

29   WHITROW, G. J. 'Why Physical Space Has Three Dimensions,' *British Journal for the Philosophy of Science,* Vol. 6 (1955), pp. 13–31; Good, I. J., 'Lattice Structure of Space-Time,' ibid., Vol. 9 (1959), pp. 317–19. On righteousness as a function of height, see Pederson-Krag, Geraldine, 'The Use of Metaphor In Analytic Thinking,' *Psychoanalytic Quarterly,* Vol. 25 (1956), p. 70. Of the opposition of right and left, Robert Hertz remarks, 'If organic asymmetry had not existed, it would have had to be invented', *Death and the Right Hand,* Glencoe, Ill.: Free Press (1960), p. 98 and Rodney Needham concludes that 'in every quarter of the world it is the right hand, and not the left, which is predominant' ('The Left Hand of the Mugwe: an Analytic Note on the Structure of Meru Symbolism,' *Africa,* Vol. 30 (1960), p. 20).

30   'Il y a une différence fondamentale dans la "façon d'être-au-monde" de l'homme et de l'animal supérieur: ce fait d'être comme englué dans l'objet, de ne pouvoir le survoler, dû . . . à l'unité que fait l'animal avec le monde. . . . L'animal ne peut transcender le réel immédiat', Filloux, Jean-C., 'La nature de l'univers chez l'animal,' *La Nature,* Vol. 85 (1957), pp. 403–7, 438–43, 490–3, on p. 493. Analogous points are made by Boulding, *Image,* p. 29; Révész, Géza, 'The Problem of Space, with Particular Emphasis on Specific Sensory Spaces,' *American Journal of Psychology,* Vol. 50 (1937), p. 434 n; Cassirer, Ernst, *An Essay on Man: Introduction to a Philosophy of Human Culture,* New Haven: Yale University Press (1944) and New York: Doubleday Anchor Books, n. d., p. 67.

31   BORN, MAX *Natural Philosophy of Cause and Chance,* Oxford: Clarendon Press (1949), p. 125. For a critique on the formation of the common world view, see McKinney, J. P., 'The Rational and the Real: Comment on a Paper by Topitsch, E., *Philosophy of Science,* Vol. 24 (1957), pp. 275–80.

32   RUSSELL, BERTRAND *Our Knowledge of the External World,* New York: Mentor (1960), p. 56. The question whether the so-called real world actually

exists lies beyond the scope of this paper. As Russell says (p. 57), 'universal scepticism, though logically irrefutable, is practically barren'. Sanity and survival depend on the 'sense of being a solid person surrounded by a solid world' (Money-Kyrle, op. cit. [see footnote 13], p. 96).

33   RUSSELL, BERTRAND *Human Knowledge: Its Scope and Limits,* New York: Simon and Schuster (1948), p. xii. But see Feibleman, J. K., 'Knowing about Semipalatinsk,' *Dialectica,* Vol. 9 (1955), pp. 3–4.

34   WRIGHT, 'Terrae Incognitae,' pp. 3–4. On the other hand, the consensual universe of discourse comprises elements from an infinite number of private worlds – not only those of existing persons, but also those that might conceivably be held. No square mile of the earth's surface has been seen from every possible perspective, but our view of the world in general is based on assumptions about such perspectives, as analogous with those that have been experienced. The Amazon basin would look different in design and detail from the top of every tree within it, but we know enough of the general character and major variations of that landscape to describe it adequately after climbing – or hovering in a helicopter over – a small fraction of its trees.

35   HUXLEY, A. *Heaven and Hell,* New York: Harper (1955), pp. 1–3.

36   'Man's spatialization of his world . . . never appears to be exclusively limited to the pragmatic level of action and perceptual experience. . . . Human beings in all cultures have built up a frame of spatial reference that has included the farther as well as the more proximal, the spiritual as well as the mundane, regions of their universe' (Hallowell, A. Irving, *Culture and Experience,* Philadelphia: University of Pennsylvania Press (1955), pp. 187–8). The genesis of these mental maps is explained in Money-Kyrle, R. E., *Man's Picture of His World: a Psycho-analytic Study,* London: Duckworth (1960); see p. 171. For instances of theological location, see Erich, Isaac, 'Religion, Landscape and Space, *Landscape,* Vol. 9 (1959–60), pp. 14–18. The visionary transfiguration of the everyday world by means of gems and precious stones is a central theme in Huxley, *Heaven and Hell.*

37   ABBOTT, EDWIN A. *Flatland, a Romance of Many Dimensions,* New York: Dover (1952), London (1884), is a classic of two-dimensional life. For samples of the impossible, see Clifton Fadiman (Ed.), *Fantasia Mathematica,* New York: Simon and Schuster (1958), notably Gardner, Martin, 'The Island of Five Colors,' pp. 196–210.

38   SMYTHIES, J. R. 'The Problems of Perception,' *British Journal for the Philosophy of Science,* Vol. 11 (1960), pp. 224–38, on p. 229; see also his *Analysis of Perception,* New York: Humanities Press (1956), pp. 81–105. 'The widespread belief that a mirage is something unreal, a sort of trick played on the eyes, is wrong. The picture a mirage presents is real but never quite accurate' (Gordon, James H., 'Mirages,' in *Smithsonian Institution, Annual Report for 1959,* Washington, D.C. (1960), pp. 327–46, on p. 328). On the form and content of landscapes in dreams, mirages and hallucinations, and their relations with 'reality,' see Fisher, Charles, 'Dreams, Images and Perception: A Study of Unconscious-Preconscious Relationships,' *Journal of the American Psychoanalytic Association,* Vol. 4 (1956), pp. 5–48; Fisher, Charles and Paul, I. H., 'The Effect of Subliminal Visual Stimulation on Images and Dreams: a Validation Study,' ibid., Vol. 7 (1959), pp. 35–83; Knapp, Peter Hobart, 'Sensory Impressions in Dreams,' *Psychoanalytic Quarterly,* Vol. 25 (1956), pp. 325–47; Chari, C. T. K., 'On the "Space" and "Time" of Hallucinations,' *British Journal for the Philosophy of Science,* Vol. 8 (1958), pp. 302–6; Huxley, Aldous, *The Doors of Perception,* London: Chatto and Windus (1954).

39   BOULDING, *The Image,* p. 50.

40   SPROUT, HAROLD and SPROUT, MARGARET, *Man-Milieu Relationship Hypotheses in the Context of International Politics,* Princeton University Center of In-

ternational Studies (1956), p. 61. The essential correspondence between the perceived and the actual milieu is stressed in Gibson, James J., *The Perception of the Visual World,* Boston: Houghton Mifflin (1950).

41   Ibid., p. 10; see also Miller, 'The Magical Number Seven' (see footnote 12), pp. 88–9. We can count only a few of the stars or raindrops we see, beyond which everything becomes blurred; but our vagueness could not be rectified by looking longer or more carefully: 'the blur is just as essential a feature of sense perception as other features are. . . . Sense perception is inexact in a very different sense from that which . . . a map is inexact' (Waismann, 'Analytic-Synthetic' (see footnote 6), *Analysis,* Vol. 13 (1953), pp. 76–7). Types and ranges of perception and learning are surveyed in Vernon, M. D., *A Further Study of Visual Perception,* Cambridge: The University Press (1952); Hirst, R. J., *The Problems of Perception,* London: Allen & Unwin (1959); Drever, James, 2d., 'Perceptual Learning,' in *Annual Review of Psychology,* Vol. 11 (1960), pp. 131–60. For a concise theoretical review, see Bevan, William, 'Perception: Development of a Concept,' *Psychological Review,* Vol. 65 (1958), pp. 34–55.

42   VERNON, MAGDALEN D. 'Perception, Attention and Consciousness,' *Advancement of Science,* Vol. 16 (1959), pp. 111–23, on p. 120; Boulding, *The Image,* p. 169. The classic story illustrating the virtues of ignorance of the geographical environment is in Koffka, Kurt, *Principles of Gestalt Psychology,* New York: Harcourt-Brace (1935), pp. 27–8.

43   CHISHOLM, *Perceiving,* pp. 55–8; for the Descartes argument, from his *Meditations,* see pp. 154–6. See also Royce, Joseph R., 'The Search for Meaning,' *American Scientist,* Vol. 47 (1959), pp. 515–35; Cassirer, Ernst, *The Philosophy of Symbolic Forms; Volume Three: The Phenomonology of Knowledge,* New Haven: Yale University Press (1957), p. 25.

44   DELGADO, 'A Possible Model for Ideas' (see footnote 12), p. 255.

45   KLUCKHOHN, CLYDE and MOWRER, O. H., ' "Culture and Personality": a Conceptual Scheme,' *American Anthropologist,* Vol. 46 (1944), pp. 1–29, on p. 13; Gombrich, E. H., *Art and Illusion: A Study in the Psychology of Pictorial Representation,* New York: Pantheon Books (1960), (Bollingen Series, XXXV, No. 5), p. 276. See also Sprout, *Man-Milieu Relationship Hypotheses,* p. 18.

46   CORNISH, VAUGHAN *Geographical Essays,* London: Sifton, Praed (1946), pp. 78–9.

47   CLARK, K. G. T. 'Certain Underpinnings of Our Arguments in Human Geography,' *Transactions of the Institute of British Geographers,* No. 16 (1950), pp. 15–22, on p. 20.

48   'Only exceptionally do we react in any literal sense to stimuli. . . . Rather, we react to our interpretations of stimuli. These interpretations are derived in considerable part from our culture and from each person's specific experiences in that culture', Kluckhohn, Clyde, 'The Scientific Study of Values and Contemporary Civilization,' *Proceedings of the American Philosophical Society,* Vol. 102 (1958), pp. 469–76, on p. 469. The classic case study is Waterman, T. T. *Yurok Geography,* Berkeley: University of California Press (1920); see also Erikson, Erik H., *Childhood and Society,* New York: W. W. Norton (1950), pp. 141–60. The literature on world views is ably summarized by Kluckhohn, Clyde, 'Culture and Behavior,' in Lindzey, Gardner (Ed.), *Handbook of Social Psychology,* 2 Vols., Cambridge, Mass.: Addison-Wesley (1954), Vol. 2, pp. 921–76.

49   FIRTH, RAYMOND *We, the Tikopia: a Sociological Study of Kinship in Primitive Polynesia,* London: Allen & Unwin (1936), p. 19. For the previous examples, see the letter from Vilhjalmur Stefansson quoted in Raisz, Erwin, *General Cartography,* New York: Mc Graw-Hill (1948), p. 4; Hallowell, *Culture and*

Experience, p. 201; Waldemar Bogoras, *The Chukchee. II. – Religion.* The Jesup North Pacific Expedition, Vol. VII, Memoir of the American Museum of Natural History, Leiden: Brill; New York: Stechert (1907), pp. 303–4. The cultural and environmental contexts of orientation are considered at length in Hallowell, op. cit., pp. 184–202, and Lynch, Kevin, *The Image of the City,* Cambridge, Mass.: Technology Press and Harvard University Press (1960), pp. 123–33.

50   DANIELSSON, BENGT *Work and Life in Raroia: an Acculturation Study from the Tuamotu Group, French Oceania,* London: Allen & Unwin (1956), pp. 30–1; Bodde, Derk, 'Types of Chinese Categorical Thinking,' *Journal of the American Oriental Society,* Vol. 59 (1939), pp. 200–19, on p. 201 n; Belo, Jane, 'The Balinese Temper, '*Character and Personality,* Vol. 4 (1935), pp. 120–46, quote on pp. 126–7. Haugen, Einar, 'The Semantics of Icelandic Orientation,' *Word,* Vol. 13 (1957), pp. 447–59, shows how cardinal orientation can depend on one's location with reference to an ultimate destination; thus an Icelander heading for the southern tip of the island is going 'south' even if his coastwise route happens to be southwest or west. For other early or 'primitive' methods of pathfinding, see Adler, B. F., *Maps of Primitive Peoples* (St. Petersburg, 1910), abridged by de Hutorowicz, H., *Bulletin of the American Geographical Society,* Vol. 43 (1911), pp. 669–79; Bogoras, Waldemar, 'Ideas of Space and Time in the Conception of Primitive Religion,' *American Anthropologist,* Vol. 27 (1925), pp. 212–15; Jaccard, Pierre, *Le Sens de la direction et l'orientation lointaine chez l'homme,* Paris: Payot (1932); Gatty, Harold, *Nature Is Your Guide: How to Find Your Way on Land and Sea by Observing Nature,* New York: Dutton (1958).

51   'I hardly found an American who knew which watershed he was in, which left me, as an Englishman who is uneasy unless he knows which ocean will receive his urination, somewhat scandalized', *Potter on America,* London: Hart-Davis (1956), p. 13.

52   HERSKOVITS, MELVILLE J. 'Some Further Comments on Cultural Relativism,' *American Anthropologist,* Vol. 60 (1958), pp. 266–73, on pp. 267–8.

53   ALLPORT, GORDON W. and PETTIGREW, THOMAS F. 'Cultural Influence on the Perception of Movement: the Trapezoidal Illusion among Zulus,' *Journal of Abnormal and Social Psychology,* Vol. 55 (1957), pp. 104–13. Under 'optimal' visual conditions, however, the Zulus mistook the trapezoid for a rectangle almost as often as Americans do, perhaps because most of them recognized it as a model of a Western-type window (Slack, Charles W., 'Critique on the Interpretation of Cultural Differences in the Perception of Motion in Ames's Trapezoidal Window,' *American Journal of Psychology,* Vol. 72 (1959), pp. 127–31). Another aspect of spatial perception which varies with culture is surveyed in Michael, Donald N., 'Cross-Cultural Investigations of Closure,' in Beardslee, David C. and Wertheimer, Michael (Eds.), *Readings in Perception,* New York: Van Nostrand (1958), pp. 160–70.

54   HALL, EDWARD T. *The Silent Language,* New York: Doubleday (1959), pp. 197–200, and 'The Language of Space,' *Landscape,* Vol. 10 (1960), pp. 41–5; Gladwin, Thomas and Sarason, Seymour B., *Truk: Man in Paradise,* Viking Fund Publications in Anthropology No. 20, New York: Wenner-Gren (1953), pp. 225–6, 269–70.

55   RAY, VERNE F. 'Techniques and Problems in the Study of Human Color Perception,' *Southwestern Journal of Anthropology,* Vol. 8 (1952), pp. 251–9, quotes on pp. 258–9; see also Ray, 'Human Color Perception and Behavioral Response,' *Transactions of the New York Academy of Sciences,* Ser. II, Vol. 16 (1953), pp. 98–104.

56   CONKLIN, HAROLD C. 'Hanunóo Color Categories,' *Southwestern Journal of Anthropology,* Vol. 11 (1955), pp. 339–44.

57 The Argentine data are cited in Karl Vossler, 'Volkssprachen und Weltspra-chen,' *Welt und Wort,* Vol. 1 (1946), pp. 97–101, on p. 98, and discussed by Basilius, Harold, 'Neo-Humboldtian Ethnolinguistics,' *Word,* Vol. 8 (1952), pp. 95–105, on p. 101. For the rest, see Gladwin and Sarason, *Truk,* p. 30, and Lynch, op. cit., pp. 131–2.

58 HALL *Silent Language,* pp. 132–3. Many landscape features exist as separate entities only in our minds. As Gombrich says (*Art and Illusion,* p. 100), 'There is a fallacy in the idea that reality contains such features as mountains and that, looking at one mountain after another, we slowly learn to generalize and form the abstract idea of mountaineity.' Owing to the 19th-century popularity of Alpine climbing, the English standard of mountains changed dramatically: for Gilbert White the 800-foot Sussex Downs were 'majestic mountains'; today anything below 2,000 feet is at best a 'hill'; Cornish, Vaughan, *Scenery and the Sense of Sight,* Cambridge: University Press (1935), p. 77.

59 'Science and Linguistics' (1940), in *Language, Thought and Reality; Selected Writings of Benjamin Lee Whorf,* Carroll, John B., (Ed.), Cambridge, Mass.: Technology Press; New York: Wiley; London: Chapman and Hall (1956), p. 213.

60 HOIJER, HARRY 'The Relation of Language to Culture,' in Kroeber, A. L., et al., *Anthropology Today: an Encyclopedic Inventory,* Chicago: University of Chicago Press (1953), pp. 554–73, on p. 560.

61 HARDIN, GARRETT, 'The Threat of Clarity,' *ETC.: a Review of General Seman-tics,* Vol. 17 (1960), pp. 269–78, on p. 270. Similarly, people more readily perceive and identify colors that have widely-known specific names (like blue and green) than those that do not; Brown, Roger W. and Lenneberg, Eric H., 'A Study in Language and Cognition,' *Journal of Abnormal and Social Psychology,* Vol. 49 (1954), pp. 454–62.

62 English terms, like "sky, hill, swamp," persuade us to regard some elusive aspect of nature's endless variety as a distinct THING, almost like a table or chair'; Whorf, *Language, Thought and Reality,* p. 240; see also pp. 140–1. But Brown, Roger W., *Words and Things,* Glencoe, Ill.: Free Press (1958), pp. 248–52, maintains that the distinction between mass and specific nouns makes perceptual sense and corresponds well with perceived reality.

　　One can easily, as critics of Whorf have pointed out, make too much of such distinctions. The fact that the word for *sun* is masculine in French and feminine in German, whereas that for *moon* is feminine in French and masculine in German, cannot easily be correlated with the habits of thought or *Weltanschauung* of either people. The fact that in Algonquian languages the gender class of 'animate' nouns includes such words as *raspberry, stomach,* and *kettle,* while 'inanimate' nouns include *strawberry, thigh,* and *bowl* does not imply 'that speakers of Algonquian have a shrine to the raspberry and treat it like a spirit, while the strawberry is in the sphere of the profane'; Greenberg, Joseph H., 'Concerning Inferences from Linguistic to Nonlin-guistic Data,' in Hoijer, Harry, (Ed.), *Language in Culture,* American Anthro-pological Association, Memoir No. 79, Chicago (1954), pp. 3–19, on pp. 15–16. In short, 'If grammar itself was once founded on an unconscious metaphysic, this linkage is now so vestigial as to have no appreciable bearing on the structure of philosophic ideas'; Feuer, Lewis S., 'Sociological Aspects of the Relation Between Language and Philosophy,' *Philosophy of Science,* Vol. 20 (1953), pp. 85–100, on p. 87. This may be true of most aspects of language, and of philosophical ideas in their broadest sense. On the other hand, the fact that English-speaking mid-Victorians clad table and piano legs in ruffs and deplored direct reference to them in mixed company was not a necessary outgrowth of prudery but depended also on the metaphorical extension of the word for human limbs to furniture – a connection not made by speakers of other languages. In this respect, language certainly altered the English – and still more the American – home landscape.

63   LA BARRE, WESTON *The Human Animal,* Chicago: University of Chicago Press (1954), p. 204. See Whorf, op. cit., p. 235.

64   PIAGET, *Child's Conception of the World,* p. 249. A book has been written to tell parents how to answer a child who asks such questions as 'What does the wind do when it's not blowing?' Purcell, Ruth, 'Causality and Language Rigidity,' *ETC.,* Vol. 15 (1958), pp. 175–80, on p. 179. Whorf, *Language, Thought, and Reality,* compares Hopi language and thought with that of 'Standard Average European' in several papers, e.g., pp. 57–64, 134–59, 207–19.

Unlike most psychologists and anthropologists, geographers have tended to assume, with positivistic philosophers, that we could rid ourselves of animistic and teleological kinds of explanation and ways of looking at the world by substituting other words and phrases in our language. 'Ritter's teleological views . . . though they colour every statement he makes, yet do not affect the essence,' according to H. J. Mackinder: 'it is easy to re-state each proposition in the most modern evolutionary terms', President's address, Section E, British Association for the Advancement of Science, *Report of the 65th Annual Meeting,* Ipswich, 1895, London: Murray (1895), pp. 738–48, on p. 743; that is, Mackinder found it easy to accommodate Ritter's brand of determinism to his own. For other views on the relation between teleological language and habits of thought, see Sprout, *Man-Milieu Relationship Hypotheses,* pp. 27–8; Bernatowicz, A. J., 'Teleology in Science Teaching,' *Science,* Vol. 128 (1958), pp. 1402–5; Nagel, Ernest, Teleological 'Explanation and Teleological Systems,' in Feigl and Brodbeck, *Readings in the Philosophy of Science,* pp. 537–58; Sinnhuber, Karl A., 'Karl Ritter 1779–1859,' *Scottish Geographical Magazine,* Vol. 75 (1959), p. 160.

65   WAISMANN 'Analytic-Synthetic' (see note 6), *Analysis,* Vol. 13 (1952), p. 2. 'The fact that an ethnologist can describe in circumlocution certain distinctions in kin that are *customarily* made by the Hopi does not alter his conclusion that the Hopi name kin and behave toward them differently from us'; Hoijer, Harry, review of Brown, Roger W., *Words and Things,* in *Language,* Vol. 35 (1959), pp. 496–503, on p. 501. For a range of views on metalinguistics see Lenneberg, Eric H., 'Cognition in Ethnolinguistics,' *Language,* Vol. 29 (1953), pp. 463–71; Fearing, Franklin, 'An Examination of the Conceptions of Benjamin Whorf in the Light of Theories of Perception and Cognition,' in Hoijer, (Ed.), *Language and Culture,* pp. 47–81; Rapoport, Anatol and Horowitz, Arnold, 'The Sapir-Whorf-Korzybski Hypothesis: a Report and a Reply,' *ETC.,* Vol. 17 (1960), pp. 346–63.

66   LEWIS, C. S. *Out of the Silent Planet,* New York: Macmillan (1952), p. 40.

67   For HELMHOLTZ, see CASSIRER, *Philosophy of Symbolic Forms,* Vol. 3, pp. 131–2. On changes in apparent color and size, see Cornsweet, T. N., *et al.,* 'Changes in the Perceived Color of Very Bright Stimuli,' *Science,* Vol. 128 (1958), pp. 898–9; Jameson, Dorothea and Hurvich, Leo M., 'Perceived Color and Its Dependence on Focal, Surrounding, and Preceding Stimulus Variables,' *Journal of the Optical Society of America,* Vol. 49 (1959), pp. 890–8: Köhler, Wolfgang, *Dynamics in Psychology,* New York: Grove Press, 1960, pp. 84–6.

68   'Without the conception of the individual and his needs, a distinction between illusion and "true" cognition cannot be made'; English, Horace B., 'Illusion as a Problem in Systematic Psychology,' *Psychological Review,* Vol. 58 (1951), pp. 52–3. The size and shape of objects seem appropriately and necessarily constant, but most of us can afford to be 'fooled' by the apparent bending of a stick half-submerged in water.

69   CORNISH *Scenery and the Sense of Sight,* p. 22. On the dissimilar impressions of identical shapes and colors, see Duncker, Karl, 'The Influence of Past Experience upon Perceptual Properties,' *American Journal of Psychology,* Vol. 52 (1939), pp. 255–65; Bruner, Jerome S. and Postman, Leo, 'Expectation and the Perception of Color,' *American Journal of Psychology,* Vol. 64 (1951), pp. 216–27; Kapp, Arthur, 'Colour-Image Synthesis with Two Unorthodox Primaries,' *Nature,* Vol. 184 (1959), pp. 710–13; Land, Edwin H., 'Experiments in Color Vision,' *Scientific American,* Vol. 200, No. 5 (May 1959), pp. 84–99.

70   TOMLINSON, H. H. 'Navy Research on Color Vision,' *Naval Research Reviews* (October 1959), p. 19.

71   *The Face of the Earth; with Some Hints for Those About to Travel,* Indianapolis: Bobbs-Merrill (1951), p. 52.

72   VERNON *A Further Study of Visual Perception,* p. 255.

73   WITKIN, H. A., *et al., Personality Through Perception: an Experimental and Clinical Study,* New York: Harper (1954); Witkin, Herman A., 'The Perception of the Upright,' *Scientific American,* Vol. 200, No. 2 (February 1959), pp. 51–6.

74   POLANYI *Personal Knowledge,* pp. 19–20, recounts this and similar episodes. Eysenck, H. J., 'Personality and the Perception of Time,' *Perceptual and Motor Skills,* Vol. 9 (1959), pp. 405–6, shows that introverts and extroverts clock the passage of time at systematically different rates. See also Kirk, John R. and Talbot, George D., 'The Distortion of Information,' *ETC.,* Vol. 17 (1959), pp. 5–27; Wallace, Melvin and Rubin, Albert I., 'Temporal Experience,' *Psychological Bulletin,* Vol. 57 (1960), pp. 221–3.

75   TROWBRIDGE, C. C. 'On Fundamental Methods of Orientation and "Imaginary" Maps,' *Science,* Vol. 38 (1913), pp. 891–2; Lynch, *Image of the City,* pp. 88–9, 136–7. See also Ryan, T. A. and Ryan, M. S., 'Geographical Orientation,' *American Journal of Psychology,* Vol. 53 (1940), pp. 204–15.

76   BRUNER, JEROME S. and GOODMAN, CECILE C. 'Value and Need as Organizing Factors in Perception,' *Journal of Abnormal and Social Psychology,* Vol. 42 (1947), pp. 33–4. Further tests yielded significant differences in size estimation principally when coins were judged from memory (Carter, Launor F. and Schooler, Kermit, 'Value, Need, and Other Factors in Perception,' *Psychological Review,* Vol. 56 (1949), pp. 200–7), but the initial general hypothesis has been substantially confirmed (Bruner, J. S. and Klein, George S., 'The Functions of Perceiving: New Look Retrospect,' in Kaplan, Bernard and Wapner, Seymour (Eds.), *Perspectives in Psychological Theory: Essays in Honor of Heinz Werner,* New York: International Universities Press (1960), p. 67).

77   SANTAYANA, GEORGE *The Sense of Beauty; Being the Outline of Aesthetic Theory* (1896), New York: Dover Publications (1955), p. 3. 'I cannot,' writes Gardner Murphy, 'find an area where hedonistic perceptual theory cannot apply'; 'Affect and Perceptual Learning,' *Psychological Review,* Vol. 63 (1956), p. 7.

78   EHRENZWEIG, ANTON *The Psycho-Analysis of Artistic Hearing and Vision: an Introduction to a Theory of Unconscious Perception,* New York: Julian Press (1953), p. 170. See also Gibson, James J. and Gibson, Eleanor J., 'Perceptual Learning: Differentiation or Enrichment?' *Psychological Review,* Vol. 62 (1955), pp. 32–41. Science is more often apt to be accelerated 'by the passionate, and even the egocentric partisan bias of researchers in favour of their own chosen methods or theories' than by disinterested impartiality (Gallie, W. B., 'What Makes a Subject Scientific?' *British Journal for the Philosophy of Science,* Vol. 8 (1957), pp. 118–39, on p. 127). Metaphysical doctrines which can neither be proved nor disproved 'play regulative roles in scientific

thinking' because 'they express ways of seeing the world which in turn suggest ways of exploring it'; Watkins, J. W. N., 'Confirmable and Influential Metaphysics,' *Mind,* Vol. 67 (1958), pp. 344–65, on pp. 360, 356.

79 GREEN, GRAHAM 'The Analysis of a Journey,' *Spectator,* Vol. 155 (1935), pp. 459–60. 'Even if we remember as many facts about Bolivia as about Sweden, this has little relevance to the relative importance of these two countries in our psychological world', MacLeod, Robert B., 'The Phenomenological Approach to Social Psychology,' *Psychological Review,* Vol. 54 (1947), p. 206.

80 HADDON, JOHN 'A View of Foreign Lands,' *Geography,* Vol. 65 (1960), pp. 286–9, on p. 286. If their view of South Africa is recognizable, the students' impressions of America leave more to be desired: 'America is a country of remarkably developed, highly polished young women, and oddly garbed, criminally inclined young men travelling at great speed in monstrous cars along superhighways from one skyscraping city to the next; the very largest cars contain millionaires with crew-cuts; everyone is chewing gum' (p. 286). Such stereotypes die hard, even face to face with contrary realities, as one traveller noted among Americans in Russia; Dettering, Richard, 'An American Tourist in the Soviet Union: Some Semantic Reflections,' *ETC.,* Vol. 17 (1960), pp. 173–201.

81 BRAIN, RUSSELL *The Nature of Experience,* London: Oxford University Press (1959), p. 3.

82 ALVAREZ, A. 'Intelligence on Tour,' *Kenyon Review,* Vol. 21 (1959), pp. 23–33, on p. 29; see James, Henry, *The Art of Travel,* Morton D. Zaubel (Editor), New York: Doubleday (1958). The virtues of the personal slant on description are discussed by Stark, Freya, 'Travel Writing: Facts or Interpretation?' *Landscape,* Vol. 9 (1960), p. 34, and Wright, 'Terrae Incognitae', (see footnote 2), p. 8.

83 BLAKE, WILLIAM 'The Everlasting Gospel' (c. 1818), in *Selected Poetry and Prose,* pp. 317–28, on p. 324.

84 GREEN, PETER 'Novelists and Travelers,' *Cornhill Magazine,* Vol. 168 (1955), pp. 39–54, on p. 49. 'Man can discover and determine the universe inside him only by thinking it in mythical concepts and viewing it in mythical images', Cassirer, *Philosophy of Symbolic Forms,* Vol. 2, p. 199; see also pp. 83, 101.

85 HUTTEN, ERNEST H. '(review of) *Sigmund Freud: Life and Work,* Vol. 3, by Jones, Ernest,' *British Journal for the Philosophy of Science,* Vol. 10 (1959), p. 81. Experience always influences the severest logic: no matter how convinced a man is that heads and tails have exactly equal prospects, he is not likely to bet on tails if heads has come up the previous fifty times; Popper, *Logic of Scientific Discovery,* pp. 408, 415; Cohen, John, *Chance, Skill, and Luck: the Psychology of Guessing and Gambling,* Baltimore: Penguin Books (1960), pp. 29, 191. See also Topitsch, Ernst, 'World Interpretation and Self-Interpretation: Some Basic Patterns,' *Daedalus,* Vol. 88 (Spring, 1959), p. 312.

86 HUTTEN, op. cit., p. 79; Quine, 'Scope and Language of Science', (see footnote 5), p. 2.

87 PORTMANN, 'The Seeing Eye,' *Landscape,* Vol. 9 (1959), p. 18. See also Hebb, D. O., *The Organization of Behavior; a Neuropsychological Theory,* New York: Wiley (1949), p. 109; Deutsch, Felix, 'Body, Mind, and Art,' in *The Visual Arts Today,* special issue of *Daedalus,* Vol. 89 (Winter, 1960), pp. 34–45, on p. 38; Tauber, Edward S. and Green, Maurice R., *Prelogical Experience: an Inquiry into Dreams and Other Creative Processes,* New York: Basic Books (1959), p. 33.

88 MARUYAMA, MAGORAH 'Communicable and Incommunicable Realities,' *British Journal for the Philosophy of Science,* Vol. 10 (1959), pp. 50–4; Mead, Margaret, 'The Implications of Culture Change for Personality Develop-

ment,' *American Journal of Orthopsychiatry,* Vol. 17 (1947), pp. 633–46, on p. 639.

89   BOULDING *The Image,* p. 6; Hering, Ewald, *Grundzüge der Lehre vom Lichtsinn,* Berlin: Springer (1920), pp. 6 ff.; quoted in Cassirer, *Philosophy of Symbolic Forms,* Vol. 3, pp. 132–3. See also Bruner, J. S. and Postman, Leo, 'On the Perception of Incongruity: a Paradigm,' in Beardslee and Wertheimer, *Readings in Psychology,* pp. 662–3; Bruner and Klein, 'Functions of Perceiving' (see footnote 78), p. 63.

90   GROSSER, MAURICE *The Painter's Eye,* New York: Rinehart (1951), p. 232.

91   HUME, DAVID *A Treatise of Human Nature* (1739), Book I, part iv, sec. vi; Money-Kyrle, *Man's Picture of His World,* p. 107; Pratolini, Vasco, *The Naked Streets,* New York: A. A. Wyn (1952), p. 204.

92   KLUCKHOHN 'Culture and Behavior', (see footnote 48), p. 939; Lynch, *Image of the City,* p. 126.

93   As natives of places we acquire and assimilate information differently from what we do we do as travellers; and personal observation, whether sustained or casual, yields impressions different in quality and impact from those we build out of lectures, books, pictures, or wholly imaginary visions. The climates of each of these modes of geographical experience, and the kind of information they tend to yield about the world, will be considered in a series of essays to which this one is meant to be introductory.

94   Quoted in WEDGWOOD, C. V. *Truth and Opinion: Historical Essays,* London: Collins (1960), p. 11.

5 The Marxist approach to the geographical environment[1]

Ian M. Matley
Michigan State University, East Lansing, U.S.A.

ABSTRACT   The geographical environment and its influence on the development of society has remained a dominant theme in Soviet geography for
several decades. The ideas of the nineteenth century Russian historians, such
as Mechnikov, on the role of the geographical environment influenced early
Marxist scholars, in particular Plekhanov, whose theories formed the basis for
Baranskiy's early philosophy of geography. Attempts by Baranskiy and others
to develop a balanced assessment of the relationship between man and the
physical environment were negated in the 1930's by a dogmatic pronouncement by Stalin which denied any environmental influence on the development of society. However, since the end of World War II a group of
geographers led by Baranskiy, Saushkin and Anuchin have attempted to rein-
troduce the theme of the geographical environment as an object of study for
geography, while opposing the Stalinist underemphasis on the natural environment. The arguments of this group have been strengthened by a pronouncement by the Communist Party in 1963, rejecting Stalin's earlier ruling
and recognizing that the geographical environment, although not a determining factor, does exercise a certain influence on the development of society.
The environmentalist debate is thus far from dead in the Soviet Union and its
outcome may result in significant changes in Soviet geographical methodology.

No other single concept in Marxist geographical methodology has excited
more controversy and attention than that of the geographical environment and
its influence on the development of human society. On two separate occasions
the Communist Party has made ex cathedra pronouncements on the nature
of the geographical environment, with a profound effect on the development
of Soviet geographical thought, and in particular on the great Soviet debate
on the unity of the science. For many years the Soviet outlook on Western
geography was colored by the idea that geographical determinism was the
basic philosophy on which the Western science was founded. An understanding, therefore, of Soviet theories of the role of the geographical environment
in the development of society is necessary before some of the other major
issues in Soviet geographical thought become clear.

Before the revolution several Russian scholars, mainly historians, had advanced environmentalist views. The famous historian Sergey Solov'yev (1820–1879) introduced the environmentalist theme in his major work on the history of Russia. He stated that[2]

> the nature of a country has important significance in history in the influence which it has on the national character. A bountiful nature, rewarding lavishly even a slight effort by man, dulls his activity, both physical and mental. . . . However, with a nature which is relatively not rich, which is monotonous and therefore not cheerful, in a climate relatively severe, among a people always active, busy and practical, a feeling of refinement cannot develop with progress; in such circumstances the character of a nation is more severe, inclined more to the useful than the pleasant; the inclination towards art, towards the embellishment of life is weaker, social pleasures more material, and all this together, without other outside influences, results in the exclusion of the woman from the society of men, which, it stands to reason, leads further to greater austerity of customs.

Vasiliy Klyuchevskiy (1841–1911) expressed environmentalist views similar to those of Solov'yev. A chapter of his history of Russia was devoted to a discussion of the effect of the physical environment on the history of the country, and in particular he drew attention to the effect of the forest, steppe, and rivers on the history of the Russian people. 'Each of these separately by itself took a lively and original part in the formation of the life and ideas of Russian man.'[3] He felt that the river had played the greatest role in the life of the Russian people and that[4]

> the river is even a kind of mentor of feeling for order and social consciousness in the people. . . . The Russian river has trained the inhabitants on its banks for communal life and sociability. . . . The river has bred a spirit of enterprise, a habit of association for common work, has compelled meditation and inventiveness, has brought together scattered parts of the population, has trained one to feel oneself a member of society, to deal with strange people, to observe their customs and interests, to exchange merchandise and experience, to know ways of life. Such was the varied nature of the historical work of the Russian river.

Klyuchevskiy's views on the importance of the role of the river in history were also reflected in the work of Lev Mechnikov (1838–1888), one of the most famous of pre-revolutionary Russian historians. The title of his best-known work *Tsivilizatsiya i velikiye istoricheskiye reki* (Civilization and the Great Historical Rivers) indicates his purpose which was to demonstrate that the ancient civilizations of the Eastern world had been born and had developed along rivers, and that the river as a synthesis of all physical-geographical conditions was one of the factors which determined the development of society. In spite of his environmentalist tendencies, however, Mechnikov was a great historian who attempted to give a balanced view of human history and who was willing to recognize also that man had played a part in the formation of the geographical environment. As he himself stated he was[5]

> far from *geographical fatalism*, which the theory of the influence of the environment is often accused of being. In my estimation one should seek the principle of the rise and character of primitive institutions and their subsequent evolution not in the environment itself but in the relations between the environment and the capacity of the people inhabiting a given environment for co-operation and solidarity.

The work and ideas of the pre-revolutionary historians are of interest mainly because of their effect on early Marxist scholars and hence on a later generation of Soviet geographers. It is interesting that most pre-revolutionary Russian geographers seem to have been relatively free from environmentalist thinking. A. Voyeykov (1841–1916), the great climatologist, showed few traces of environmentalism in his work, and D. Anuchin (1843–1923), perhaps the greatest geographer of the Czarist period, saw man as one of the productive forces of nature and regretted that the development of geography in the past had led to the 'placing in the foreground of the influence of geographical conditions on man and his culture – influences for the exact study of which there were neither means nor methods.'[6] It was the historians and not the geographers who were largely responsible in influencing the thought of the most eminent of early Russian Marxist philosophers, G. Plekhanov (1856–1918).

Plekhanov was influenced to a considerable extent by Mechnikov, whose work he valued highly. In a review of Mechnikov's book *Tsivilizatsiya i velikiye istoricheskiye reki* Plekhanov pointed out that Mechnikov saw the difference in the historical roles played by different groups of human beings as being caused by the influence of the geographical environment. This, Plekhanov said, was not a novel idea, as many earlier thinkers had arrived at the same conclusion. However, many of them had erred in their interpretation of the effect of the environment on the development of human society because they had looked for traces of the influence of the environment in the psychology or even the physiology of various human groups rather than in their social life. 'In order to assess correctly the influence of the geographical environment on the historical destiny of mankind it is necessary to trace how the natural environment acts upon the type and nature of the *social environment* which in the closest way determines the character and proclivities of man.'[7] This Mechnikov had done in his account of the influence of nature on the rise of the ancient civilizations of the East.

Adapting Mechnikov's basic ideas Plekhanov attempted to develop the theory of geographical determinism into something more than the earlier scholars had made of it. His idea was to introduce the theory into the framework of the Marxist analysis of society and by so doing to take the analysis a stage further than Marx himself did. Marx had claimed that the basis of social life and of historical development lay in the mode of production. This mode of production included productive relations and productive forces. Plekhanov, taking this argument a stage further, stated that[8]

> ... with this answer of Marx the whole question of the development of the economy therefore boils down to the question, by what principle is the development of the productive forces, which are found in ordered society, explained. In its final form it is determined above all by the nature of the geographic environment.

He further claimed that[9]

> ... the peculiarities of the geographical environment determine the development of the productive forces, the development of the productive forces determines the development of the economic forces and directly after them also all the other social relations.

Although attacked by later Soviet scholars for adopting this theory, Plekhanov's eminence and reputation as a Marxist saved him from the scorn poured on the head of others guilty of the sin of 'geographical deviation.' Criticism is restrained, and it is carefully pointed out that in some of his other works he did not commit the same error.[10] However, there is little doubt that Plekhanov saw in the geographical environment one of the major influences in the development of society. As Wetter points out, 'Marxism for him is the application to social development of the Darwinian theory of the adaptation of biological species to the conditions of the environment.'[11]

Plekhanov's ideas on the role of the geographical environment in the development of society influenced several Marxists, including the unfortunate Bukharin who, along with Trotskiy, became one of the chief scapegoats of the late 1920's and early 1930's. Bukharin saw the development of society determined not by the direct influence of the geographical environment on society but by its indirect influence through the productive forces on the technical base of society.[12] This was, of course, directly taken from Plekhanov. However, Bukharin's environmentalism did not seem to be very strong; he emphasized the influence of man upon nature, and he saw the physical environment chiefly as a source of raw materials and energy.[13] Although Bukharin has been attacked for many heretical views, he is in fact rarely accused of environmentalism in Soviet literature.

A more interesting subscriber to the ideas of Plekhanov was N. Baranskiy, who became perhaps the greatest and best known of modern Soviet economic geographers. In 1926 he made the following statement, in which the influence of Plekhanov can be clearly seen:[14]

> The influence of natural conditions on man is taken into account in the Marxist scheme of social development to the extent in which these natural conditions form a natural basis for the 'material productive forces,' which determine the 'productive relations' and through them the 'legal and political superstructure' and finally, the 'forms of social conscience.'

As support for his views Baranskiy further quoted one of the most interesting passages from Marx on the subject of the geographical environment, one which he revived again in later years. In *Capital* Marx stated that[15]

> ... where nature is too lavish she keeps man in hand, like a child in leading strings. She does not impose on him any necessity to develop himself. It is not the tropics with their luxuriant vegetation but the temperate zone that is the mother-country of capital. It is not the mere fertility of the soil, but the differentiation of the soil, the variety of its natural products ... which forms the real basis of social division of labour.

Marx followed this with examples of the necessity placed on man to control his environment before developing production and cited the cases of Egypt, Lombardy and Holland, each having to control water in order to obtain soil for agriculture. Baranskiy commented that[16]

> ... it is particularly characteristic that Marx chooses just these cases in which the influence of natural conditions affects the acceleration of the formation of the very 'economic stuff' of human society. Marx's point of view on the 'geographical factor' is nevertheless far from exaggerating its significance, a fault of several non-Marxist geographers, as it is also from any minimizing of it. In explaining the differences in the historical development of various countries, the difference in natural conditions must, of course, be taken into account in the first place. However, acccording to Marx, the geographical environment acts on man not directly but through the agency of the social environment; in this neither the degree nor the direction itself of this action remain constant but change according to their subordination to the changes in the social environment itself.

This mixture of geographical and economic determinism is lifted straight from Plekhanov, with little addition on the part of Baranskiy. He partially admitted this by giving Plekhanov and Bukharin in the bibliography of his article. Plekhanov had made use of the same quotation from Marx and had drawn the conclusion from it that[17]

> ... the character of the surrounding natural environment determines the character of man's productive activities, the character of his means of production. However, the means of production determine just as inevitably the mutual relations of people in the proceess of production. . . .

In turn, these productive relations determine the structure of society, and thus one can say that 'the character of the natural environment determines the character of the social environment.'[18] This is in effect the explanation given by Baranskiy, though he seems to hesitate to carry the argument quite as far as Plekhanov does, and is cautious about overemphasizing the role of the natural environment. Through the years Baranskiy's position on this subject has remained remarkably consistent.

The acceptance of the doctrine of environmentalism by many Marxist and other scholars in the 1920's led to an article in three parts by Wittfogel, published in *Pod znamenem marksizma*, in 1929.[19] It took the form of an attempt to study the relationship between man and nature, with special reference to labor, and was illustrated with quotations from Marx. Wittfogel attacked geographical determinism as a scientific weapon of the bourgeois revolution, and in particular singled out Montesquieu, Herder, Hegel, and Ritter for condemnation. The article was partially intended as an attack on the environmentalist views of the left-wing scholar Graf, who thought that 'the mistake of Karl Marx and many of his pupils consists in the fact that they have transferred the whole center of gravity to economic and social facts and have lost sight of the primary facts of nature.'[20] Graf elsewhere criticized Marx's inability to think in geographical terms and his neglect of 'questions of relationship between earth space and the development of culture.'[21] In fact, neither environmentalists nor anti-environmentalists have found much in the works of either Marx or Engels to support their views either way.

Although Wittfogel was one of the earliest critics of Marxist environmentalism, Soviet scholars point to the pre-Marxist revolutionary scholar Chernyshevskiy (1828–89) as an early opponent of geographical determinism as an explanation of the development of society.[22] In an article written in 1887–8 he discussed races, the difference between peoples and their national character, and the general character of elements promoting progress without once raising the bogy of the geographical environment.[23]

The 'geographical deviation' died hard in the Soviet Union. Pokrovskiy, the famous Marxist historian, who fell into disfavor shortly before his death in 1932, claimed that[24]

> ... in basic outline the development of all the countries in the world is similar. If one country is not similar to another, if the Eskimos up to the present have not emerged from the stone age and the people of the little prominence of Asia, called 'Western Europe,' live in the machine age, the cause of this is above all climate and other geographical peculiarities.

The case against Pokrovskiy as an environmentalist is not, however, completely straightforward. At the beginning of the 10th edition of his *Brief History of Russia*, published in 1931, he stated that[25]

> ... man ... depends on nature, and the rate of progress of a given nation depends to a considerable degree on its natural environment. But this power of nature over man is not unlimited. Man can master nature and it is not nature that is the foundation of his economic activity. Nature is only the material for this activity. The foundation of the economic activity of man is man's labor. ... It is easy to foresee that in the future, when science and technique have attained to a perfection which we are as yet unable to visualize, nature will become soft wax in his hands which he will be able to cast into whatever form he chooses.

There is, in fact, little evidence to show that Pokrovskiy was guilty to any great degree of environmentalism. His view seemed to be that at an early stage of human development the environment has a more decisive influence than at a later stage, when improved tools and techniques enable man to control nature to an increasing degree. A similar view is expressed by Chernyshevskiy who is extolled as an anti-environmentalist: 'Especially at the beginning of national existence, geographical conditions exercise all their power on national activities.'[26] At a later stage of development, however, geographical and climatic factors are no longer of such significance. This was essentially the theory put forward by the English historian Buckle (1821–62), author of the *History of Civilization in England*, who saw the history of man as a progressive liberation of himself from the control of nature, ending in the ultimate control of nature itself by man. Buckle's views have been consistently attacked by Soviet scholars, who resent his claims to have discovered the scientific laws of history.

Pokrovskiy had the misfortune to fall foul of Stalin, and his timely demise saved him from a complete fall from grace. The crime of 'geographical deviation' is just another one in a list of heresies, real and imaginary, of which he has been accused before and after his death.

Although writers such as Wittfogel were critical of environmentalist views, no hard and fast line on the subject developed until the late 1930's, by which time the Party had formulated its ideas on the forces determining the development of society.

The first official pronouncement on the subject was contained in an article written by Stalin in September 1938, which condensed the Marxist approach to environmentalism into a few sentences. This statement has appeared in the original form, or with only slightly different wording, in various publications as the official Soviet view.[27]

The key sentences in Stalin's article stated that[28]

> ... the geographical environment indisputably is one of the constant and necessary conditions of society and, of course, influences the development of society; it accelerates or retards the speed of development of society. However, its influence is not a *determining* influence, inasmuch as the changes and development of society proceed incomparably faster than the changes and development of the geographical environment. ... Geographical environment cannot be the chief cause of development of that which undergoes fundamental changes in the course of a few hundred years.

Stalin added that the force determining the material life of society, the character of the social system, and its development is the[29]

> ... method of procuring the means of life necessary for human existence, the mode of production of material values – food, clothing, footwear, houses, fuel, instruments of production, etc., – which are indispensable for the life and development of society.

The publication of this statement by Stalin effectively silenced any Soviet scholar who might still harbor any environmentalist ideas and unleashed a flood of attacks on Western geographers and historians for their erroneous ideas on the subject. The fact that most Western geographers were by this time abandoning the crude environmentalism of the 1920's did not seem to be recognized.

Along with these attacks on environmentalism went a new approach to the physical environment which was revealed in grandiose plans for the alteration of the natural environment, personally backed by Stalin. The most famous of these was the 'Stalin Plan for the Transformation of Nature' (Stalinskiy plan preobrazovaniya prirody) first publicized in *Pravda* in October 1948, which aimed at changing the natural conditions in the steppe and forest steppe lands of European Russia by the planting of forest belts. By the end of 1953 little was heard of this plan, but the attitude of mind encouraged by this type of planning remained and manifested itself as a tendency on the part of some Soviet geographers to regard themselves as the ultimate masters of nature.

Typical of Soviet geographical literature of the Stalin period is the work of Voskanyan on the Armenian Academy of Sciences, published in 1956 and translated into Russian.[30] It is highly propagandistic in tone and, in the words of the author, criticized the 'anti-scientific and reactionary doctrines, following the line of idolization of the role of the geographical environment, the foundation of racism and the justification of aggressive wars.'[31] He attempted to link the then current Soviet approach to the physical environment with the original teachings of Marx and Engels with little success. It is difficult to find any specific statement that might be applied to environmentalism in the works of Marx, although Engels, with his greater interest in the natural sciences, is more quotable. For example, Voskanyan quoted Engels in *Dialectics of Nature* on the activities of man in changing the natural features of Germany.[32] There is little doubt that the founders of Marxism looked on man as the modifier of nature but never felt strongly enough to make a special point of it.

Many Soviet textbooks on geography, and especially on economic geography, of the 1950's include a paragraph or two on the subject of environmentalism, stating the Marxist view and attacking the bourgeois approach. In the introduction to the first volume of his *Economic Geography of the U.S.S.R.*[33] Breyterman discussed the Soviet and Western views on the subject in a succinct fashion. After quoting almost verbatim from Stalin's definition of the influence of environment on society, he went on to state that all changes in the social structure are caused by changes in the means of production, which in the final reckoning are explained by the development of the productive powers of society. He did not deny that natural conditions in some parts of the Soviet Union were more favorable to rapid economic development than in others, but claimed that this had no significance until society was in a suitable economic position to exploit these natural conditions and resources. Changes in techniques, e.g., the use of coal instead of wood as a fuel, can change man's utilization of his natural environment completely.[34] However, he did not deny that the natural environment influences economic development, especially that of agriculture and in particular in the southeast and north of the Soviet Union. Likewise, weather affects such varying industries as hydro-electric power, the peat industry, and transportation. Natural conditions can, however, change under the influence of economic activity. Capitalist countries allow their soil resources to deteriorate, e.g., soil erosion in the United States, whereas the Soviet Union is carrying out a 'planned reformation of the physical geographical environment on a large scale.'[35]

Contrasting it to the Soviet approach, Breyterman discussed the development of environmentalism in the West along the same lines as Voskanyan. He traced its history, beginning with Herodotus and Aristotle. He saw geographical determinism as a step forward from the explanation of social phenomena as the will of the gods or of kings. The theory was revived again in the eighteenth century, especially by Montesquieu, and in the nineteenth century it was still upheld by such scholars as List, Buckle, Ratzel, and others as an[36]

> ... anti-scientific reactionary trend. ... The reason for its tenacity of life is the tendency of bourgeois authors to appeal to geographical environment or biological laws for proof of the suitability or unsuitability (in accordance with the interests of the bourgeoisie at a given moment) of historically complex social relations.

Bourgeois economists, for example, attempt to show that manufacturing is more suitable for temperate countries and that tropical countries should confine themselves to producing colonial goods. These reactionary ideas on the subject of the geographical environment have led to the development of theories of geopolitics.

In conclusion, Breyterman points out that the Soviet geographer must not, however, ignore the role of the physical geographical environment. Too often 'general sociological schemes, equally suitable for any country and any region'[37] have been put forward in place of a deeper and more detailed study of the local conditions in a particular region or country.

It might be thought that such a strong reaction to geographical determinism might lead Soviet geographers along the road to possibilism. This was in fact claimed by a Canadian geographer, Sebor,[38] but his article was reviewed by I. Gerasimov, for many years the ideological spokesman for Soviet geography, and its claims rejected on the grounds that possibilism is simply a form of geographical determinism.[39] In fact, Sebor's claim that Soviet geographers are possibilists is not borne out by the evidence. Saushkin has also demonstrated that possibilism is just as incompatible with Marxism as is environmentalism, although he admits that 'the idea that geographers should be interested in nature only from the point of view of the possibilities that it offers to the economy is rather widely held among Soviet economists.'[40] He agreed that possibilism completely rejects determinism, but nevertheless considered that it errs in regarding nature as only an object for exploitation. His explanation of the origin of possibilism was, to say the least of it, novel. After pointing out that the relations between man and nature are not similar to the relations between exploiter and exploited, he went on to say that 'in the capitalist world it frequently happens that bourgeois scientists try to apply the social relationships of their system, based on exploitation, to natural phenomena; hence, the idea of geographical "possibilism."'[41]

Leszczycki, director of the Institute of Geography of the Polish Academy of Sciences, has also commented on the Marxist approach to possibilism. He pointed out that[42]

> ... possibilists in general do not reckon with the laws of social development and explain this development rather by the progress of civilization and human thought (inventions, technology, the genius of individuals) and therefore differ significantly from Marxists. In time various views developed among the followers of possibilism. Some of them looked for the causes of changes, arising from the relation between man and the geographical environment, with the manner of man's activity and even in changes in the structure of society and the state. Nevertheless, most continued to assign the chief role to human genius and the development of technology.

He also attacked views held mainly by economists and statisticians engaged in planning, who believe in the power of an able society to transform the geographical environment in an unlimited and free manner. These 'economic' views, ascribing an insignificant role to the geographical environment in the development of society, he called 'geographical nihilism.'[43] (The 'nihilists' referred to are presumably allied to the 'leftists' who attempted in the 1930's in the Soviet Union to annex economic geography as a branch of political economy, and traces of whose influence still seems to linger in the Soviet bloc countries. They also undoubtedly included geographers of the Stalin school of thought.)

Modern Marxist scholars are thus opposed to the belief in the absolute will of man, who is able to alter the material world when and how he wishes and in the power of strong personalities who are capable of changing nature and society. This type of free will does not exist according to the Marxist, as man must take the laws of both nature and social development into account. Viewed in this light it is clear that possibilism in its more extreme forms cannot be acceptable to the Soviet geographer any more than is geographical determinism.

Baranskiy also attacked 'geographical nihilism' as being 'theoretically incorrect,' as, denying any significance for natural conditions and removing human society from the material environment of its existence and development, it inevitably leads to Men'shevik idealism.[44] The Marxist-Leninist position is as far from this view as it is from that of environmentalism, or 'geographical fatalism.'[45]

The Bol'shevik approach to nature is above all an 'engineering' approach: it does not bow before nature, like Buddhists, but studies as a whole its uses for the welfare of mankind. This is the mountain which it is necessary to conquer. He who is accustomed to bow before nature will advise its circumvention. He who is accustomed to ignore nature will scale it without looking; for him it is best if it is covered with clouds. The sensible man begins by inspecting and studying the mountain and then on the basis of this study draws up a plan of conquest of the difficulties presented by it and conquers them.

We have already seen Baranskiy's reference to Marx in 1926 on the subject of the geographical environment. In 1960 he used the same quotation in an attack on the 'leftists' who deny any correlation between nature and the structure of society.[46] This group consisted mainly of economists and economic geographers with 'nihilist' tendencies, the main body of which was routed before the war. However, in spite of the assertion that 'we are completely convinced that on the question of the role of the geographical environment the views of the leftists have absolutely nothing in common with the position of Marx and Engels,'[47] one feels that Baranskiy was not only attacking the 'leftists' but a large proportion of conventional Marxist geographers of the Stalin school as well. Although Baranskiy did not interpret the quotation from Marx in such a deterministic fashion in 1960 as he did in 1926, he still demonstrated a very positive approach to the geographical environment.

Baranskiy has had more success with 'geographical' quotations from Marx than most other Soviet geographers. There is no doubt that Marx and Engels now and again mentioned the natural environment as one of the factors to be taken into account in history or the study of a region or country, but without particular emphasis. Baranskiy, however, managed to build up within the Marxist framework a strong case for a more realistic approach to the natural environment by Soviet geographers. His attitude is best summed up by his warning that in taking account of man's activity in controlling and changing nature one must remember:[48]

1   that this activity is far from limitless (there can be no talk of any 'escaping from nature' or of any miracles);
2   that with the development of human society, not only its power increases but also its requirements; and
3   that with the development and complication of technique and the 'dominance of man over nature' this link with nature not only does not diminish but on the contrary becomes more intense and complicated, as the increase of the 'dominance of man over nature' in the scientific meaning of this process does not mean the freeing of man from nature, but only a wider, fuller and expedient utilization of this nature. Thus, for example, when man did not use coal or oil in his economy, his economy could in no way depend on deposits of these minerals. Now when these deposits have entered the picture, their distribution has

begun to affect in the strongest way the distribution of industry and transportation, and from this also the distribution of the human economy as a whole.

These ideas of Baranskiy contain an echo of Ratzel as well as a rejection of the ideas of Buckle, Chernyshevskiy, and Pokrovskiy.

Baranskiy's views on the relationship between man and the geographical environment have influenced other Soviet geographers, especially those at Moscow University. One of the latter, V. Anuchin, has carried Baranskiy's ideas a stage further and has successfully challenged the Stalinist approach to the natural environment.

Anuchin's first attack was launched in 1957 when he discussed the nature of the geographical environment along novel lines, claiming that it included 'not only "pure" nature but also man and the results of his activities.'[49] By 'geographical environment' Anuchin means that part of the 'landscape envelope' of the earth in which direct links are found between human society and the rest of nature. (The 'landscape envelope' of the earth is seen as the surface of the earth, including the bottom of the seas and oceans, along with the hydrosphere and atmosphere, this 'landscape envelope' providing conditions for the rise and development of life, including its highest form, human society.) In other words it is the part of nature which has been changed by man's activities and which in turn exercises certain influences on the development of human society. This is in contrast with the use of the term 'geographical environment' by Stalin, which referred to the whole of nature, or 'pure' or 'dehumanized' nature as Anuchin calls it. Anuchin at the same time accused Soviet geographers of 'indeterminism' or an underestimation of the influence of the geographical environment on the life of society.

In 1960 Anuchin broadened the basis of his attack in his book *Teoreticheskiye problemy geografii* (Theoretical problems of geography), where he elaborated his ideas with the aim of demonstrating the unity of geography.[50] He devoted considerable space to a review of the history of geographical determinism, and analyzed the theory from a Marxist point of view.[51] He realized that 'the influence of geographical conditions on the life of society is beyond all doubt. It would be completely wrong to see in the assertion of this influence a manifestation of geographical determinism: the scientific worthlessness of the latter does not rest in the fact that it recognizes the influence of the geographical factor on the development of society as a basic condition of this development, but in the fact that this influence was elevated to the position of a basic principle of development.'[52] Anuchin saw the main fault of geographical determinism as lying in its mechanical nature. The environmentalists saw certain prime influences as producing certain consequences, whereas in fact the problem of cause and effect is a very complex one. It is this mechanical aspect of determinism, according to Anuchin, that Marxism rejects and not determinism in the broader sense of the word.[53]

Anuchin was more worried about indeterminism in Soviet geography than he was about determinism. He saw the former as an antiscientific, idealistic view of the world, arising from subjective idealism.[54]

> In our time in many capitalist countries, the indeterminists present 'theories' of the absolute will of man who can, as if at his own discretion, alter the material world without taking any factor into account (voluntarism). They attempt theoretically to prove the necessity for the rule of powerful personalities, 'heroes,' capable, as of their own will, of changing nature and society.

This point of view has also considerable support in the Soviet Union and must be firmly rejected.

These 'indeterminists' are obviously the 'geographical nihilists' and 'leftists' attacked by Baranskiy and are presumably to be found mainly among the ranks of economic geographers and economists and others who 'erect an insurmountable wall between human society and the rest of nature.'[55] It is in essence a continuation of Baranskiy's plea for a balanced appraisal by economic geographers of the natural environment, adapted by Anuchin for special purposes, namely, a demonstration of the unity of nature and society in the geographical environment and thus the emergence of a single object of study for all geographers.

Early in 1961 Anuchin's book was rejected as a doctoral thesis in the Geography Faculty of Leningrad University, the author being severely taken to task by such prominent Soviet geographers as Kalesnik, Konstantinov, and Semevskiy for his unorthodox views.[56] On the other hand Anuchin had the full support of Baranskiy and Saushkin. The former was full of praise for Anuchin's book and pointed out that[57]

> Anuchin comes out, on the one hand, against geographical determinism and on the other hand against a nihilistic underestimation of the environment, which he aptly calls geographical indeterminism. The geographical environment in itself does not determine the historical process of social production, but in a number of cases it exerts, in intermediary form, a decisive influence on the development of the economy.

He claimed that Anuchin is the first to have formulated this proposition in such a clear form, but presumably is only referring to Soviet geographers. Baranskiy took the opportunity to point out that a 'nihilistic attitude' to the natural environment has led to a stereotyped approach to economic management, the ignoring of local peculiarities of natural and economic conditions, and the inadequate participation of geographers in economic construction and planning.[58]

The second of Anuchin's influential supporters, Saushkin, had, for some years before Anuchin's first methodological article appeared, been a supporter of a unified geography and of a more realistic appraisal of the role of the natural environment in economic geography. In 1958 Saushkin published his *Vvedeniye v ekonomicheskuyu geografiyu* (Introduction to economic geography), which showed his thinking on the major issues of Soviet geography as being very close to those of Baranskiy and Anuchin. He emphasized that[59]

> the geographical environment is not a passive and uniform witness of human history. The history of nature and the history of society are linked in the closest manner and influence the development of one another. This becomes even clearer the more that society changes the surrounding geographical environment.

He went on to say that although man at every stage in his history has become more capable of changing nature in his own interests[60]

> he can never change the laws of nature or formulate new laws of nature. The whole mastery of man over nature is confined to understanding her laws and being able to apply them in his interests.

These natural laws are thus in no way dependent on man's will and the incorrect comprehension of this basic tenet of Marxism leads to serious mistakes in theory and practice. In particular the concept of man as the master and exploiter of nature and of nature as only an object of exploitation leads to recklessness and finally to nature taking its revenge. These erroneous views of the relationship between man and the natural environment are reflected in the adoption of the concept of 'anthropocentrism' by some Soviet scholars and scientists.[61] These 'anthropocentrists' presumably belong to the same group as the 'geographical nihilists,' the 'leftists,' and the 'indeterminists.'

It must be remembered that while Saushkin was airing the above views and Anuchin was preparing his doctoral thesis, Stalin's pronouncement on the role of the natural environment in the development of society was still the official view held by the Communist Party and without some modification of the Stalinist line it was impossible for these new ideas to receive any real recognition or even for them to be regarded as truly Marxist. However, in October 1963, one of the most significant events in the history of Soviet geographical methodology took place when the Stalinist interpretation of the geographical environment was rejected by the Party. Ilyichev, a prominent ideological spokesman of the Party, spoke before the Presidium of the Academy of Sciences of the U.S.S.R., his speech being published in the *Vestnik Akademii Nauk S.S.S.R.* and in a shorter form in *Voprosy Filosofii*, journals used for the reporting of items of ideological interest and importance to a larger audience than a specialized group of scholars.

Ilyichev supports the concept of a unified geography, although he does not accept the view that human society is part of the geographical environment as Anuchin claims. This would mean that sociology would be replaced by geography. Most significant is his rejection of Stalin's 'one-sided definition of the relationship between the geographical environment and human society.'[62] Ilyichev points out that in studying the laws of social life we cannot ignore the influence of nature on society, just as in studying the natural environment we cannot ignore the effect of man's activities on it. The geographical environment is not a determining factor of historical progress, but it exerts a certain influence on it. 'And, although, obviously, historical development cannot be reduced to that factor, one cannot completely ignore the interaction between man and the surrounding environment and the changes he produces in nature.'[63] Stalin erred by presenting the geographical environment as being purely external to society and as something relatively unchanging. The adoption of Stalin's views has led to a 'rift between society and nature, particularly ignoring the significance of the geographic environment in the life of society,' which has resulted on occasions in a restricted approach to certain practical problems.[64] The erection of 'an insurmountable wall between natural science and the social sciences' by some Soviet scholars is due to their acceptance of Stalin's definition of the natural environment as a theoretical basis for their ideas.[65]

The importance of this statement by Ilyichev on behalf of the Party cannot be overemphasized for its effect on the future development of Soviet geographical philosophy. Not only has Anuchin received support for many of his ideas, but Saushkin and the supporters of a unified geography have been placed in a strong position vis-à-vis Gerasimov, Kalesnik, and others, mainly physical geographers, who favored a segregated science. Anuchin celebrated the partial vindication of his theories with a further discussion of the relations between society and nature in the journal *Voprosy Filosofii* in which he expanded his views on the nature of the geographical environment. He points out that[66]

> although the mode of production is the determining factor of social development, the geographical environment, as a condition of social development, will introduce corrections in that process. The results of the operation of social laws under a given mode of production will differ in various geographical environments although these differences will not change the direction of the social laws.

Although the geographical environment does not determine the transition between social or economic systems, the environment, acting through the process of production, determines many characteristics of different economic systems, as material goods are produced in different ways depending on the environmental conditions, in spite of common features in the system.

In this article Anuchin also reveals that he has accepted Ilyichev's pronouncement and admits that 'it would ... be wrong to regard society as part of the geographical environment,' giving Ilyichev's reasons. Anuchin has thus significantly modified his original thesis in this regard.[67]

Anuchin has utilized the columns of *Literaturnaya Gazeta* for the further propagation of his ideas to a wide audience and to attack 'antigeographism,' 'voluntarism,' and the Stalinist view of the geographical environment. He selects in particular Gerasimov for attack for presiding over the virtual liquidation of geography as a discipline.[68] This article has led to further exchanges between geographers in the columns of the newspaper, but little of a methodological nature has been discussed.

Saushkin has also followed Ilyichev's statement with an article on the geographical environment and human society in which he specifically criticizes the Stalinist viewpoint. He supports Anuchin's concept of the geographical environment as not being simply 'external nature' which is relatively unchanging, but that part of nature transformed by man, which can change not only over a long period but also over a comparatively short one. Saushkin thinks that Stalin took his ideas on the nature of the environment not from the classics of Marxism but from the French geographer Réclus (1830–1905), always a favorite with Soviet geographers, partly because of his political beliefs (although not a Marxist he was an anarchist and a participant in the Paris Commune), and partly because of his lack of environmentalist tendencies. Whether Réclus was the original source of his inspiration or not, Stalin's views are, according to Saushkin, not in line with those of the early Marxists, who in fact emphasized the principle that[69]

> in the process of struggle against nature, man not only changes the character of nature (transforming it into a historically conditioned geographical environment), but also himself, by acquiring new qualities, habits and experience.

Saushkin has also taken the opportunity of reaching a wide audience of schoolteachers and others by publishing a short article in *Geografiya v shkole* (Geography in school) on the role of the geographical environment in geography, reviewing the role of Stalin, emphasizing Baranskiy's contributions, and explaining the significance of Ilyichev's pronouncement.[70]

These articles by Anuchin and Saushkin can be regarded as representative of the 'new look' in Soviet geography, which has received the seal of approval of the Party. It must be realized, however, that it was not Ilyichev's speech which gave birth to new ideas about the role of the natural environment in geography, but rather that the gradual evolution of these ideas among a group of geographers led to the Party pronouncement. The phrasing of Ilyichev's speech, and especially the remarks about the 'insurmountable wall' erected by some Soviet scientists between the social and natural sciences, suggest the direct influence of Anuchin's thesis and of the writings of Baranskiy and Saushkin, rather than any independent thinking on the part of Party ideologists. There seems little doubt that the influence of such prominent physical geographers as Gerasimov and Kalesnik, both supporters of a segregated geography, will be reduced in the future and that the economic geographers led by Saushkin will find their star in the ascendancy.[71]

The withdrawal by Anuchin of his earlier statements on society as part of the geographical environment was no doubt owing to the opposition to this theory which had developed in ideological circles. An article by Konstantinov, a prominent Marxist philosopher and Party spokesman on ideological matters, stressed the traditional Marxist opposition to a unity of nature and society on the grounds of the impossibility of mixing laws of different qualities, i.e., natural laws and social laws.[72] Konstantinov joins Anuchin and Saushkin with an attack on the Stalinist 'one-sided' and 'negative' approach to the physical environment. However, he rejects Anuchin's claim that society forms part of the geographical environment and invokes Ilyichev's reservations on this point. He cannot accept the idea that nature forms part of society even though 'all material–productive activities of man are based on cognition and utilization of the forces of nature and its laws.'[73] Likewise, although society, in the broadest sense of the word, exists within nature as part of the material world, it is wrong to include society in the composition of nature 'because this would lead to erasure of the qualitative boundary between the two, to a search for social laws in nature, and to the transfer of natural laws to society.'[74] With the opposition of two such important figures as Konstantinov and Ilyichev it seems unlikely that Anuchin's earlier views on society and nature will be revived.

In reviewing the history of Marxist geographical methodology it is important for the Western geographer to realize that Anuchin and Saushkin have been seeking a single object of study for both physical and human geographers alike. Having rejected the chorological concept of Hettner, a synthesis of physical and human phenomena could be most conveniently arrived at in the concept of the geographical environment, suitably defined. Without this single object of study geography could not be considered as a unified discipline. Hence the important place occupied by the geographical environment in recent Soviet methodology.

It is difficult to assess how far the ideas of Western geographers have influenced the latest Soviet theories. Anuchin himself pointed out that Humboldt and Ritter saw geography as a unified science, studying the environment surrounding man, including also man himself and the result of his works.[75] Perhaps even closer to the ideas of the modern Soviet school than they are willing to admit are the ideas of Vidal de la Blache and Brunhes. The latter developed a view of the geographical environment very close to that of Anuchin, pointing out that human beings live in contact with only a limited part of the total natural environment, i.e., the surface of the earth and a slice of the atmosphere. In this limited part of the environment man is active as a modifier. Brunhes put it that[76]

> The *ensemble* of all these facts in which human activity has a part forms a truly special group of surface phenomena – a complex group of facts infinitely variable and varied, always contained within the limits of physical geography, but having always the easily discernible characteristic of being related more or less directly to man.

This specific group of geographical phenomena forms the object of study of human geography.

Not only did Brunhes state a concept very close to the concept later stated by Anuchin, of the geographical environment as the portion of nature which has direct links with human society, but he also voiced an opinion on the relationships between man and nature which Saushkin closely paralleled, as quoted above. Brunhes pointed out that:[77]

> the power and means which man has at his disposal are limited and he meets in nature bounds which he cannot cross. Human activity can

within certain limits vary its play and its movement; but it cannot do away with its environment; it can often modify it, but it can never suppress it, and will always be conditioned by it.

Man's adaptation of the environment must be preceded by scientific investigations, which should help us to moderate our ambitions and not undertake tasks[78]

> that would mean such bold opposition to the forces of nature that man would run the risk of seeing sooner or later his patient work annihilated at a single stroke. The more imposing and glorious man's conquest, the more cruel the revenge of the thwarted physical facts.

It would certainly be difficult for a Soviet geographer to find any traces of 'geographical nihilism' or 'indeterminism' here. One is led to suspect that the theory of possibilism has not been properly understood in the Soviet Union and that a one-sided view of it has been fostered.

Whether any of the ideas developed by Anuchin or Saushkin are close to those of Western geography or not, it must be remembered that Soviet geographers have consistently rejected practically all aspects of Western geographical methodology and are interested only in demonstrating the superiority of the Marxist dialectical approach. Anuchin, for example, denied any attempt on his part to bring Soviet and Western geography closer and claims a completely Marxist line of reasoning for his theories.[79]

The great debate on the geographical environment and the unity of geography has demonstrated the fact that open debate on controversial issues is now possible between the members of the geographical profession in the Soviet Union without serious repercussions for the losing side except, perhaps, loss of influence within their profession, in contrast with the situation during Stalin's lifetime. As a result of this many articles now appearing in geographical journals cannot be said to represent any 'official' Soviet point of view on geography; this is, for example, also true of many of the chapters of *Sovetskaya geografiya: itogi i zadachii* (Soviet geography: accomplishments and tasks), where the ideas of the individual authors have been expressed and no attempt has been made to present a completely unified Soviet view of geography.

The existence of open debate on geographical problems does not, of course, negate the existence of an officially approved Marxist view of geography, based on Marxist ideas on the nature of science and society. An analysis of these ideas made it possible to state a few years ago that[80]

> in spite of the Marxist arguments put forward in favor of geographical monism, there are few signs of any serious attempts at synthesis in Soviet geography, and a major pronouncement on the nature of the physical and social sciences must first be awaited before any radical change can take place.

It is such a major pronouncement which has recently been made and which will form the basis for a revised Marxist approach to geography. Saushkin has attacked Western scholars for regarding Marxism as a rigid, dogmatic philosophy, although in general his evaluation of Western views on Soviet geography are more objective than those of any other Soviet geographer.[81] He stated that[82]

> the 'wall' between the natural and social sciences was forcibly built in the period of the personality cult to provide a 'theoretical' justification for voluntarism in the solution of problems in the development of society and in projects for the transformation of nature. Now this 'wall' is being torn down . . .

However, the tearing down of this 'wall' is not the result of flexibility within the Marxist framework but of the restatement of a basic dogma, just as the erection of the 'wall' was not a whim of Stalin's but the result of following certain selected Marxist principles to their logical conclusion. That it is possible to find quotations in the Marxist classics to support many different points of view is not denied; the devil can quote Scripture for his own ends. No doubt from now on much use will be made of Marx's statement that[83]

> history is itself a real part of natural history, of the transformation of nature and man. However, the natural sciences will eventually include the science of man, just as the science of man will include the natural sciences: there will be only one science.

It should be noted that in 1957 Saushkin used Marx's views on the unity of science to argue the unity of geography, whereas in 1960 Pokshishevskiy argued that Marx on the contrary meant a unity of development, i.e., a unity of science in time, and not of object of study. Perhaps the flexibility of Marxism lies in the inability of its followers to agree on what Marx really meant.

Perhaps the most important lesson of all arising from the debate on the role of the geographical environment in the development of society is that the concept still exists as a central theme of Soviet geography at a time when many American geographers can feel, and state, that geography, to the satisfaction of much of the profession, has finally rid itself of what is thought to be the last traces of geographical determinism and has exorcized its ghost. There are, of course, American geographers who share preferences for some degree of environmentalism or possibilism, but they now are less outspoken. Indeed, so thoroughly has this exorcism taken place that many geographers experience a strong feeling of discomfort when any suggestion is made that the natural environment may play any role other than a purely passive one; any mention at all of the natural environment is embarrassing to some. Perhaps some of us have leaned too far backward to avoid the stigma of environmentalism. At any rate, wherever the truth lies, and each will have his own ideas on this, it behoves American geographers to give more than a passing glance at their own theoretical foundations. At a time when schools of environmental studies and biometeorology are being founded and when anthropologists and ecologists are increasing their research into environmental problems perhaps geographers are busily cutting off another supposedly dead limb from their tree to have it grafted on to that of another discipline more interested and more capable of making it blossom.

References

1   Reprinted from *Annals of the Association of American Geographers,* Vol. 56, No. 1 (March 1966), Washington, D.C.
2   SOLOV'YEV, S. *Istoriya Rossii s drevneyshikh vremen* (History of Russia from ancient times), Moscow: Universitetskaya tipografiya (1883), Vol. I, pp. 32–3.
3   KLYUCHEVSKIY, V. *Kurs russkoy istorii* (Course of Russian history), Moscow: Gos. sotsial'no-ekonomi-cheskoye izd-vo (1937), Part 1, p. 57.
4   Ibid., pp. 60–1.
5   MECHNIKOV, L. I. *Tsivilizatsiya i velikiye istoricheskiye reki* (Civilization and the great historical rivers), Moscow: Gosudarstvennoye izd-vo (1924), p. 69.
6   ANUCHIN, D. N. 'Geografiya' in *Entsiklopedicheskiy Slovar*; Vol. VIII, St. Petersburg: Brokgauz and Yefron (1892), p. 385.
7   PLEKHANOV, G. V. 'O knige L. I. Mechnikova: "Tsivilizatsiya i velikiye istoricheskiye reki" ' (The Book of L. I. Mechnikov: "Civilization and the great historical rivers"), in *Sochineniya,* Vol. VII, Moscow: Gosudarstvennoye izd-vo (1923), pp. 20–1.
8   PLEKHANOV, G. V. *Osnovnyye voprosy marksizma* (1908) (Basic problems of Marxism), in *Isbrannyye filosofskiye protzvedeniya,* Vol. III, Moscow: Gos. izd-vo polit. literatury (1957), p. 151.
9   PLEKHANOV, op. cit., footnote 8, p. 153.
10   Ibid., footnote 8, introduction by Maslin, A., p. 17.
11   WETTER, G. A. *Dialectical Materialism,* London: Routledge and Kegan Paul (1958), p. 107.
12   BUKHARIN, N. *Teoriya istoricheskogo materializma* (The theory of historical materialism), Moscow and Petrograd: Gosudarstvennoye izd-vo (1923), pp. 131–2.
13   Ibid., pp. 116–21.
14   See 'Antropogeografiya' in *Bol'shaya Sovetskaya Entsiklopediya,* Vol. III, Moscow: Gos. nauchnoye izd-vo (1926), p. 112. Appendix by Baranskiy, N.
15   MARX, K. *Capital,* Vol. I, Moscow: Foreign Languages Publishing House (1954), p. 513.
16   *Bol'shaya Sovetskaya Entsiklopediya,* Vol. III, Moscow: Gos. nauchnoye izd-vo (1926), p. 113.
17   PLEKHANOV, G. V. *Ocherki po istorii materializma* (Essays on historical materialism), *Izbrannyye filosofskiye proizvedeniya,* Vol. II, Moscow: Gos. izd-vo polit. literatury (1956), p. 155.
18   Ibid., p. 155.
19   WITTFOGEL, K. A. 'Geopolitika, geograficheskiy materializm i marksizm' (Geopolitics, geographical materialism and Marxism), *Pod znamenem marksizma,* Part I, Nos. 2–3 (February–March); Part II, No. 6 (June); Part III, Nos. 7–8 (July–August), (Moscow, 1929).
20   GRAF, G. E. *Die Landkarte Europas 'gestern und morgen,* Berlin: Cassirer (1919), p. 29.
21   GRAF, G. E. 'Geographie und materialistische Geschichtsauffassung,' in *Der lebendige Marxismus,* Festgabe z. 70 Geburtstage v. Karl Kautsky. Edited by O. Jenssen, Jena: Thüringer Verlagsanstalt u. Druckerei (1924), p. 563.

22   VOSKANYAN, A. M. *O roli geograficheskoy sredy v razvitii obshchestva* (The role of the geographical environment in the development of society), translated from Armenian, Yerevan: Izd-vo A. N., Armyanskoy, S.S.R. (1956), pp. 16–19.

23   CHERNYSHEVSKIY, N. G. *Ocherk nauchnykh ponyatiy po nekotorym voprosam vseobshchey istorii* (An outline of the scientific concepts of several problems of general history), in *Izbrannyye filosofskiye sochineniya*, Vol. III, Moscow: Gos. izd-vo polit. literatury (1951), pp. 557–62.

24   POKROVSKIY, M. N. *Ob Ukraine* (The Ukraine) (Kiev, 1936), p. 19, quoted by Breyterman, A. D. in *Ekonomicheskaya Geografiya S.S.S.R.*, Vol. I, Leningrad: Izd-vo Leningradskogo Universiteta (1958), p. 10.

25   POKROVSKIY, M. N. *Russkaya istoriya v samom szhatom ocherke* (A brief outline of Russian history), Moscow: Partiynoye izd-vo (1931). Published in English as *Brief History of Russia*, New York: International Publishers (1933), pp. 33–4.

26   CHERNYSHEVSKIY, N. G. *O nekotorykh usloviyakh, sposobstvuyushchikh umnozheniya narodnogo kapitala* (Several conditions which assist the increase of national capital), in *Izbrannyye filosofskiye sochineniya*, Vol. II, Moscow: Gos. izd-vo polit. literatury (1950), p. 187.

27*a.*   STALIN, I. V. 'O dialekticheskom i istoricheskom materializme' (Dialectical and historical materialism), in *Voprosy leninizma*, 11th ed.; Moscow: Gos. izd-vo polit. literatury (1952), pp. 587–9.

  b.   *Istoriya vsesoyuznoy kommunisticheskoy partii* (The history of the all-union Communist Party), Moscow: Gos. izd-vo polit. literatury (1951), chap. iv, p. 113.

  c.   'Geograficheskaya sreda' (Geographical environment), in *Bol'shaya Sovetskaya Entsiklopediya*, Vol. X, Moscow: Gos. nauchnoye izd-vo (1952).

28   STALIN, op. cit., footnote 27a, p. 588.

29   Ibid., p. 589.

30   VOSKANYAN, A. M. *O roli geograficheskoy sredy v razvitii obshchestva* (The role of the geographical environment in the development of society), translated from Armenian, Yerevan: Izd-vo Armyanskoy, A. N., S.S.R. (1956).

31   Ibid., p. 8.

32   Ibid., p. 61.

33   BREYTERMAN, A. D. *Ekonomicheskaya geografiya S.S.S.R., chast' I* (*geografiya tyazheloy promyshlennosti*), (Economic geography of the U.S.S.R., Part I, Geography of heavy industry), Leningrad: Izd-vo Leningradskogo Universiteta (1958), Introduction, pp. 5–12.

34   It is interesting that a similar point was made by Isaiah Bowman in *Geography in Relation to the Social Sciences,* New York: C. Scribner's Sons (1934), p. 115.

35   BREYTERMAN, op. cit., p. 8.

36   Ibid., p. 10.

37   Ibid., p. 12.

38   SEBOR, M. M. 'La géographie soviétique: sur quels principes est-elle actuellement basée?' *Revue Canadienne de Géographie*, Vol. X, No. 1 (January–March 1956), pp. 41–4.

39   GERASIMOV, I. P. 'Nauchnyye printsipy sovetskoy geografii v osveshchenii kanadskogo geografa' (The scientific principles of Soviet geography in the view of a Canadian geographer), *Izvestiya Vsesoyuznogo geograficheskogo obshchestva*, Vol. LXXXIX, No. 4 (July–August, 1957), pp. 370–1.

40   SAUSHKIN, YU. G. 'On "Environmentalism" and "Possibilism",' *Vestnik Moskovskogo Universiteta, seriya geografiya*, No. 3 (1960), in translation in *Soviet Geography*, Vol. II, No. 2 (February 1961), p. 74.

41   Ibid., p. 74.

42 LESZCZYCKI, S. 'Nowsze kierunki i prady w geografii' (The latest directions and trends in geography), *Przeglad Geograficzny,* Vol. XXX, No. 4 (1958), p. 549. Translated into Russian as 'Novyye napravleniya i techeniya v geografii,' in *Izvestiya Vsesoyuznogo geograficheskogo obshchestva,* Vol. XCI, No. 1 (1959), pp. 51–9.

43 Ibid., *passim.*

44 'Men'shevik idealism' was a deviation of the early 1930's, consisting mainly of an idealistic interpretation of the dialectic, the separation of theory from practice and an overestimation of Plekhanov. See Wetter, G. A., op. cit., pp. 154–74.

45 BARANSKIY, N. N. 'Uchet prirodnoy sredy v ekonomicheskoy geografii' (An estimation of the natural environment in economic geography), in *Ekonomicheskaya geografiya – ekonomicheskaya kartografiya,* Moscow: Gos. izd-vo geog. literatury (1960), p. 40.

46 Ibid., p. 45.

47 Ibid., p. 48.

48 Ibid., p. 54.

49 ANUCHIN, V. A. 'O sushchnosti geograficheskoy sredy i proyavlenii indeterminizma v sovetskoy geografii' (The nature of the geographical environment and the appearance of indeterminism in Soviet geography), *Voprosy geografii,* Vol. XLI (1957), p. 52.

50 For a discussion of Anuchin's book and reactions to it see Hooson, D. J. M., 'Methodological Clashes in Moscow,' *Annals of the Association of American Geographers,* Vol. 52 (1962), pp. 469–75.

51 ANUCHIN, V. A. *Teoreticheskiye problemy geografii* (Theoretical problems of geography), Moscow: Gos. izd-vo geog. literatury (1960), pp. 37–50.

52 Ibid., p. 150.

53 Ibid., p. 150.

54 Ibid., p. 151.

55 Ibid., p. 152.

56 See Hooson, D. J. M., op. cit., footnote 49, for details of the attacks on Anuchin.

57 BARANSKIY, N. N. review of Anuchin, V. A., *Teoreticheskiye problemy geografii, Vestnik Moskovskogo Universiteta, seriya geografiya,* No. 2 (1961), pp. 76–7, in translation in *Soviet Geography,* Vol. II, No. 8 (October 1961), p. 83.

58 Ibid., p. 83.

59 SAUSHKIN, YU. G. *Vvedeniye v ekonomicheskuyu geografiyu* (Introduction to economic geography), Moscow: Izd-vo Moskovskogo Universiteta (1958), p. 118.

60 Ibid., p. 118.

61 Ibid., p. 119.

62 'L. F. Ilyichev's Remarks about a Unified Geography,' from *Vestnik Akademii Nauk S.S.S.R.,* No. 11 (1963), pp. 14–15, in translation in *Soviet Geography,* Vol. V, No. 4 (April 1964), pp. 32–3.

63 Ibid., p. 32.

64 Ibid., p. 33.

65 Ibid., p. 33.

66 ANUCHIN, V. A. 'The Problem of Synthesis in Geographic Science,' *Voprosy Filosofii,* No. 2 (1964), pp. 35–45, in translation in *Soviet Geography,* Vol. V, No. 4 (April 1964), p. 39.

67 Ibid., p. 41.

68 ANUCHIN, V. 'Istoriya s geografiyey' (The sad story of geography), *Literaturnaya Gazeta* (18 February 1965), p. 2.

69 SAUSHKIN, YU. G. 'The Geographical Environment of Human Society,' *Geografiya i Khozyaystvo,* No. 12 (1963), pp. 67–77, in translation in *Soviet*

Geography, Vol. IV, No. 10 (December 1963), p. 14. The principle referred to was emphasized by Plekhanov in 'K voprosu o razvitii monistichekogo vzglyada na istoriyu' (On the question of the development of a monistic view of history) in *Sochineniya*, Vol. VII, Moscow: Gosudarstvennoye izd-vo (1923), p. 162.

70   SAUSHKIN, YU. G. 'Vzaimodeystviye prirody i obshchestva' (The interaction of nature and society), *Geografiya v Shkole,* No. 4 (July–August 1964), pp. 10–13.

71   Baranskiy died on 29 November 1963.

72   This is Konstantinov, F. V. and not the Konstantinov, O. A., a geographer, whose criticisms were answered by Anuchin in his *O kritike yedinstva geografii* (On the criticism of the unity of geography) (Moscow, 1961), in translation in *Soviet Geography,* Vol. III, No. 7 (September 1962), pp. 22–39.

73   KONSTANTINOV, F. V. 'Vzaimodeystviye prirody i obshchestva i sovremennaya geografiya' (The interaction of nature and society and modern geography), *Izvestiya Akademii Nauk S.S.S.R., seriya geograficheskaya,* No. 4 (July–August 1964), pp. 12–22, in translation in *Soviet Geography,* Vol. V, No. 10 (December 1964), p. 68.

74   Ibid., p. 68.

75   ANUCHIN, op. cit., footnote 50, p. 60.

76   BRUNHES, J. *Human Geography*, Chicago and New York: Rand Mc Nally and Co. (1920), p. 4.

77   Ibid., p. 603.

78   Ibid., p. 611.

79   ANUCHIN, op. cit., footnote 50, pp. 109–10.

80   MATLEY, I. M. *The Soviet Approach to Geography,* unpublished Ph.D. dissertation, University of Michigan (1961), p. 54.

81   SAUSHKIN, YU. G. 'Methodological Problems of Soviet Geography as Interpreted by Some Foreign Geographers,' *Vestnik Moskovskogo Universiteta,* No. 4 (1964), pp. 28–41, in translation in *Soviet Geography,* Vol. V, No. 8 (October 1964), p. 7.

82   Ibid., p. 7.

83   MARX, K. *L'Economie politique et la philosophie,* in *Oeuvres philosophiques,* Vol. VI, Paris: A. Costes (1937), p. 36.

II Geography and the Methods of Modern Science

'. . . Just at this moment, somehow or other, they began to run. Alice never could quite make out, in thinking it over afterwards how it was that they began: all she remembers is that they were running hand in hand, and the Queen went so fast that it was all she could do to keep up with her. . . .'

'The most curious part of the thing was that the trees and other things around them never changed their places at all: however fast they went, they never seemed to pass anything.

"Here, you see", said the Queen, "It takes all the running you can do, to keep in the same place. If you want to get somewhere else, you must run at least twice as fast as that." '

Lewis Carroll *Alice through the Looking Glass*

In contrast to the epistemological emphasis of the previous section, this set of six essays charts the development of interest in the methods and concepts of modern science, in particular with the use of theories, models and laws and the more rigorous application of scientific logic and quantification to geography. However, all the essays, whatever their primary concern, pay particular attention to the need to develop an adequate body of theory in geography. All agree that theory should represent the core of the discipline. In addition, each essay deals explicitly with the methodological pitfalls that must be avoided if the procedures of modern science are to be applied correctly. Hence it is not surprising that the epistemological issues of the first section of this book are constantly relevant, though it must be emphasized that they represent the background, or the framework, upon which the individual arguments of these essays are based.

Before dealing with each individual paper, it is worth noting that all the authors echo Burton's assertion in Chapter 6 ('The Quantitative Revolution and Theoretical Geography') that geography has been a 'following' rather than a 'leading' discipline in terms of the general body of scientific knowledge. Examples of this point have already been provided by Stoddart (Chapter 3) and Matley (Chapter 5) in the preceding section. Here, not only is additional evidence forthcoming, but all the essays look at the situation in

contemporary terms by showing how the scientific methodology of neighbouring disciplines has provided the stimulus for similar developments in
geography. This alone would make it important to continue to preserve and
foster interdisciplinary contacts, but this is shown to be imperative when one
realizes the unusual slowness of the influence of modern scientific methodology
on geography. No doubt it is part of the price paid for an overzealous
concern with the empirical approach, though Lewis's paper (Chapter 7)
shows that it may also have been due to the persistence of outmoded ideas
about the nature of scientific concepts. Yet it must be remembered that not
all geographers rejected these developing ideas of modern science. It is a
salutary experience to find that many of the modern concepts appeared quite
early in the literature, particularly good examples being referred to by
Chorley in his study of models (Chapter 9); Burton in relation to quantification (Chapter 6); and Grigg in his survey of regional procedures (Chapter
10). However an early appearance in the literature does not ensure acceptance: most geographers seemed to be unaware of the importance of such
developments and the general consensus failed to incorporate these concepts.
The explanation of this situation is still the subject of debate, though it might
be suggested that part of the reason lies in the fact that these scientific concepts were derived primarily as a by-product of some empirical enquiry,
rather than being recognized as an interlocking part of a scientific methodology. Hence the methodological implications of these ideas were rarely
precisely explored or explicitly formulated as a body of scientific principles.

Without exception, all the authors in the section are careful to point
out that their acceptance of the modern scientific methods does not mean that
all that has gone before is wasted, nor that the methods represent any
universal panacea. The experience of the several workers leads all of them to
believe that the new methods have been found to be more productive than
the established lines of investigation. Hence, by providing a more rigorous
organization and statement of the existing state of knowledge, and by
pointing out the problems and gaps of existing schemes, these articles are
designed to help geographers to find new directions of understanding –
directions that are particularly well illustrated in the essays by Grigg
(Chapter 10) and Berry (Chapter 11) on regionalization.

In the course of his review of the growth of quantification in geography,
Burton (Chapter 6, 'The Quantitative Revolution and Theoretical Geography')
maintains that the so-called quantitative revolution was over by the early
1960's. Given the continuing debate on this question, such an opinion may
seem to be rather optimistic at first sight, particularly in the light of a recent
review by Lavalle, McConnell and Brown.[1] Yet one must remember that
Burton was judging the issue primarily on the acceptance of the need for
quantification in geography among the leading universities of North America,
and on the published debate in the American journals. Although the timing
of this change may be disputed, it is important to note that it is the change in
general attitude that is all important. Indeed, Burton shows that quantification
has long antecedents in geography, but it has only recently become part of

the conventional wisdom. He suggests that the vigorous but misinformed opposition of leading geographers and the acceptance of alternative lines of thought constituted the major blocking mechanisms, arguing that the change only took place when mathematical ideas spread to all the natural and social sciences.

The review of quantification is of far less significance, however, than Burton's reasoned insistence that quantification alone is not enough. He maintains that it is useful only if used as an adjunct to the development of theory in geography. A valuable example of this point is provided by his discussion of the dangers of depending upon statistical correlation alone: in isolation it could lead to a new kind of determinism. Basically, therefore, the essay provides a balanced and still timely reminder of the priorities of geographical enquiry.

The second paper of the section analyses one of the major barriers to the growth of a scientific methodology in geography, the rejection of the possibility of developing laws. Lewis's small but valuable paper (Chapter 7: 'Three Related Problems in the Formulation of Laws in Geography') illuminates the inadequacy of the arguments used to justify this rejection so, in view of the importance of these points, these issues must be looked at in some detail.

Lewis clarifies the general/unique debate by pointing out that it does not matter that everything is unique – logically this must be accepted. Yet even if scientists accept this point, it does not prevent anybody from searching for similarities. If geographers are going to follow the accepted standards of scientific methodology they should recognize this fact and base their discipline on this principle. Once this is accepted, the similarities within the world (particularly the functional similarities) should be investigated and the results formulated as laws.

Another concept that seems to have gained a great deal of currency in geographical circles is the attitude already referred to by Stoddart in Chapter 3, namely that a law implies some sort of cause-effect relationship. This opinion, probably derived from nineteenth century attitudes towards causality, is carefully debunked. Lewis observes that it is the sameness of relationships that lie behind a law, not causality. This adds emphasis to Davies's view (Chapter 1) that the retreat from determination into possibilism and microscopic scale studies may have proved to be counter-productive. Indeed, the discovery of similarities between the environment and the human response in different areas may represent important and useful generalizations; it is the unthinking connection of these associations as a causal law that represents the danger.

Recognition of these two points makes it much easier to resolve the third problem raised by Lewis, the controversy over 'free will', and again it is shown to rest on another misconception relating to causality. When sets of individual actions or sequences of events are aggregated to provide generalizations, it does not make one iota of difference if these actions take place in accordance with the wishes of man: the similarities recognized are the same and may be formulated as a law. If it is accepted that the same processes are likely to operate in the future, these similarities may be used to calculate possible future states of the system without any deterministic connection being implied. The detailed implications of such a point of view are, of course, still being investigated and provide the subject-matter for Section IV.

The acceptance of the argument in the third paper of this section (Chapter 8) by Golledge and Amedeo, depends not only on a prior commitment to Lewis's three points (Chapter 7) but also to the principle of nonrandomness in the universe subjected to study and to a nomothetic role for geography. The authors maintain that statements of individual fact can provide a useful enquiry but have limited value, mainly because they do not possess any explanatory or predictive ability. By connecting facts together into structured relationships or 'laws', the authors believe it may prove possible to extend knowledge beyond what was originally observed, to predict from what is known about a part of a class of objects to the whole class. Moreover, if the results of enquiries are formulated in this way, their relevance to other studies would be more explicit and their utility for theory building would be improved. The only caveat introduced is that this applies only when a research worker obtains an answer to 'why' any set of relationships exists, rather than 'how' it exists. In methodological terms the latter investigation leads to the discovery of classifications, not laws, and is illustrated by Griggs' paper (Chapter 10).

After a necessary introduction to the difference between analytic and synthetic statements, the authors point out the difference between an individual fact and a law: the former relating to an individual instance, the latter to all instances of a kind. One must, however, be careful; not all instances of any situation or event can be dealt with, so laws cannot be conclusively confirmed. Instead one accepts the ideal of universality, while admitting that this ideal is not entirely feasible in practice. This concern with generalizing from the particular (or particulars) means that some type of inference must be made if laws are to be established, and Golledge and Amedeo survey the six major types of inference used to derive laws, namely: inductive, deductive, probabilistic, genetic, extensional and intermediary explanations. Although necessarily brief, this survey does touch upon part of the wider problem of scientific inference, in particular with the respective roles of induction and deduction, a point already raised in Chapter 2.

The substantive part of the paper is concerned with a discussion of the most important types of law in scientific enquiry, providing, where possible, examples derived from the geographical literature. The framework used in this discussion derives from the work of a former member of the Vienna Circle, Gustav Bergman, a philosopher long resident at Iowa City.[2] As the authors were graduates at Iowa State University, they join a long list of geographers who seem to have been considerably influenced by Bergman's work, in particular Schaefer, McCarty and Bunge.[3] When the history of the conceptual revolution in geography is written, Bergman's influence may be seen to be considerable.

In their analysis of various types of law, Golledge and Amedeo draw a major distinction between static and dynamic laws. In the former, examples of cross sectional and equilibrium laws are provided, in the latter historical and developmental laws. Two major problems connected with the discovery of laws, namely 'imperfection' and 'limitation or scope' are also investigated. In the first case the problem relates to the extent to which an individual law covers all conditions and to the number of exceptions required before the 'law' is discredited. This problem leads to probability notions and to the discussion of a fifth type of law, statistical law. Again it is worth emphasizing with the authors that not all quantitatively expressed work can be classed as 'lawful': it is the search for generalities, not quantified case studies, that is the all important task, confirming once again that quantification is not enough (cf. Burton, Chapter 6). The second problem relating to the discovery of laws concerns the 'limitations' of a law and its alternative, the 'scope' of a law. In this case the importance of a law is judged to be the extent to which different systems can be deduced from the original law. Hence the so-called 'composition rules' are important, for they show how other laws can be obtained from the original statements. Again it is necessary to be wary here; one does not imply that a set of complete universal laws applicable to all cases can be forthcoming – even the best or most useful of laws break down at certain levels of complexity. However, if adequate 'composition rules' are forthcoming, individual laws may be successively connected until everything is known about the relevant variables. At this stage, a sixth type of law, a 'process law' is obtained. Golledge and Amedeo conclude that at this stage in scientific geography process laws have still to be formulated, but that this is the goal towards which geographers should work.

Since the appearance of the fourth paper in this section 'Geography and Analogue Theory' in 1964 (Chapter 9), the term 'model' has become one of the most frequently heard 'new' words of modern geography. In this succinct and well documented review of the way models can be applied in scientific study, Chorley is careful to point out that the history of geographical enquiry is littered with examples of models. Although models have been used, the methodological implications of such an analysis in geography have never been thoroughly analysed. Chorley redresses this balance by a rigorous examination of model-building, and in doing so outlines the possibilities as well as the dangers of this method. Thus the essay provides a vivid demonstration of the fact that models do not represent any universal salvation for the problems of geography: they are merely more productive ways of organizing and interpreting the similarities for which all sciences search. They provide, therefore, alternatives to the more usual method of reaching conclusions about the real world by direct reasoning.

In the course of his paper Chorley outlines the basic procedure followed by any model-builder. It consists of abstracting those parts of the complex reality that the research worker considers to be significant to his problem. These are sifted to form a simplified pattern or model of reality and it must be noted that creative ability and vision are crucially important elements at this stage of the enquiry (cf. Lowenthal's comments in Chapter 4 about the conditioning of individual thought processes). Once this simplified interpretation of the world has been made, it is possible to explore this structure in order to arrive at predictive statements. Three major types of enquiry lead to three types of model, namely mathematical, experimental and natural models. All produce results that need to be carefully checked against the real world, although once a satisfactory number of checks are made the hypothesis may be called a theory. Once again, this is the crucial test. Model-building, like quantification, is a means to an end.

The fifth and sixth papers in this section are both concerned with the application of scientific methodology to one of the basic procedures of geographic enquiry – regionalization. Grigg, in a lucid paper entitled 'The Logic of Regional Systems' (Chapter 10), illuminates one of the major themes of this section: namely, that geographers are not grappling with operational problems that are necessarily unique to their discipline: they share many of these problems with other scientists. His major objective is to show that there is sucifficient identity between the procedures of regionalism and classification to treat them as parallel lines of enquiry. In carrying out this task the logic behind the process of regionalization is clearly exposed and the major problems that have to be faced by would-be region-builders are identified.

In his paper Grigg quotes Bunge's view that geographers seem to have independently rediscovered the entire logic of systems of classification. Although it is worth emphasizing the waste of effort that this independent discovery may have involved, this point does seem to represent a somewhat unreal view of the situation. Certainly it may be relevant if applied to the opinions of the leading methodologists, but the majority of geographers seem to have remained comparatively indifferent to the issues involved. They were more concerned with producing a set of descriptions about the real world, forgetting that the usefulness of the results depends very largely upon the logical construction of the vehicles used in this description, namely the regions. Moreover, the tendency to venerate certain regional schemes has meant that they have been perpetuated long after their usefulness has been exposed.

Grigg emphasizes these points by showing that many regional schemes of classifications are possible: there is unlikely to be any single classification that serves every purpose. Moreover, he maintains that there is little point in classification per se, (cf. similar views held by Chorley in relation to models, and Burton with reference to quantification); instead it should be used for a particular purpose, and this purpose should be made clear.

One of the exceptionally valuable parts of this paper is Grigg's survey of the problem of the 'individual' in geography, the factor that produces the major point of difference between regionalization and classification. He shows that the problem is not peculiar to geography but is shared by pedologists and ecologists. All these disciplines are unable to deal with any single discrete individual which can provide a basis for classification. Indeed, the introduction of a scientific methodology emphasizing the search for similarities may intensify the problems, because the similarities that are of interest extend over areas, and it is difficult to break into such a continuum. No final solution to this problem is provided. However, it is implied that the effect of the problem is diminished once geographers stop searching for an all-purpose classification or set of regions, and remember that they, like other scientists, always deal with operationally defined individuals. Perhaps a more important limitation upon regionalization as a device for describing similarities lies in the fact that a set of similarities may occur among a set of individuals located in space, but that unless these individuals are contiguous, a region cannot be identified.

A major part of Grigg's paper is devoted towards an explicit statement of the issues that have to be resolved when any regional scheme is formulated. Ten principles derived from the work of taxonomists and logicians are presented, and the applicability of each principle to the work of geographers is carefully documented. Hence the article provides a valuable case study of the productivity of scientific methodology in clarifying and identifying the issues lost or disregarded by the subjectivity of so many regional classifications.[4]

The final essay of the section, 'Approaches to Regional Analysis: A Synthesis' by Berry (Chapter 11), begins with a discussion of the role of geography among the sciences. It seems appropriate at the end of this set of essays to emphasize that the distinctiveness of geography (like all scientific disciplines) does not arise from the study of particular phenomena, but from the series of interconnected concepts and conceptual schemes that have been developed by any discipline. In the case of geography the object of study is the world-wide ecosystem in which man is the dominant part, and the relationships involved are viewed in spatial perspective.

The major part of the paper involves the construction of a conceptual scheme or model that depicts the variety of approaches to regional analysis. The basis of this model is derived from the fact that the spatial character of objects can be studied in two different ways, either by looking at the same character in different locations, or by looking at different characters at the same place. Berry shows that it is possible to manipulate this data by constructing a data matrix in which 'locations' are represented by the columns, and 'characteristics' by the rows. Addition of the 'time' dimension gives ten possible approaches for study. It is notable that Berry does not regard this framework as rigid in any way, and he explicitly observes that it should not be allowed to obscure the fact that it is nothing more than a clarification of approaches to analysis. The interdependence of variables in particular should always be kept in mind. Moreover he demonstrates that the study of particular topics is frequently influenced by the traditional study of the columns, for example the literature of urban geography has been primarily concerned with the Western city.

Although Berry uses the model to illuminate the work already carried out in the economic geography of the U.S.A., he concludes by pointing out the limitations of existing knowledge, as well as the problems of the model he uses. In particular Berry is conscious of the concentration of regional studies at the static structure level. Comparatively few investigations of the connectivity existing between the parts of a system or studies of dynamic processes have been made (cf. Chapter 20), though he maintains that the developing literature of central place systems does represent a breakthrough in the former case. Such points really summarize the primary concern of all the essays in this section, namely the need to embrace modern scientific methodology, not because it is the only path to understanding, but because it is the most orderly and productive line of investigation so far developed. In the process of organizing and clarifying present knowledge by the constant search for generalization, not only are individual facts integrated, but the gaps existing in any field are uncovered and the path of future progress is made easier.

References

1   LAVALLE, P., MC CONNELL, H. and BROWN, R. G. 'Certain aspects of the expansion of quantitative methodology in American geography', *Annals of the Association of American Geographers,* Vol. 57 (1967), pp. 423–6.
2   BERGMAN, G. *Philosophy of Science,* University of Wisconsin Press (1958).
3   Examples of the work of these geographers are:
  a.   SCHAEFER, F. 'Exceptionalism in Geography', *Annals of the Association of American Geographers,* Vol. 43 (1953), pp. 226–49.
  b.   MCCARTY, P. and LINDBURG, J. 'A Preface to Economic Geography', Prentice Hall (1966).
  c.   BUNGE, W. *Theoretical Geography,* Lund: Gleerups, (2nd edition) (1966).
4   See also: (*a*) the discussion between Bunge, W. and Grigg, D.:
      BUNGE, W. 'Locations are not Unique', *Annals of the Association of American Geographers,* Vol. 56 (1966), pp. 375–7.
  b.   JOHNSON, R. J. 'Choice in Classification: The Subjectivity of Objective Methods', *Annals of the Association of American Geographers,* Vol. 58 (1968), pp. 575–89.

Selected bibliography for further reading

BOOKS

BERRY, B. J. L. and MARBLE, D. (Eds.), *Spatial Analysis: a Reader,* Englewood Cliffs: Prentice Hall (1968).
BLALOCK, H. M. *Causal Inferences in Non-Experimental Research,* Chapel Hill (1961).
CASETTI, E. 'Classificatory and Regional Analysis by Discriminant Iterations', *Northern University Office of Naval Research Task No. 389-135,* Technical Report No. 12 (1964).
CHURCHMAN, C. W. *Prediction and Optional Decision,* Englewood Cliffs: Prentice Hall (1961).
CHURCHMAN, C. W. and RATOOSH, P. (Ed.), *Measurement: definitions and theories,* New York (1959).

COLE, J. and KING, C. *Quantitative Geography*, London: J. Wiley (1968).

FISHER, R. A. *Design of Experiments*, Edinburgh: Oliver and Boyd (1966).

GARRISON, W. L. and MARBLE, D. F. (Eds.) 'Quantitative Geography Parts I and II', *Northwestern University Studies in Geography*, Nos. 13 and 14 (1967).

GARRISON, W. L., ALEXANDER, R., BAILEY, W., DACEY, M. F. and MARBLE, D. F. 'Data Systems Requirements for Geographic Research', Evanston: *Northwestern University Geographic Information System* (1965).

HAGGETT, P. and CHORLEY, R. J. *Network Analysis in Geography,* London: Arnold (1967).

HESSE, M. B. *Models and Analogies in Science*, London: Sheed and Ward (1963).

HUDSON, J. C. *Theoretical Settlement Geography,* University of Iowa Ph.D. dissertation (1967).

KING, L. J. *Statistical Analysis in Geography*, New Jersey: Prentice Hall (1969).

KRUMBEIN, W. C. and GARRISON, W. L. 'Computer Applications in the Earth and Environmental Sciences: Reports and Manuals', *Office of Naval Research, Geography Branch*, Evanston: Northwestern University (1963–present).

MERRIAM, D. F. (Ed.), 'Computer Applications in the Earth Sciences No. 7', *Colloquium on Classification Procedures, Kansas State Geological Survey* (1966).

MORGENSTERN, O. *On the Accuracy of Economic Observations*, Princeton, N.J. (1965).

NEFT, D. S. 'Statistical Analysis for Areal Distributions', *Regional Science Research Institute Monograph Series,* No. 2, p. 173 (1966).

OLSSON, G. 'Distance and Human Interaction', *Regional Science Research Institute*, Philadelphia (1965).

YOUNG, M. (Ed.), *Forecasting and the Social Sciences,* Heinemann, London (1968).

ARTICLES

ACHINSTEIN, P. 'Models, Analogies and Theories', *Philosophy of Science,* Vol. 31 (1964), pp. 328–50.

CURRY, L. 'A note on Spatial Association', *Professional Geographer,* Vol. 18 (2) (1966), pp. 97–9.

CURRY, L. 'Quantitative Geography 1967', *Canadian Geographer,* Vol. 11, No. 4, (1967), pp. 265–79.

HARVEY, D. W. 'Some Methodological Problems in the Use of the Neyman Type A and the Negative Binomial Probability Distribution for the Analysis of Spatial Point Patterns', *Transactions of the Institute of British Geographers*, Vol. 44 (1967), pp. 85–95.

KAO, R. 'The Use of Computers in the Processing and Analysis of Geographic Information', *Geographical Review*, Vol. 53 (1963), pp. 530–47.

KING, L. J. 'The Analysis of Spatial Form and its Relation to Geographic Theory', *Annals of the Association of American Geographers*, Vol. 59 (1969), pp. 573–95.

LAVALLE, P., MC CONNELL, H. and BROWN, R. G. 'Certain Aspects of the Expansion of Quantitative Methodology in American Geography', *Annals of the Association of American Geographers*, Vol. 57 (1967), pp. 423–6.

MANDELBAUM, M. 'Historical Explanation: The Problem of Covering Laws', *History and Theory,* Vol. 1 (1961), pp. 229–42.

NEYMAN, J. 'Indeterminism in Science', *Journal of the American Statistical Association,* Vol. 55 (1960), pp. 625–39.

OLSSON, G. 'Inductive and Deductive Approaches to Model Formulation', *Proceedings of the 1st Scandinavian–Polish Regional Science Seminar*, Szczcin (1967), pp. 173–83.

SWANSON, J. W. 'On Models', *British Journal of the Philosophy Science,* Vol. 17 (1967), pp. 297–311.

6 The quantitative revolution and theoretical geography[1]

Ian Burton
University of Toronto, Canada

In the past decade geography has undergone a radical transformation of spirit and purpose, best described as the 'quantitative revolution.' The consequences of the revolution have yet to be worked out and are likely to involve the 'mathematization' of much of our discipline, with an attendant emphasis on the construction and testing of theoretical models. Although the future changes will far outrun the initial expectations of the revolutionaries, the revolution itself is now over. It has come largely as the result of the impact of work by non-geographers upon geography, a process shared by many other disciplines where an established order has been overthrown by a rapid conversion to a mathematical approach.

Geographers may look with the wisdom of hindsight on a recent statement by Douglas C. North who points out that in the field of economic history 'a revolution is taking place. . . . It is being initiated by a new generation of economic historians who are both skeptical of traditional interpretations of U.S. economic history, and convinced that a new economic history must be firmly grounded in sound statistical data.'[2] North's paper has a familiar ring in geographical ears, but is not primarily concerned with where the revolution is likely to lead. If the example of other social sciences is any criterion, it will lead to a more mathematical, not solely statistical, economic history.

The movement which led to the revolution in geography was begun by physicists and mathematicians, and has expanded to transform first the physical and then the biological science. It is now strongly represented in most of the social sciences including economics, psychology, and sociology. The movement is not yet strongly represented in anthropology or political science, and has scarcely been felt in history, although early rumblings may perhaps be heard from a new journal devoted to history and theory.[3]

This paper presents a discussion of the general characteristics of the quantitative movement; describes in somewhat greater detail the coming of the quantitative revolution to geography; and attempts an assessment of the value of quantitative techniques in the development of theory. Some scholars have chosen to regard the revolution in terms of a qualitative-quantitative dichotomy. It does not help to cast the debate in this form. For 'what is philo-sophically distinctive about contemporary science is its disinterest in dubious dichotomies or disabling dilemmas,'[4] which fascinate and ensnare the mind because they give the illusion of coming close to the essential nature of things. O. H. K. Spate, in his paper on 'Quantity and Quality in Geography,' goes so far as to cry 'down with dichotomies,'[5] but fails to heed his own advice and apply it to the title of his paper. Furthermore, to specify the presence or absence of an attribute or quality is merely to begin the process of measure-ment at its lowest level on a nominal scale. Viewed in this manner, observa-tions of qualitative differences are but the prelude to measurements of a higher order on ordinal, interval, or ratio scales.

The quantity-quality dichotomy has also been allowed to embrace and perhaps conceal a number of related but distinct questions. These include measurement by instruments versus direct sense-data; rational analysis versus intuitive perception; cold and barren scientific constructs versus the rich variety of daily sense experience; continuously varying phenomena versus discrete cases, nomothetic versus ideographic, and the like.

The desire to avoid this confusion reinforces my inclination to side-step the quality-quantity issue, and to view the movement toward quantification as a part of the general spread and growth of scientific analysis into a world formerly dominated by a concern with the exceptional and unique.

Quantification as indeterminism

Geography has long been a 'following' rather than a 'leading' discipline. The main currents of thought have had their origins in other fields. The mechani-stic approach of much nineteenth-century science was represented to some extent among the environmental determinists from Ratzel (if he *was* a deter-minist) to Semple, Huntington, and Griffith Taylor. They were preoccupied by the notion of cause and effect, and were constantly seeking 'laws'. A similar mechanistic flavour is present in much of the recent work by the 'quantifiers.' It is as if geography is re-emerging after the lapse into ideo-graphy which followed the retreat from environmental determinism. The quantitative revolution is taking us back much closer to environmental determinism. It is surely not coincidental that the quantitative revolution is contemporaneous with the appearance of neo-determinism in geography.[6]

It seems clear that a strong reaction to environmental determinism has served to delay the coming of the quantitative movement to geography, and has postponed the establishment of a scientific basis for our discipline that the quantifiers hope to provide (and which the determinists were seeking, al-though for the most part did not find).

It is not surprising, therefore, that the quantitative revolution was resisted most strongly by American geographers, for it was in the United States that the reaction to environmental determinism was strongest. Charac-teristically, the source of strongest opposition is now the source of greatest support, and the United States has achieved a very favourable balance of trade in the commodity of quantitative techniques.

Although quantification in geography has been mechanistic, new techniques being used and others being developed are in line with the contemporary trend in science in that they are probabilistic. The probabilistic approach as exemplified in Curry's work on climatic change,[7] and Hägerstrand's simulation of diffusion[8] offers a most promising vista for future research. As Bronowski notes, statistics 'is the method to which modern science is moving. ... This is the revolutionary thought in modern science. It replaces the concept of inevitable effect by that of probable trend.'[9] It is more accurate, therefore, to refer to some of the later examples of quantification in geography as indeterministic. With Jerzy Neyman, 'One may hazard the assertion that every serious contemporary study is a study of the chance mechanism behind some phenomena. The statistical and probabilistic tool in such studies is the theory of stochastic processes, now involving many unsolved problems.'[10]

Of great significance in the development of laws in the social sciences is the scale of analysis. As Emrys Jones explains, 'The lack of stringency lies in the finite numbers dealt with in the social sciences as opposed to the infinite numbers dealt with in the physical sciences. At this latter extreme, statistical regularity is such that it suggests extreme stringency or absolute validity; while at the other end statistical variations and exceptions are much higher, and deviations themselves warrant study.'[11]

The end of a revolution

Although its antecedents can be traced far back, the quantitative revolution in geography began in the late 1940's or early 1950's; it reached its culmination in the period from 1957 to 1960, and is now over. Ackerman remarks that, 'Although the simpler forms of statistical aids have characterized geographic distribution analysis in the past, the discipline is commencing to turn to more complex statistical methods – an entirely logical development. The use of explanatory models and regression, correlation, variance and co-variance analysis may be expected to be increasingly more frequent in the field. In the need for and value of these methods geography does not differ from other social sciences.'[12]

Similarly, Hartshorne says that 'to raise ... thinking to the level of scientific knowing, it is necessary to establish generic concepts that can be applied with the maximum degree of objectivity and accuracy and to determine correlations of phenomena with the maximum degree of certainty. Both purposes can best be accomplished if the phenomena can be fully and correctly described by quantitative measurements and these can be subjected to statistical comparisons through the logic of mathematics.[13]

Spate, although somewhat sceptical about quantitative methods, concedes that 'increasingly young geographers will feel that they are not properly equipped without some statistical nous,'[14] and adds parenthetically that he is relieved not to be a young geographer.

An intellectual revolution is over when accepted ideas have been overthrown or have been modified to include new ideas. *An intellectual revolution is over* when the revolutionary ideas themselves become a part of the conventional wisdom. When Ackerman, Hartshorne, and Spate are in substantial agreement about something, then we are talking about the conventional wisdom. Hence, my belief that the quantitative revolution is over and has been for some time. Further evidence may be found in the rate at which schools of geography in North America are adding courses in quantitative methods to their requirements for graduate degrees.

Many would concur with Mackay's comment that 'the marginal return on arguing for the need of quantitative methods is now virtually nil.'[15] This does not deny that many ramifications of the revolution remain to be worked out. Nor does it mean that the ramifications will be painless. It is not easy to agree with Spate's argument that the need for statistical nous applies only to young geographers. Is the field to progress only as rapidly as the turnover in generations? The impact of cybernation is already creating unemployment at the white collar level. Its impact on the managerial and professional strata is likely to mean more work, not less. It is no flight of fancy to foresee the day when geographers, if they are to remain abreast of developments, must relearn their craft anew every decade. Nor is it difficult to see that the present generation of quantifiers may rapidly be replaced by younger men more thoroughly versed in mathematics.

Although the quantitative revolution is over, it is instructive to examine its course because to do so tells us something about the sociology of our profession, and because it provides a background for the question, 'quantification for what?' considered below.

The course of the quantitative revolution in geography

Although the origins of the revolution lie in the fields of mathematics and physics, the direct invasion came from closer to home. A list of the more important antecedents, having a direct or indirect impact on geography, would include Von Neuman (a mathematician) and Morgenstern (an economist) for their *Theory of Games and Economic Behavior*,[16] first published in 1944; Norbert Wiener, whose 1948 volume on cybernetics[17] emphasized the necessity of crossing academic boundaries; and Zipf, who published *Human Behavior and the Principle of Least Effort*[18] in 1949.

Geographers began to look for quantitative techniques that could be applied to their problems, and some non-geographers began to bring new methods to bear on old geographic questions. One example is physicist J. Q. Stewart's paper, 'Empirical Mathematical Rules Concerning the Distribution and Equilibrium of Population,' published in the *Geographical Review*[19] as early as 1947.

Stewart has been a leader in the development of social physics, and the declaration of interdependence signed by a group of physical and social scientists at the Princeton conference in 1949 is a landmark in the growth of the application of mathematics to the social sciences.[20] That economists were engaging in methodological debate at this time, in a way that geographers were to do five years later, is evidenced by the Vining and Koopmans controversy in the *Review of Economics and Statistics for 1949.*[21]

The impact of quantification began to be felt in geography almost immediately. It was initiated by a number of statements calling for quantification. Such calls had been issued earlier. For example, in 1936 John Kerr Rose, in his paper on corn yields and climate argued that 'The methods of correlation analysis would seem especially promising tools for geographical investigation.'[22] This call went largely unheeded. Similar statements in 1950, however, were followed up. An outstanding early plea was made by Strahler in his attack on the Davisian explanatory-descriptive system of geomorphology,[23] and his endorsement of G. K. Gilbert's dynamic-quantitative system.[24]

Quantitative geomorphology and climatology

If Gilbert's 1914 paper was as sound as Strahler seems to think, why was it not adopted as a signpost to future work in geomorphology, instead of being largely forgotten and ignored for thirty years? The answer may be, as Strahler himself seems to imply, that geomorphology was a part of geography. Hydrologists and geologists did not direct their major interest towards such matters, or when they did they followed Davis. The followers included Douglas Johnson, C. A. Cotton, N. M. Fenneman, and A. K. Lobeck. Strahler held that they made 'splendid contributions to descriptive and regional geomorphology,' and 'have provided a sound base for studies in human geography,'[25] but they did not greatly advance the scientific study of geomorphological process. This is not to say that there was no quantitative work in geomorphology prior to Strahler.[26]

One immediate response to Strahler's attack on Davis came from Quam, who wondered whether mathematical formulae and statistical analysis might not give a false impression of objectivity and accuracy.[27] A more violent response, however, came from S. W. Wooldridge, who notes that:

> There has been a recent attempt in certain quarters to devise a 'new' quasi-mathematical geomorphology. At its worst this is hardly more than a ponderous sort of cant. The processes and results of rock sculpture are not usefully amenable to treatment by mathematics at higher certificate level. If any 'best' is to result from the movement, we have yet to see it; it will be time enough to incorporate it in the subject when it has discovered or expressed something which cannot be expressed in plain English. For ourselves we continue to regard W. M. Davis as the

founder of our craft and regret the murmurings of dispraise heard occasionally from his native land.[28]

Lester King is inclined to support Strahler.

Statistical analysis is essentially the method of the bulk sample, and is admirable for the study of complex phenomena and processes into which enter a large number of variables. As yet few geomorphic topics provide data suited directly to statistical treatment, and methods may have to be adapted to the new field of enquiry, so that too facile results should not be expected. The net result must be, however, a greater precision in geomorphic thinking.[29]

Several geomorphologists, including Chorley,[30] Dury,[31] Mackay,[32] Wolman,[33] and others, in addition to Strahler, are using quantitative methods and the practice seems likely to spread.

There has been little argument about the application of quantitative techniques to climatology. This branch of our subject embraces the most apparently manageable and quantifiable continuum that geographers have been concerned to study. Thornthwaite and Mather,[34] Hare,[35] Bryson,[36] and others have been applying quantitative techniques to climatic problems for some time, and with great effect. The quality of their work has virtually silenced the potential critics.

Quantification in human and economic geography

By far the greatest struggle for the acceptance of quantitative methods has been in human and economic geography. This is not surprising in view of the possibilist tradition.[37] It is here that the revolution runs up against notions of freewill and the unpredictability of human behaviour. Here the comparison with physical science is helpful. Physicists working on a microcosmic level encounter the same kinds of problems with quanta and energy that social scientists do with people. The recognition of such parallels is cause for rejoicing, not for despair. To be accepted and accorded an honoured place in our society, social science needs to acquire demonstrable value as a predictive science without a corresponding need to control, restrict, or regiment the individual. A social science which recognizes random behaviour at the microcosmic level and predictable order at the macrocosmic level is a logical outgrowth of the quantitative revolution.

The catalogue of claim and counter-claim, charge and counter-charge that appeared in the literature in the 1950's is a long one. It includes Garrison's[38] comment on Nelson's[39] service classification of American cities; the Reynolds[40]–Garrison[41] exchange of 1956 on the (then) little use of statistical methods in geography; the Spate–Berry editorial exchange in *Economic Geography* in which the former reminds us that 'Statistics are at best but half of life. The other half is understanding and imaginative interpretation,'[42] and the latter defends the quantifiers for their clear distinction between facts, theories and methods, and in turn accuses his critics of creating a quantitative bogeyman and tilting at windmills;[43] Dacey's[44] criticism of Burghardt's[45] conclusions on the spacing of river towns, and Porter's defence with the fable of 'Earnest and the Orephagians';[46] the Zobler[47]–Mackay[48] exchange on the use of chi-square in regional geography; Arthur Robinson's classification of geographers into 'Perks and Pokes;'[49] the debate between Lukermann[50] and Berry[51] on a 'geographic' economic geography, and so on.

By 1956, the quantifiers were arguing with each other through the medium of the professional journals as well as with their opponents. In so doing, they occupied an increasing amount of attention and space. In 1956 also the Regional Science Association was established and gave further impetus to quantification in geography.

The erstwhile revolutionaries are now part of the geographic 'establishment,' and their work is an accepted and highly valued part of the field.

The opposition to quantification

The opposition to the quantitative revolution can be grouped into five broad classes. There were those who thought that the whole idea was a bad one and that quantification would mislead geography in a wrong and fruitless direction. If such critics are still among us they have not made themselves heard for some time. There were those like Stamp who argued that geographers had spent too long perfecting their tools (maps, cartograms, and other diagrammatic representations) and should get on with some real building. Stamp was 'a little alarmed by the view that the geographer must add to his training a considerable knowledge of statistics and statistical method, of theoretical economics and of modern sociology. Sufficient perhaps to appreciate what his colleagues are doing so that team work may be based on mutual appreciation seems to me the right attitude.'[52] This seems to be another dubious dichotomy. The notion that geographers either improve their tools or engage in research with available tools seems false. Surely advances in technology are most likely to occur at the moment when we are grappling with our toughest problems. Furthermore, to argue that geographers should not use statistical methods comes close to defining geography in terms of one research tool – namely the map. One weakness of this position has been well demonstrated by McCarty and Salisbury who have shown that visual comparison of isopleth maps is not an adequate means of determining correlations between spatially distributed phenomena.[53]

A third kind of opposition holds that statistical techniques are suitable for some kinds of geography, but not all geography, because there are certain things that cannot be measured. This may be true for some variables. However, even with qualitative characteristics, nominal observations can be made and there is an expanding body of literature on the analysis of qualitative data.[54] A variant of this argument is that the variables with which geography is concerned are too numerous and complex for statistical analysis. Quantifiers claim that it is precisely because of the number and complexity of the variables that statistical techniques are being employed.

Another class of objections is that although quantitative techniques are suitable and their application to geographic problems is desirable, they are nevertheless being incorrectly applied; ends are confused with means; quantitative analysis has failed on occasion to distinguish the significant from the trivial; the alleged discoveries of the quantifiers are not very novel; and so on. That these criticisms have a grain of truth cannot be denied, but to the valid, correct use of quantitative methods (and this is surely what we are concerned with) they are merely irrelevant. Incorrect applications have been and no doubt will continue to be made, and in some cases for the wrong reason such as fashion, fad, or snobbery. More often, however, they are genuine and honest attempts to gain new knowledge and new understandings.

A final kind of criticism to note is in the *ad hominem* that quantification is all right but quantifiers are not. They are perky, suffer from over-enthusiasm, vaulting ambition, or just plain arrogance. To this charge also perhaps a plea of guilty with extenuating circumstances (and a request for leniency) is the most appropriate response. When you are involved in a revolution, it is difficult not to be a little cocky.

The consequences of the revolution

The revolution is over, in that once-revolutionary ideas are now conventional. Clearly this is only the beginning. There is a purpose other than the establishment of a new order. If the revolution had been inspired by belief in quantification for its own sake, or by fad and fashion, then it would have rapidly run its course and quickly died. But the revolution had a different purpose. It was inspired by a genuine need to make geography more scientific, and by a concern to develop a body of theory. Dissatisfaction with ideographic geography lies at the root of the quantitative revolution. The development of theoretical, model-building geography is likely to be the major consequence of the quantitative revolution.

Description, or as some have said, 'mere description,'[55] may be an art or at least call for the exercise of certain talents best described as artistic. Nevertheless, description is an essential part of the scientific method. In examining the real world, our first task is to describe what we see, and to classify our observations into meaningful groups for the sake of convenience in handling. The moment that a geographer begins to describe an area, however, he becomes selective (for it is not possible to describe everything), and in the very act of selection demonstrates a conscious or unconscious theory or hypothesis concerning what is significant.

In his examination of significance in geography, Hartshorne rejects the notion that significance should be judged in terms of appearance, that is, as in objects in a landscape, and establishes as an alternative the criterion that observations should express 'the variable character from place to place of the earth as the world of man.'[56] In many geographic pursuits, man is the measure of significance, and spatial variations the focus. But how else can significance to man be measured except in terms of some theory of inter-relationships?

In this connection there is reason to question Strahler's assertion, quoted above, that the Davisian geomorphologists 'provided a sound basis for studies in human geography.' The genetic and morphological landform classifications they produced may have provided a sound basis for most studies in human geography prior to 1950, but they are not truly anthropocentric. No attempts to assess significance to man were made until the work was substantially completed. This can be contrasted with Sheaffer's recent stream classification,[57] based on flood-to-peak interval, a variable known to be of significance for human adjustment.

The observation and description of regularities, such as these in the spatial arrangement of cultural features, human activities, or physical variables are first steps in the development of theory. Theory provides the sieve through which myriads of facts are sorted, and without it the facts remain a meaningless jumble. Theory provides the measure against which exceptional and unusual events can be recognized. In a world without theory there are no exceptions; everything is unique. This is why theory is so important. As Braithwaite puts it, 'The function of a science is to establish general laws covering the behavior of empirical events as objects with which the science in question is concerned . . . to enable us to correct together our knowledge of the separately known events, and to make reliable predictions of events yet unknown.'[58]

The need to develop theory precedes the quantitative revolution, but quantification adds point to the need, and offers a technique whereby theory may be developed and improved. It is not certain that the early quantifiers were consciously motivated to develop theory, but it is now clear to geographers that quantification is inextricably intertwined with theory. The core of scientific method is the organization of facts into theories, and the testing and refinement of theory by its application to the prediction of unknown facts. Prediction is not only a valuable by-product of theory building, it is also a test by which the validity of theory can be demonstrated. Scientific inquiry may or may not be motivated by the desire to make more accurate predictions. Whatever the motivation, the ability to predict correctly is a sound test of the depth of our understanding.

Given the need to comply with the rigorous dictates of the scientific method, the need to develop theory, and to test theory with prediction, then mathematics is the best tool available to us for the purpose. Other tools – language, maps, symbolic logic – are also useful and in some instances quite adequate. But none so well fulfils our requirements as mathematics.

The quantification of theory, the use of mathematics to express relationships, can be supported on two main grounds. First, it is more rigorous. Second and more important, it is a considerable aid in the avoidance of self-deception.

These points may be illustrated by reference to a paper by Robinson,

Lindberg, and Brinkman on rural farm population densities in the Great Plains.[59] The authors point out that the statistical-cartographic techniques which they use may be properly employed after the establishment of 'tentative descriptive hypotheses regarding the mutuality that may exist among the distributions of an area, inferred through the study of individual maps and other sorts of data. Coefficients of correlation and related indices provide general quantitative statements of the degree to which each hypothesis is valid.'[60]

My submission is that the testing of hypotheses does not make much sense unless these hypotheses are related to a developing body of theory. High correlation does not necessarily confirm a hypothesis, and it is well known that nonsense correlations are possible. The authors propose rural farm population density as a dependent variable and proceed to examine spatial variations using average annual precipitation, distance from urban centres, and percentage of cropland in the total land area as explanatory variables. Having calculated correlation coefficients, the authors conclude that the general hypothesis concerning the association of spatial variations of these variables is confirmed.[61] This use of quantitative techniques demonstrates rigour to the extent that precise measurements of association are made. It also demonstrates the need and possibility of avoiding self-deception.

Nowhere in the paper is it possible to find an explicit statement of theory. Nowhere are we told why rural farm population density is highly correlated with average annual precipitation. Perhaps the explanation lies in the fact that as precipitation decreases, larger farm units are required to support a farm family, owing to lower yields of the same crops, or the cultivation of less remunerative crops. This is a theory, and a test of it would be to examine rural farm population density and farm size. It is conceivable that these two variables are not closely correlated. If this is the case, the theory will need revision. It is surely not much of an explanation, however, to correlate rural farm population density with precipitation. If there is a causal relationship here, it is an indirect one and several links have been omitted.

A more logical treatment would relate farm population to farm size, farm size to yields and land use, yields and land use to precipitation; but it is by no means certain that the causal chain of relationships could be carried so far. The correlations which John K. Rose[62] obtained between corn yields and July precipitation are not as high as Robinson, Lindberg, and Brinkman obtained for average annual precipitation and rural farm population. Admittedly, the two studies were concerned with different measurements, in different areas, at a different point in time. Nevertheless, it is significant that the Robinson group was able to show higher correlation between remotely connected variables than Rose could show between much more closely connected variables.

Robinson's study is deficient because it is not related to an explicit statement of theory. Quantitative analysis of variables cannot be justified for its own sake. The mere restatement of accepted ideas in numerical form instead of in 'plain English' is not what the quantitative revolution is about.

Examination of spatial variables of rural farm population of the Great Plains in terms of an explicit theory would have led Robinson *et al.* to select other, or at least additional, variables than those considered. Some might argue that the hypothesis relating rural farm population and average annual precipitation is a theory. If so, it sounds dangerously like the old deterministic hypotheses and has the same quality of inferring a causal relationship without any explanation or testing of a connecting process leading from cause to effect.

Conclusion

Quantitative techniques are a most appropriate method for the development of theory in geography. The quantitative era will last as long as its methods can be shown to be aiding in the development of theory, and there can be no end to the need for more and better theory. It follows that any branch of geography claiming to be scientific has need for the development of theory, and any branch of geography that has need for theory has need for quantitative techniques.

Not all statements of theory need to be expressed quantitatively in their initial form. Firey, for example, has developed a general theory of resource use[63] without resort to hypothesis testing in a formal sense. Such statements of theory are extremely valuable, and many more of them are needed in geography. Once formulated they should not long remain untested, but the testing need not be undertaken by the same person, or even by persons in the same discipline.

The development and testing of theory is the only way to obtain new and verifiable knowledge and new and verifiable understandings. As Curry points out, 'Methods of representing various phenomena of nature and speculation about their inter-relationships are closely tied together. It is too often forgotten that geographical studies are not descriptions of the real world, but rather perceptions passed through the double filter of the author's mind and his available tools of argument and representation. We cannot know reality: we can have only an abstract picture of aspects of it. All our descriptions of relations or processes are theories or, when formalized, better called models.'[64]

Curry relates model-building to another element in recent geographical work – the problem of perception which may soon come to merit a place alongside the quantitative revolution in terms of significant new viewpoints.[65]

Our literature is replete with ideographic studies. There is a strong urge to get something into the literature because it has not been described before. If these ideographic studies and new descriptions are to have lasting value, their theoretical implications must be shown. In an increasing number of cases, the relationship to theory can best be shown in quantitative terms. In some instances a simple description of an exceptional case may serve to highlight defects in theory. The theory can then be revised or modified to take account of another kind of variation not previously noted, or the theory may have to be abandoned. Theories are not usually abandoned, however, because a few uncomfortable facts do not happen to fit. Theories are abandoned when newer and better theories are produced to take their place. Although observation and description of exceptional cases may be achieved without quantification,[66] the eventual incorporation of modifications into a theory will normally require the rigour of statistical techniques to demonstrate their validity.

There is not a very large literature in theoretical geography. Our discipline has remained predominantly ideographic.[67] A small proportion of the large volume of central place literature can be described as theoretical.[68] It is appropriate to speak of central place theory as one relatively well-developed branch of theoretical economic geography. A recent volume by Scheidegger has emphasized the theoretical aspects of geomorphology.[69] Wolman comments that 'the emphasis on principles that Scheidegger stresses directs attention to inter-relationships and hopefully lessens the tendency to observe, measure, and record everything because it's there.'[70] This remark can be applied with equal value to the development of theory in other branches of geography.

Geographers are now making a conscious effort to develop more theory. A recent volume on theoretical geography[71] attempts to develop theory basic to some areas of the subject. In particular, the author presents a measurement of shape and discusses a general theory of movement and central place theory. This volume will help to focus the attention of geographers on the need for theory. Perhaps a rash of attempts to develop geographic theory will begin. Such a development seems unlikely, however. For while the use of quantitative methods is a technique that can be learned by most, few seem to have that gift of insight which leads to new theory. North comments that a difficult problem is 'the development of the theoretical hypotheses necessary for shaping the direction of quantitative research.'[72]

Attempts to develop theory in geography need not mean a wholesale shift in emphasis. Many an ideographic study could be of greater value if it contained but two paragraphs showing the theoretical implications of the work. This is often easier or at least possible for the author, while it is more difficult or even impossible for others who try to use the work at a subsequent time to develop or test theory. Of course, if case studies are designed with a theory in mind, it is likely that they will differ considerably from studies unrelated to a conscious statement of theory.

Theoretical geography does not mean the development of an entirely new body of theory exclusive to geography. Scheidegger has not attempted to develop new laws of physics, but has merely refined and adapted these laws to the study of geomorphological phenomena and processes. Central place theory is in keeping with some schools of economic theory. One role of an economic geographer is to refine and adapt available economic theory. In doing so he will improve the theory he borrows. If the Anglo-Saxon bias in economics has been to ignore the spatial aspects of economic activity, the geographer is one of those to whom we should look for the remedy. It need not be thought that the growth of regional science completely fills the gap. Those geographers who study drainage networks, highway networks, power distribution systems, flood problems, airline routes, social organization, and the venation of leaves all have in common a concern for a 'flow' between 'points' over a network of links arranged in a particular pattern. Graph theory is a branch of mathematics concerned with networks and may be adapted to fit all manner of collection, distribution, and communications systems. It is conceivable that a body of useful theory could be built up around the application of graph theory to geographical problems.[73] This is an example of what is meant by theoretical geography. It is a direction that an increasing number of geographers are likely to follow. Let us hope that the effort will meet success.

References

1   Reprinted from *The Canadian Geographer,* Vol. 7, No. 4 (1963), Department of Geography, Toronto, Canada.
      A shorter version of this paper was presented at the 13th annual meeting of the Canadian Association of Geographers, Quebec City, June 1963.

      ACKNOWLEDGMENTS During the preparation and revision of this paper. I have benefited from discussions with Brian J. L. Berry, J. W. Birch, W. C. Calef, Michael Church, John Fraser Hart, Robert W. Kates, Leslie King and Jacob Spelt.

2   NORTH, D. C. 'Quantitative Research in American Economic History,' *American Economic Review,* Vol. 53 (1961), 128–30.

3   See, for example, BERLIN, I. 'History and Theory, the Concept of Scientific History,' *History and Theory,* Vol. 1 (1960), pp. 1–3.

4   From the editor's introduction, LERNER, DANIEL (Ed.), *Quality and Quantity,* New York, The Free Press (1961), p. 22.

5   SPATE, O. H. K. 'Quantity and Quality in Geography,' *Annals of the Association of American Geographers,* Vol. 50 (1960), pp. 377–94.

6   See, for example, SPATE, TOYNBEE and HUNTINGTON, 'A Study in Determinism', *Geographical Journal,* Vol. 118 (1952); also Spate, *The Compass of Geography,* Canberra (1953), pp. 14–15. 'There are signs of at least a neodeterminism, more subtle than the old, less inclined to think of environment as exercising an almost dictatorial power over human societies, but convinced that it is far more influential than the current view admits; and with this trend I would identify myself.' Quoted in JONES, EMRYS, 'Cause and Effect in Human Geography,' *Annals of the Association of American Geographers,* Vol. 46 (1956), pp. 369–77 (see p. 370). See also Martin, A. F., 'The Necessity for Determinism,' *Transactions and Papers, Institute of British Geographers,* Vol. 17 (1951), 1–11.

7   CURRY, LESLIE 'Climatic Change as a Random Series,' *Annals of the Association of American Geographers,* Vol. 52 (1962), pp. 21–31.

8   HÄGERSTRAND, TORSTEN 'On Monte Carlo Simulation of Diffusion.' Unpublished paper; and 'The Propagation of Innovation Waves,' *Lund Studies in Geography,* B, 4 (1952), Department of Geography, Royal University of Lund.

9 BRONOWSKI, J. *The Common Sense of Science*, New York: Random House (1959).

10 NEYMAN, J. 'Indeterminism in Science and New Demands on Statisticians.' *Journal of the American Statistical Association*, Vol. 55 (1960), pp. 625–39.

11 JONES, op. cit., p. 373.

11 JONES Cause and Effect in Human Geography, p. 373.

12 ACKERMAN, EDWARD A. 'Geography as a Fundamental Research Discipline'. *University of Chicago, Department of Geography, Research paper* No. 53 (1958), p. 11.

13 HARTSHORNE, RICHARD *Perspective on the Nature of Geography*, Publication for the Association of American Geographers, Chicago: Rand Mc Nally and Co. (1959), p. 161. (Monograph Service of Association of American Geographers, No. 1).

14 SPATE, op. cit., p. 386. Spate makes a similar statement in 'Lord Kelvin Rides Again.' Guest editorial, *Economic Geography*, Vol. 36 (1960), preceding p. 95.

15 Personal communication, 30 March 1963.

16 VON NEUMAN, JOHN, and MORGENSTERN, OSKAR *Theory of Games and Economic Behavior,* Princeton: Princeton University Press (1944).

17 WIENER, NORBERT *Cybernetics*, New York: John Wiley and Sons (1948).

18 ZIPF, G. K. *Human Behavior and the Principle of Least Effort*, Cambridge: Addison-Wesley Press (1949).

19 STEWART, J. Q. 'Empirical Mathematical Rules Concerning the Distribution and Equilibrium of Population,' *Geographical Review*, Vol. 37 (1947), 461–85.

20 STEWART, J. Q. 'The Development of Social Physics,' *American Journal of Physics*, Vol. 18 (1950), pp. 239–53.

21 VINING, RUTLEDGE 'Methodological Issues in Quantitative Economics,' *Review of Economics and Statistics*, Vol. 31 (1949), pp. 77–86. See also T. C. Koopman's reply and Vining's rejoinder, pp. 86–94.

22 ROSE, J. K. 'Corn Yield and Climate in the Corn Belt,' *Geographical Review*, Vol. 26 (1936), pp. 88–102. For a much earlier paper on a similar topic see Hooker, R. H., 'Correlation of the Weather and Crops,' *Journal of the Royal Statistical Society*, Vol. 70 (1907), pp. 1–51.

23 STRAHLER, A. N. 'Davis's Concepts of Slope Development Viewed in the Light of Recent Quantitative Investigations', *Annals of the Association of American Geographers*, Vol. 40 (1950), pp. 209–13.

24 GILBERT, G. K. *The Transportation of Debris by Running Water*, U.S. Geographical Survey, Professional Paper No. 86, Washington, G.P.O. (1914).

25 STRAHLER, op. cit., p. 210.

26 Strahler notes that important work was initiated in the Soil Conservation Service in the middle and late 1930's. In addition, in 1945 there is R. E. Horton's classical paper on quantitative morphology, 'Erosional Development of Streams and their Drainage Basins: Hydrophysical Approach to Quantitative Morphology,' *Bulletin of the Geological Society of America*, Vol. 56 (1945), pp. 275–370.

27 See QUAM, LOUIS O. 'Remarks on Strahler's paper,' *Annals of the Association of American Geographers,* Vol. 40 (1950), p. 213.

28 WOOLDRIDGE, S. W. and MORGAN, R. S. *An Outline of Geomorphology*, London: Longmans, Green, and Co. (1959). The quotation is from the preface to the 2nd ed., p. v.

29 KING, LESTER *Morphology of the Earth*, Edinburgh and London: Oliver and Boyd Ltd (1962), p. 231.

30 See, for example, CHORLEY, R. J. 'Climate and Morphometry,' *Journal of Geology*, Vol. 65 (1957), pp. 628–38.

31 See, for example, DURY, G. H. 'Contribution to a General Theory of Meandering Valleys,' *American Journal of Science*, Vol. 252 (1954), pp. 193–224; also Tests of a General Theory of Misfit Streams, *Transactions and Papers, Institute*

of British Geographers, Vol. 25 (1958), pp. 105–18; and 'Misfit Streams: Problems in Interpretation, Discharge and Distribution,' *Geographical Review*, Vol. 50 (1960), pp. 219–42.

32 MACKAY, 'Pingos of the Pleistocene Mackenzie Delta Area', *Geographical Bulletin*, No. 18 (1962), 21–63; and *The Mackenzie Delta Area*, N.W.T. Department of Mines and Technical Surveys, Geography Branch, Members No. 8, Ottawa (1963).

33 See, for example, WOLMAN, M. G. *The Natural Channel of Brandywine Creek, Pa.* U.S. Geological Survey, Professional Paper No. 271, Washington, G.P.O. (1955).

34 Much of the work of THORNTHWAITE, C. W. and MATHER, J. R. has appeared in the Thornthwaite Association Laboratory of Climatology, *Publications in Climatology*, Centerton, N.J.

35 See, for example, HARE, F. K. 'Dynamic and Synoptic Climatology,' *Annals of the Association of American Geographers*, Vol. 45 (1955), pp. 152–62; also The Westerlies, *Geographical Review*, Vol. 50 (1960), pp. 345–67.

36 See, for example, HORN, L. H. and BRYSON, R. A. 'Harmonic Analysis of the Annual March Precipitation over the United States,' *Annals of the Association of American Geographers*, Vol. 50 (1960), pp. 157–71. Also SABBAGH, M. E., and BRYSON, R. A., 'Aspects of the Precipitation Climatology of Canada Investigated by the Method of Harmonic Analysis,' *Annals of the Association of American Geographers*, Vol. 52 (1962), pp. 426–40.

37 A useful summary is provided by TATHAM, G. 'Environmentalism and Possibilism'. In Taylor, G. (Ed.), *Geography in the Twentieth Century*, New York (1933), pp. 128–62.

38 GARRISON, WILLIAM L. 'Some Confusing Aspects of Common Measurements,' *Professional Geography*, Vol. 8 (1956), pp. 4–5.

39 NELSON, H. J. 'A Service Classification of American Cities,' *Economic Geography*, Vol. 31 (1955), pp. 189–210.

40 REYNOLDS, R. B. 'Statistical Methods in Geographical Research,' *Geographical Review*, Vol. 46 (1956), pp. 129–32.

41 GARRISON 'Applicability of Statistical Inference to Geographical Research,' *Geographical Review*, Vol. 46 (1956), pp. 427–9.

42 SPATE op. cit., 'Lord Kelvin Rides again'.

43 BERRY, B. J. L. 'The Quantitative Bogey-Man.' Guest editorial, *Economic Geography*, Vol. 36 (1960), preceding p. 283.

44 DACEY, M. F. 'The Spacing of River Towns,' *Annals of the Association of American Geographers*, Vol. 50 (1960), pp. 59–61.

45 BURGHARDT, A. F. 'The Location of River Towns in the Central Lowland of the United States,' *Annals of the Association of American Geographers*, Vol. 49 (1959), pp. 205–23.

46 PORTER, P. W. 'Earnest and the Orephagians: A Fable for the Instruction of Young Geographers,' *Annals of the Association of American Geographers*, Vol. 50 (1960), pp. 297–9.

47 ZOBLER, L. 'Decision-Making in Regional Construction,' *Annals of the Association of American Geographers*, Vol. 48 (1958), pp. 140–8.

48 MACKAY, J. ROSS 'Chi-Square as a Tool for Regional Studies,' *Annals of the Association of American Geographers*, Vol. 48 (1958), p. 164. See also Zobler, 'The Distinction between Relative and Absolute Frequencies in using Chi-Square for Regional Analysis'. Ibid., pp. 456–7, and Mackay and Berry, 'Comments on the Use of Chi-Square.' Ibid., Vol. 49 (1957), p. 89.

49 ROBINSON, ARTHUR H. 'On Perks and Pokes,' Guest editorial, *Economic Geography*, Vol. 37 (1961), pp. 181–3.

50 LUKERMANN, F. 'Toward a More Geographic Economic Geography', *The Professional Geographer*, Vol. 10 (1958), pp. 2–10.

51   BERRY 'Further Comments Concerning "Geographic" and "Economic" Economic Geography', *The Professional Geographer,* Vol. 11 (1959), pp. 11–12, Part I.

52   STAMP, L. DUDLEY 'Geographical Agenda: A Review of some Tasks Awaiting Geographical Attention,' President's address, *Transactions of the Institute of British Geographers,* Vol. 23 (1957), pp. 1–17 (see p. 2).

53   MC CARTY, HAROLD H., and SALISBURY, NEIL E. 'Visual Comparison of Isopleth Maps as a Means of Determining Correlations between Spatially Distributed Phenomena,' *University of Iowa, Department of Geography, Publication* No. 3, (1961).

54   MAXWELL, A. E. *Analyzing Qualitative Data.* London: Methuen and Co. (1961).

55   SHAEFFER, FRED K. 'Exceptionalism in Geography: A Methodological Examination,' *Annals of the Association of American Geographers,* Vol. 43 (1953), pp. 226–49.

56   HARTSHORNE *Perspective on the Nature of Geography,* Chap. 5, pp. 36–47.

57   SHEAFFER, J. R. 'Flood-to-Peak Interval,' in White, G. F. (Ed.), Papers on Flood Problems, *University of Chicago, Department of Geography, Research Paper* No. 70 (1961), pp. 95–113. Also Burton, I. 'Types of Agricultural Occupance of Flood Plains in the United States', *University of Chicago, Department of Geography, Research Paper* No. 75 (1962), represents a similar attempt to classify flood plains on the basis of characteristics significant for agricultural occupance.

58   BRAITHWAITE, R. B. *Scientific Explanation,* Cambridge: Cambridge University Press (1955).

59   ROBINSON, ARTHUR H., LINDBERG, JAMES B., and BRINKMAN, LEONARD W. 'A Correlation and Regression Analysis Applied to Rural Farm Population Densities in the Great Plains,' *Annals of the Association of American Geographers,* Vol. 51 (1961), pp. 211–21.

60   Ibid., p. 211.

61   Ibid., p. 215.

62   ROSE, op. cit., pp. 95–7, Figs. 7 and 8.

63   FIREY, WALTER *Man, Mind and Land: A Theory of Resource Use,* New York: The Free Press (1960).

64   CURRY, op. cit., p. 21.

65   One recent publication in this newly developing field of geography is KATES, ROBERT W. 'Hazard and Choice Perception in Flood Plain Management' 5, *University of Chicago, Department of Geography, Research Paper* No. 78 (1962). See also papers by Ian Burton and Robert Kates, and by Robert Lucas and Dean Quinney in a forthcoming issue of the *Natural Resources Journal.*

66   See, for example, my description of a dispersed city as an exception to the classical central place theory in 'A Restatement of the Dispersed City Hypothesis,' *Annals of the Association of American Geographers,* 53 (1963).

67   SIDDALL, WILLIAM R. 'Two Kinds of Geography,' Guest editorial, *Economic Geography,* Vol. 37 (1961), preceding p. 189.

68   BERRY, B. J. L., and PRED, ALLAN 'Central Place Studies: A Bibliography of Theory and Applications,' *Bibliographical Series* No. 1, Philadelphia: Regional Science Research Institute (1961).

69   SCHEIDEGGER, ADRIAN E. *Theoretical Geomorphology,* Berlin: Springer-Verlag (1961).

70   SOLMAN Review of Ibid., *Geographical Review,* Vol. 53 (1963), pp. 331–3.

71   BUNGE, WILLIAM 'Theoretical Geography', *Lund Studies in Geography,* C, 1 (1962), Department of Geography, Royal University of Lund.

72   NORTH, op. cit., p. 129.

73   Some recent work has been done in this direction. See GARRISON, 'Connectivity of the Interstate Highway System,' *Papers—Regional Science Association*, Vol. 6 (1960), pp. 121–37; Nystuen, John D., and Dacey, Michael F., A Graph Theory Interpretation of Nodal Regions. Ibid., 7 (1961), pp. 29–42; Kansky, Karl, 'Structure of Transportation Networks: Relationships between Network Geometry and Regional Characteristics, *University of Chicago, Department of Geography, Research Paper,* No. 84 (1963); Burton, Ian, Accessibility in Northern Ontario: 'An Application of Graph Theory to a Regional Highway Network,' unpublished report to Ontario Department of Highways, Toronto (1962).

7 Three related problems in the formulation of laws in geography[1]

Peter W. Lewis
University of Hull, U.K.

In the past twenty years the use of mathematics by geographers has increased in an attempt to give greater precision to geographic descriptions, to explanatory statements concerning relationships between data and, recently, to prediction. These attempts have raised fundamental issues about the nature of geography and particularly about the three related concepts of whether geographic data are unique or general, of the nature of scientific laws and of human volition. This paper attempts to introduce briefly some additional comments into the discussions which are already available.[2]

Hartshorne considered that:

> Of all the problems of current concern in the thinking of geographers, the most disturbing appears to be the question whether geography 'like the sciences' can develop 'the knowledge of principles, laws and general truths' – and thus lay claim to the name of science – or whether its function is merely to describe innumerable unique things.[3]

The fundamental premise of Hartshorne and many other geographers when discussing this issue is the assumption that there are two classes of object, one of which consists of unique things and the other of general things. The subsequent inference is that the ability of a discipline to establish scientific laws depends upon which type of object is studied. It is then suggested that geographic data are unique, and, as a necessary consequence of this, that geographers study individual cases rather than construct scientific laws.

This argument hinders the progress of geography as a science, and in an attempt to remedy the difficulties, Bunge suggests that 'whether a phenomenon is unique or general can be considered to be a matter of point of view or of the inherent property of the phenomenon itself.'[4] But this seems to add to the confusion by introducing a third possibility: namely that things may be considered unique at will. Bunge overcame these difficulties by discarding uniqueness as a premise and by assuming that things are general. He stated that science assumes things to be general, not unique, and proposes that geography should adopt the same assumption. He emphasized this point by affirming that '. . . only by the complete rejection of uniqueness can geography resolve its contradictions.'[5]

It seems more rational to discard the possibility of such choice and to assert that all things either are or are not unique and that which is true can be resolved logically and by observation. The term unique is synonymous with singular, for as no two objects are one then, if only in this respect, each is unique. This is comparable with saying, in geography, that regions of whatever size are unique because they occupy different parts of the earth's surface. This truism can be extended to include those objects studied by the physical and natural sciences and contradicts Hartshorne's statement that their 'ability to establish scientific laws is dependent on the number of identical . . . cases available for examination.'[6]

But the essential point is that uniqueness does not preclude similarity, and from these similarities relationships can be elicited to provide the basis for scientific generalization. The moon, the earth and an apple are unique and yet they share certain features: in trivial descriptive terms they each have an inside and outside, whereas Newton recognized a major functional similarity when he propounded his laws of gravitation.

By analogy settlements vary greatly in size: many have been described in considerable detail, and certain functional relationships have been recognized. Reilly investigated the relationships between towns and their market areas in 1929. Zipf attempted to measure the exchange of information and people between paired cities, and Stewart considered the distance-density ratio of people from urban centres. Stewart also recognized that these three indexes were related to each other and to Newton's laws of gravitation. These concepts have recently been extended by Warntz with his use of the potential of population.[7]

The task is to search for similarities, and especially functional similarities between unique things, something that geographers have done for a long time, but whose importance and implications have been accentuated by the attempts to formalize these relationships mathematically. Formalization is too often taken to mean cause-effect relations and this is seen as directly opposing free will. It would appear that the unique and individual case study in geography was looked upon as a refuge from this conclusion.

If generalizations can be made from unique things, such general statements are most useful when they are expressed formally. The most efficient method of expressing statements formally is in mathematical symbols, which give a succinct expression of previous findings and a precise statement of what is to be tested in subsequent work. Geographers may be suspicious of such laws because the association seems to be made between laws in general and 'causal laws' in particular as illustrated by a quotation from Hartshorne: 'Thus in order to explain fully by scientific laws of cause and effect a single decision of any single human being we should need . . . far more data than we could ever hope to secure.'[8] This association may reflect the statements made in the only complete theory of geography to be developed, and commonly referred to as environmental determinism. Laws were fashionable in the late nineteenth century, and cause and effect were common coin to many law-makers. The disrepute attached to traditional geographic determinism lingers, and so does the association of law with the notion of cause. Bunge has shown that science does not presume to explain fully.[9] But if scientific laws are not synonymous with causal laws, the general distaste for laws may be dispelled and the proper place of mathematics in geography be recognized.

The problems introduced by determinism are not peculiar to geography, but are important in philosophy and general scientific enquiry, and it is from a philosopher and scientist that acceptable statements can be found concerning the nature of causation and of scientific laws.[10] Baldwin's dictionary defines cause and effect as '... correlative terms denoting any two indistinguishable things, phases, or aspects of reality, which are so related to each other that whenever the first ceases to exist the second comes into existence immediately after, and whenever the second comes into existence the first has ceased to exist immediately before.'[11] Temporal contiguity is asserted, but this is clearly impossible as the time-series is a continuum and event A and event B must be separated by a time-interval. Event A lasts a finite length of time and is very likely to suffer change itself, and thus only the 'last part' of it could be considered as the 'cause', and as the 'last part' can be diminished until it is infinitely small there must be plurality of causes, which is not contained in the definition, and an impasse is reached. No matter how short the interval between events A and B may be, the sequence could be interrupted: the depression of a light switch causes the light to come on, but the electricity could fail in the interval between depression and illumination. The supposed isolated cause is inadequate; as the conditions are extended to give precision to the event, the whole environment is included, the antecedents become impossibly complex, and the probability of repetition is reduced drastically.

But geographic literature clearly contains many observations of fairly reliable regularities of sequence and it is such regularities that have suggested the supposed law of causality. When exceptions appear it is assumed that a superior statement could be found to which there would be no exceptions. It was the regularity of the exceptions that prevented the general acceptance of causal laws in geography, but, nevertheless, the recognition and statement of such regularities are useful in the early stages of enquiry even if there are exceptions. All such uniformities depend on a considerable vagueness of definition of the events, both the earlier event or cause and the later event or effect. The progress of a discipline is reflected by increasing precision, and it has already been noted that the more precisely the events are defined the less likely is the repetition of the sequence. In the traditional concept of the law of causality based upon a sequence that has been observed many times there is a strong inductive probability that it will hold in future cases. But this makes the later event, the effect, no more than probable, whereas objections to the traditional cause and effect are based upon the belief that the relationship is necessary. This does not mean that in a particular case a relationship is not causal, but only that the effect is not inevitable. For example, in the Maidstone area of Kent the early paper mills which made white paper after 1850 were all located adjacent to springs of water from the Hythe Bed aquifer, but not all such sites were occupied by paper mills. A, the presence of a particular type of springwater, may be the cause of B, a paper mill, even if there are cases where B does not occur. There is no need to assume that every event has some antecedent which is its cause in this sense.

Here, then, is some indication for the sentiment for the unique whose study is believed to be incompatible with generalizations and laws. This impediment is equally applicable to scientific enquiry and would be equally restrictive. As such restrictions do not seem to have operated, it is useful to enquire what science searches for and how the results are expressed. Causal laws tend to be displaced by different ones as a science progresses. In Newton's law of gravity there is nothing that is 'cause' or 'effect': there is simply a formula which states one thing as a function of a number of determinants. Russell clearly stated that in science, 'There is no question of the "same" cause producing the "same" effect; it is not in any sameness of causes and effects that the constancy of scientific law consists, but in sameness of relations. And even "sameness of relations" is too simple a phrase; "sameness of differential equations" is the only correct phrase.'[12] The word 'determine' can be introduced here in its purely logical sense in that a number of variables 'determine' another variable if that other variable is a function of them. There is no implicit distinction between past and future, so that the future 'determines' the past in just the same sense that the past 'determines' the future. That is to say, the relationship between the variables is constant, assuming the uniformity of nature and as long as time is not integral to the relationships. (Lapse of time may be included in the relationship.) Thus if the relationships mentioned earlier as stated by Reilly, Zipf and Stewart are accurate, then they are true regardless of time and of place, '. . . or, that, if it does not itself hold, there is some other law, agreeing with the supposed law as regards the past, which will hold for the future.'[13]

The considerable dissatisfaction with the traditional concept of cause and effect was expressed by geographers initially in the ideas of possibilism and subsequently by an increasing concentration on microscopic studies. The concepts contained in scientific laws as outlined above avoid those features of causality which were repulsive and offer considerable scope for theory; theory is fundamental to science and to scientific progress. But this sort of functional relationship, which has been suggested as symptomatic of the more advanced states of scientific progress, involves considerable use of mathematics and of formal statements of generalizations based upon observed facts and their analysis. A strong case can therefore be constructed for ordering as precisely as possible the basic descriptive facts and for analyzing them mathematically so that comparisons can be made between studies of similar topics.

'Can the actions of individual human beings be considered as theoretically determined by inexorable, unvarying scientific laws?'[14] This is clearly a *cri de coeur* and the emotion is related not to actual scientific law but to the supposed identity of such laws with cause and effect relationships. The problem of free will seems to be associated with man's supposed ability to change the future, but it is quite clear that the future will simply be what it will be. And therefore the nexus between the notion of cause and the loss of free will must be compulsion because it is erroneously assumed that causes compel their effects in some way in which effects do not compel their causes. This idea is eliminated in terms of functional relationships, but even so it should be recognized that compulsion involves thwarted desire and so long as a person does what he wishes to do there is no compulsion involved, however much his wishes may be calculable by the help of earlier events. In this sense determinism is not in opposition to volition and the introspective sense of freedom of will is not affected.

There is implicit in many discussions of determinism the idea of a formula by which volitions can be predicted. The complexity of such formulae, if indeed they are possible, would seem to preclude any practical application. The associated problem of whether volitions are part of such a deterministic system is a question of fact, but this fact cannot yet be demonstrated to be either true or false. On the other hand there are certain observed uniformities with regard to volitions which are empirical evidence that they are determined. Much of the geographer's work is concerned with such evidence, of which an early example is found in Mary Somerville's book *Physical Geography*.[15] Recent work in the U.S.A. and Sweden deals with similar material. 'It is important to observe, however, that even if volitions are part of a mechanical system, this by no means implies any supremacy of matter over mind ... thus a mechanical system may be determined by sets of volitions, as well as by sets of material facts.'[16]

Conclusion   It was agreed that, logically, everything is unique. But similarities can be found among unique things and general statements can be made from these similarities. These statements are then expressed as formal laws. These laws do not imply that one event A is always followed by another event B, but they state a functional relationship between given determinants. Finally, it was seen that determinism in this sense involves no conflict with introspective free will.

References

1   Reprinted from *The Professional Geographer*, Vol. 17, No. 5 (September 1965).

2   ACKERMAN, E. A. 'Geography as a Fundamental Research Discipline,' *University of Chicago, Department of Geography, Research Paper* No. 53, Chicago (1958); Bunge, W., 'Theoretical Geography,' *Lund Studies in Geography, Series C*, No. 1 (1962). Hartshorne, R. *Perspective on the Nature of Geography*, Chicago: Rand McNally & Co. (1959); Schaefer, F. K., 'Exceptionalism in Geography: A Methodological Examination,' *Annals of the Association of American Geographers*, Vol. 43 (1953), pp. 226–49.

3   HARTSHORNE, op. cit., p. 146.
4   BUNGE, op. cit., p. 7.
5   Ibid., p. 13.
6   HARTSHORNE, op. cit., pp. 147–50.
7   WARNTZ, W. *Toward a Geography of Price*, Philadelphia: University of Pennsylvania Press (1959), and *Geography, Geometry and Graphics* (1963).
8   HARTSHORNE, op. cit., p. 155.
9   BUNGE, op. cit., p. 12.
10   RUSSELL, B. *Mysticism and Logic*, New York: Longmans, Green & Co. (First published 1918, reprinted 1954.)
11   Quoted in RUSSELL, op. cit., p. 172.
12   RUSSELL, op. cit., p. 184.
13   Ibid., p. 185.
14   HARTSHORNE, op. cit., p. 153.
15   SOMERVILLE, M. *Physical Geography II*, London: J. Murray (1849), pp. 382–4.
16   RUSSELL, op. cit., p. 195.

8   On laws in geography[1]

Reginald Golledge
Ohio State University, Columbus, U.S.A.

Douglas Amedeo
University of California, Irvine, U.S.A.

ABSTRACT   Although many geographers have discussed philosophical problems related to 'explanation' and 'theories' in geography, few have attempted to examine the question of the types of laws and their occurrence in the discipline. It is suggested here that at least five different types of laws may be formulated by geographers, such as Cross-Section, Equilibrium, Historical, Developmental, and Statistical. By following principles of logical reasoning about geographic facts, geographers may discover laws from statistical to process types which emphasize spatial relationships between these facts.

In 1954 Schaefer argued that 'the present conditions of the field (of geography) indicate a stage of development, well known from other sciences which finds most geographers still busy with classifications rather than laws.'[2]

Current geographical thinking has advanced well beyond this stage to one where considerable emphasis is being placed on the discovery and formulation of laws and theories.[3] This is a logical outcome of the continued search for scientific explanation of the spatial relations of phenomena and is an attempt to increase the scope and reliability of geographic knowledge. That is, instead of concentrating on the question of how phenomena are located and arranged, the emphasis is now directed toward the general question of why are phenomena arranged and located the way they are in a certain place. Reasons are being sought to answer the question why the facts are as alleged.

If this assessment of the current emphasis in geographic research is correct, then it follows that geographical thinking is also involved with the search for and development of laws concerning spatial relations. Since geography counts itself among the sciences, then such an emphasis is in the right direction; for as Nagel suggested, explanation is an integral part of any science:[4]

To explain, to establish some relation of dependence between propositions superficially unrelated, to exhibit systematically connections

163

between apparently miscellaneous items of information are distinctive marks of scientific inquiry.

Whereas this nomothetic aim is accepted by many geographers, not all are aware of the nature, types, and relationships of scientific laws developed by philosophers of science. It should be of value, therefore, to examine some of these developments and try to assess their application to work in geography. The aim of this paper is to provide some information on lawfulness, to relate this information, when possible, to geographic research, and to offer suggestions regarding the interpretation of this information for geographic purposes. The discussion on lawfulness, however, necessarily depends on how utilized terms are defined. Hence, prior to this discussion, space will be allotted to an exposition of concepts, facts, inference, and explanation.

Our approach is naturally biased from the point of view of terminology and classification towards those philosophies of science with which we are most familiar and which we happen to support, but it is hoped that the material covered is presented objectively enough to allow for interpretations different from our own and to provide a basis for further discussion of the topic.

Concepts and facts

Within the general field of knowledge a number of fields of inquiry are identified, geography being one such field. Each field of inquiry uses a variety of technical terms. Sometimes different fields use the same words. These words may be defined in the same way in each field or they may be given a different definition in each field. The precise meanings given to words used in a field of inquiry are used to help distinguish the content of each field.[5]

The language of each field of inquiry consists of descriptive words and logical words. The members of one subclass of descriptive words are called 'proper' words, or words which name physical objects.[6] Members of another subclass of descriptive words are called 'character' words or words which name *properties* of physical objects. The term 'concept' is frequently used as a synonym for 'character word.' The language of geography, like the language of any other science, consists wholly of declarative sentences. Through the use of these sentences geographers talk about concepts such as location, spatial distribution, spatial interaction, and areal differentiation. Some concepts used in geography are borrowed directly from other fields of inquiry. Others, such as central place and regionalism, are technical terms which geographers have added to their language and have endowed with precise meanings. The distinguishing common dimension or fundamental implication of all these concepts is a spatial emphasis, an emphasis that places geography as a distinct field of specialization within the broad area of social sciences. However, as with other fields of inquiry, the concepts of geography vary in their degree of significance. One measure of the degree of significance of a concept is its frequency of use, together with other concepts, in lawful statements.

Connecting descriptive terms leads to the production of statements of individual fact. 'Town A is a central place' or 'in 1960, factory X was located in the northeast corner of Township J' are examples of statements of individual fact resulting from connections made between descriptive words. These statements of fact may be true or false. The critical point here is that the truth or falsity of such statements must be determined by empirical investigation.

Strictly speaking, every sentence in a language is either synthetic or analytic. Conceivably, there are sentences which have elements of both in their content. However, the analytic-synthetic distinction is significant because it provides the first clue to the concept of lawfulness. A true analytic sentence is called a *tautology*. A statement is said to be tautological when it is logically true.[7] Arithmetic statements, such as $2 + 2 = 4$, are analytic statements. In fact, every arithmetical proposition is analytic – and is either analytically true or analytically false. Analytic statements in effect say nothing 'new' – their truth depends on the order and arrangements of the logical words (or arithmetic signs) occurring in them.

Statements which are not analytic are synthetic. Synthetic statements are produced by combining statements of fact, and their truth depends on whether these facts are true or false. Whereas an analytic statement can be logically true and is thus automatically verified, a synthetic statement has to be empirically verified before its truth or falsehood can be established. For example, the statement of fact 'Seven people are in this office,' is a synthetic one which, in this case, can be verified by counting. A law is another type of synthetic statement whose validity must be tested by controlled experimentation and/or observation. It must be remembered, however, that not all synthetic statements are lawful. Both true and false statements of individual fact can be found in the language of geography. A true statement of fact describes the state of things as they are.

The sentence 'Town A is a central place' was cited as an example of a statement of fact. If geographical knowledge consisted of nothing but individual statements like this, it would be difficult to achieve explanation or prediction within the framework of the discipline, for a statement of fact achieves neither explanation nor prediction. This is because such statements are not designed to explain and/or predict.

Since statements of fact neither explain nor predict, the accumulation of individual facts leads to little understanding of what is going on in one's universe. Even when single facts are connected with other facts the result may not be a meaningful statement (even if it is grammatically correct) but rather may be a nonsense statement. However, the act of connecting or combining facts is a step towards the achievement of understanding. In other words a statement of fact may become significant when its connections with other statements of fact are known. Significance, however, is not achieved merely by connection or synthesis.

Whereas a statement of fact says something about an individual instance, laws state something about all instances of a kind. Thus, for example only, the statement of fact above 'Town *A* is a central place' ideally becomes a very simple law when stated as all towns of *A*'s character are central places. This is not just a matter of definition because presumably in order for a town to be a central place, it has to 'behave' in a certain way. This law might have been determined by a process of induction in which numerous observations over a long period of time and in many places established that towns of *A*'s character behave in a manner typical of central places and therefore justify the formulation of the law. Here, an invariable connection is made between towns of *A*'s character and the concept of central place. In pointing out that certain features of a universe are always connected with certain other features of the same universe, laws state the order or regularity that is found in a universe. Thus any (presumably) invariable associations known by man are usually stated by him as laws, and these laws are usually restricted to a specific set of variables, relations, concepts, and definitions.

An important distinctive difference between statement of fact and law statements is that laws cannot be conclusively confirmed whereas facts can be (i.e., if an instance of the fact can be found). Evidence for the existence of a particular fact can be obtained by observation. Evidence for the existence of laws is not as easily obtained. Because laws state something about all instances of a kind and it is generally not possible to observe all conceivable instances, direct evidence for the existence of a law is necessarily incomplete. Whereas the notion that all laws must be strict universals is not always adhered to by scientists, it is generally agreed that if an exception to a stated law is found within the conditions set down as prerequisites for the law to exist, then considerable doubt is thrown on the law's validity and it is generally revised to account for the new evidence. Although all instances of a law can rarely be observed, a lawful statement implies that if it were possible to examine all instances of the law, then the proposed connection would invariably hold.

The use of the terms 'all' and 'invariable' in the definition of a law may be objectionable to some and, indeed, we, too, object to their use in any strict sense; for in this case one may justifiably ask 'Is a statistical generalization a law, even though it cannot be thought of as invariable?' Such a question and others like it lead inevitably to the basic question of 'What is the accepted definition of a law?' The answer appears to be that no single definition of a law exists which is acceptable to all. Nagel made this point clear when he stated that:[8]

The label 'law of nature' (or similar labels such as 'scientific law,' 'natural law,' or simply 'law') is not a technical term defined in any empirical science and it is often used, especially in common discourse, with a strong honorific intent but without a precise import. There undoubtedly are many statements that are unhesitatingly characterized as 'laws' by most members of the scientific community, just as there is an even larger class of statements to which the label is rarely if ever applied. On the other hand, scientists disagree about the eligibility of many statements for the title of 'law of nature,' and the opinion of even one individual will often fluctuate on whether a given statement is to count as a law. This is patently the case for various theoretical statements . . . which are sometimes construed to be at bottom only procedural rules and therefore neither true nor false, although viewed by others as examples par excellence of laws of nature. Divergent opinions also exist as to whether statements of 'regularities' containing any reference to particular individuals (or groups of such individuals) deserve the label of 'law.' For example, some writers have disputed the propriety of the designation for the statement that the planets move on elliptic orbits around the sun, since the statement mentions a particular body.

Similar disagreements occur over the use of the label for the statements of statistical regularities; and doubts have been expressed whether any information of uniformities in human social behavior (e.g., those studied in economics or linguistics) can properly be called 'law.' The term 'law of nature' is undoubtedly vague. In consequence, any explication of its meaning which proposes a sharp demarcation between lawlike and nonlawlike statements is bound to be arbitrary.

In what sense, then, do we write about laws? We write about laws in the sense that we accept the adjectives 'universal,' 'invariable,' and 'all' as ideals toward which to strive, but we admit at the same time, in company with writers on the Philosophy of Science, that something other than this ideal is feasible or achievable in actuality. Such a situation is mainly true in the Social Sciences and to a degree in Physics. This 'other than ideal sense' will be made more explicit later in this paper when consideration is given to types of law.

Types of inference and explanation

The principal method of obtaining laws is by inference. Inference is a process of arriving at some conclusion by reasoning (generally in a logical way) from facts or evidence and is frequently utilized in an attempt to obtain explanations. Several major types of inference are commonly identified.

For example, let us assume we have two spatial distributions A and B. Let A be the distribution of X's and B be the distribution of Y's. Suppose that upon repeated observation under different conditions and in various locations the occurrence of an element of A is invariably related to occurrence of an element of B. Suppose further that after a great number of observations it is possible to conclude that the distribution A is related to the distribution B. Then the generalization becomes an example of our first type of inference, that is, *induction*. Using this type of reasoning, a conclusion of a general nature was reached about the two distributions based on repeated observations, under varying circumstances and in many places, of the behavior of specific occurrences of X and Y.

Many philosophers of science argue that true inductively derived generalizations are rarely produced. For example Popper says: 'I do not believe that we ever make inductive generalizations in the sense that we start with observations and try to derive our theories from them. . . .'[9] However, he also goes on to say: '. . . it is irrelevant from the point of view of science whether we have obtained our theories by jumping to unwarranted conclusions or merely stumbling over them (i.e., by 'intuition') or else by some inductive procedure.'[10]

Philosophers such as Nagel do not explicitly include inductive inference as a means of obtaining 'explanations.'[11] Whereas it may be true that most researchers have, prior to conducting research, either precise or informal notions about the expected outcomes of an experiment, their procedures and statements of results often follow a pattern of inference which has historically been identified as an inductive process. Geographers frequently proceed in this manner when they compare known patterns of phenomena with the pattern of their problem phenomenon either by map correspondence or some more rigorous means; generally, however, such procedure is restricted to the initial stages of investigation and to the analysis of the distribution of some phenomenon where little or no theoretical framework exists. This is, in a sense, the geographic equivalent to the chemist's experiment where substances are mixed to obtain some knowledge of reaction which cannot ordinarily be deduced. Such a process involves reasoning from a subset of a class of events to a conclusion about the whole class of events. Generalizations resulting from the use of an inductive process state more than is implied by the individual facts or premises that are given. This is necessarily true because it is unlikely that all of the relevant facts about which the generalizations are concerned have been or can be observed.

Another method of deriving laws or generalizations is by the *deductive* process of reasoning. In this process laws are used as axioms in a deductive framework for the purpose of deducing other laws. In this case the explicandum or 'fact to be explained' is a logically necessary consequence of the explanatory premises. No 'new' laws really result, but the deductive process makes explicit laws that are only implicit in the original formulation. This raises the question of why we bother with deduction if the conclusion arrived at says nothing 'new.' In answer to this question Bergmann argued that 'our minds are not so constituted that when we grasp what a sentence or group of sentences *says*, we also know what they *imply*. To know that, we must 'reason' deductively.'[12]

Perhaps the best example of the *deductive process, per se*, can be drawn from Euclidean geometry. For the most part, its axioms are simple and well known. However, many of the theorems derived from those axioms are quite complicated. Yet all those theorems are deduced from the axioms and simply make explicit what is implicit (but not at once obvious) in the axioms. Thus the geometric 'proofs' of Euclidean geometry are examples of deductive reasoning, and although the theorems produced are analytic[13] rather than synthetic statements, they provide an example of the power of deduction as a means of unpacking the implications of a set of axioms.

An excellent example of deductive reasoning in geography is provided in Central Place Theory.[14] Here, axioms concerning economic behaviour of consumers and entrepreneurs, threshold sizes and ranges of goods, and the competitive market structures are deductively related to each other and to the facts and assumptions about the distribution and socio-economic structure of the population and conditions of the chosen physical environment in a logical manner such that the distribution, the number, and the sizes of central places on the plain in question are directly deducible.

On a smaller scale, predicting and explaining the pattern of climatic variabilities over the surface of the earth is yet another example of deductive reasoning. Axioms on the behavior of gases, thermodynamics, pressure systems, winds, and movements of bodies are again deductively connected with each other and with the facts about the distributions of land, water, and barriers in a manner provoking logical conclusions about the distribution of climatic types on the surface of the earth.

A final example of the *deductive process* can be developed using a migration problem. We choose this example not for any light it may throw on migration studies, but solely for its usefulness in illustrating deductive reasoning. Suppose that we have a situation in which there are two countries *A*, *B* and we desire to determine whether migratory movements will take place from *A* to *B* or from *B* to *A*. Suppose also that we are given the following:

1   The two countries *A* and *B* are separated by a distance that is feasible to traverse in an acceptable amount of time at a reasonable cost;

2   no barriers of a social, political, economic, or physical nature exist between these two countries to suppress the movement of population;

3   there are no countries closer to *A* than *B* and vice versa;

4   The following conditions exist in country *A*:

a.   *A* has surplus population and lacks sufficient resources to fulfil the needs and goals of this population;

b.   *A*'s birth rate exceeds its death rate and a large proportion of its population is young in age.

5   The following conditions describe country *B*:

a.   the population of *B* is not sufficient to attain full economic growth for its society;

b.   the population of *B* is small enough in size so that there exists a labor shortage;

c.   country *B*'s demand for specialized skills exceeds its supply;

d.   country *B*'s population is aging and its natural increase rate can be described as declining.

Findings from researchers who have investigated the problem of population movements in other areas include such generalizations as:[15]

1   Over the long run, free international migration is tied to differential economic opportunity; sending countries have a high rate of population growth relative to economic growth; receiving countries have a need for human talent to exploit resources.

2   At least over the short run, the state of the economy in the receiving country is more influential in stimulating free migration than is that of the sending country.

Conclusion: If we have considered all the relevant variables and if the generalizations developed by other researchers have some generality, then by reasoning deductively from the given facts and the generalizations derived by other researchers, it follows that the main population movements must take place from *A* to *B* and not vice versa.

Probabilistic inference is our third example of a method used increasingly frequently to achieve explanation. According to Nagel, this occurs when: '. . . though premises are logically insufficient to secure the *truth* of the explicandum (i.e., the statement to be explained) they are said to make the latter "probable".'[16] The use of probabilistic inference may result in a statistical law or a statistical generalization. The nature of such laws will be discussed later in the paper.

Functional or teleological reasoning state the instrumental role an action plays in bringing about some goal. Resulting explanations make possible 'a world comprehensively lawful and yet not "deterministic".'[17] Apparently cogent arguments exist both for the acceptance and rejection of teleological explanation, and while reporting on the suggested existence of this form of explanation we hesitate to expand further on it.

Another form of reasoning, set forward explicitly by Nagel, is *genetic* in nature. This sets out the sequence of major events through which some earlier system has been transformed into a later one. Genetic explanations are, by and large, probabilistic in nature. Often sets of developmental laws may be used as assumptions in the explanatory process.

Other subtypes of explanations include extensional and intermediary explanations. To use a trivial example: if it was established that a drumlin, an esker, a cirque, and an isthmus all had specific and peculiar shapes, and from this it was inferred that all geomorphic features could be identified by their shapes, then an extensional generalization would have been made. The veracity of this generalization would, of course, then have to be tested before any significance could be attached to it. On the other hand, if it is inferred that if two sets of phenomena are both related to a third set of phenomena then they are related to each other, then this is sometimes called an intermediary generalization.

Having accumulated some information on types of inference and explanation we turn now to a discussion of some of the types of laws which we feel either occur in geography or may be sought after by geographers.

Types of laws

The types of laws to be discussed in this paper do not exhaust the varieties of laws suggested by various philosophers of science. They do, however, represent a common core of laws which are included in one guise or another in many reputable philosophies of science. For the most part we have accepted the definitions and schema put forward by Bergmann. He used the schema or logical structure of statements to distinguish between types of laws. This bypasses the rather difficult problem of distinguishing and categorizing laws on the basis of their subject matter and results in the definition of types of laws that can be found in any field of knowledge. Thus, statements found in geography, history, economics, and physics can be identified as having a common structure or as belonging to a recognizable class of laws, regardless of the differences in jargon and/or concepts used by each discipline. The major types of laws identified by Bergmann are: cross-section, equilibrium, historical, developmental, statistical, and process laws.[18] This résumé will give the schema or form of each type, one or two principal characteristics of the law, and an example from geography.

CROSS-SECTION LAWS

A cross-section law states the functional connections between the values which several variables have at the same time. If we regard a system as consisting of an infinite number of 'states' – each state representing the condition of the system at a given time period – then a cross-section law is one relating to a specific state of a system. The schema of a cross-section law is: For every X, X is A if and only if X is B. This format is used extensively in geometry in such forms as: For any triangle, a side of the triangle is not smaller than the difference and not greater than the sum of the other two sides.[19]

The cross-section law is found in geography in the form 'where A there B.' Crude attempts at forming such laws can be found in the works of the environmentalists; however, many of these so-called 'laws' have proved to be either improperly stated (incomplete) or merely the statement of instances of a functional relationship which were in no way lawful.

An example of an attempt to formulate a cross-section law can be paraphrased from the work of Christaller:[20]

> Wherever there occurs an urban place of size A, then there will be f(A) urban functions associated with it.

Determining the exact relationship between size of places and number of functions in different countries has been one of the more popular research problems in urban geography.[21] Such investigations are essentially empirical tests, under different conditions, of this cross-sectional law.

Building on the postulates of Christaller and Lösch, Dacey has been able to further formalize the urban size-function relationship by regarding the Christaller system as a system of Dirichlet regions.[22] Under such conditions he shows that if the degree of the system is q, and there is an order of functions ranging from l (the smallest) to m (the largest), then there is at least:

$$
\begin{aligned}
&1 \text{ point with function } m \\
&q \text{ points with function } m - 1 \\
&q^2 \ldots\ldots\ldots m - 2 \\
& \cdot \qquad\qquad\qquad \cdot \\
& \cdot \qquad\qquad\qquad \cdot \\
& \cdot \qquad\qquad\qquad \cdot \\
&q^{m-r} \ldots\ldots\ldots m - r \\
&\phantom{q^{m-r}} \cdot \qquad\qquad\qquad \cdot \\
&\phantom{q^{m-r}} \cdot \qquad\qquad\qquad \cdot \\
&\phantom{q^{m-r}} \cdot \qquad\qquad\qquad \cdot \\
&q^{m-1} \ldots\ldots\ldots 1
\end{aligned}
$$

Such a system is described by Berry as the steady state of an open-state system which can be observed at a point in time.[23]

Perhaps the most publicized attempts at formulating a cross-sectional law have come from work on gravity models. This work ranges both through social and economic geography with numerous test cases commonly available to geographers.

The 'where-there' format of cross-section laws is also used extensively in studies of areal association. Such studies sometimes pose their conclusions as a form of syndromatic lawfulness in which the occurrence of certain characteristics implies the occurrence of other characteristics. For example, Ginsburg has provided sufficient empirical evidence in his *Atlas of Economic Development* for geographers to argue that, given the traits: large proportion of the work force engaged in manufacturing A, high per capita income B, low illiteracy rate C, high per capita energy consumption D, and high per capita revenue from overseas trade E, then if A,B,C occur in a country then D,E will also occur. Such a statement is syndromatically lawful in that it identifies the characteristics of the 'highly developed country' syndrome, and places these characteristics in a 'where-there' framework.

It will be noticed that there is no explicit reference to time in either example of cross-section schema; the implicit temporal idea is *simultaneity*.

Cross-section laws therefore are discovered for a certain state and, though they may apply at several time periods, they are essentially static laws and should not be used to predict the course of events over time. The information they contain, however, may well be used in the process of discovering other laws which may not be temporally confined.

EQUILIBRIUM LAWS

Whereas a cross-section law states a functional connection that exists under some given circumstances, an equilibrium law states that some change will occur if the connection the equilibrium formula states does not obtain. These laws are also static and, in a sense, they are imperfect – i.e., they say what will occur if certain conditions are fulfilled, but do not say what will occur if the conditions are not fulfilled.

An example of the schema of an equilibrium law is:

If X and Y, then if there is to be equilibrium, $X = Y$.

This type of law is exemplified by the law of the lever in physics and by the demand-supply relationships in economics. It appears in geography in research on interaction between places, and it has also been used in defining trade areas, hinterlands, and so on. For the most part the 'equilibrium' situation has been used to define boundaries where the attractive force of one center equals that of another center. Commonly such lines of indifference, or market-supply area boundaries, are defined empirically. In many cases, however, it would be possible to formulate the equilibrium situation in a manner similar to that attempted by Reilly or by Fetter.[24] A paraphrase of Fetter's Law of Market Area Boundaries would be:

If $(p_1 + TC_1)$ is the delivered-to-customer cost of a good from center A and $(p_2 + TC_2)$ is the delivered-to-customer cost from center B, then a market area boundary will exist between A and B at that point where $(p_1 + TC_1) = (p_2 + TC_2)$.

Although not formulated by a geographer, this law has considerable spatial implications. Its scope can be expanded if we remove the limiting assumption that only economic variables determine market area boundaries and concentrate on finding that combination of variables that appear to geographers to most suitably replace the *p's* and *TC's* in Fetter's Law. Many empirical studies have already contributed information that may be profitably used in producing a set of lawful statements concerning the spatial structure of, and relationships between, market areas.[25]

Both laws discussed so far are static laws. The general format of a dynamic law is somewhat different from the schemas already shown. Reduced to its simplest elements, the dynamic law states that, if a system exemplifies at a certain time a certain character, then it exemplifies at some later time a certain other character (and conversely). That is, if at time t_0, a system has the character a, then it will have at some later time t_4 the character b. Conversely if a system at time t_4 has the character b then it must have had a at a specified earlier time, namely t_0.

In dynamic laws the phrase 'and conversely' is stressed because it points to one of the main imperfections in the common interpretation of dynamic processes. This imperfection can be traced to the fact that many attempts to formulate dynamic laws contain conditions which are merely 'sufficient' or merely 'necessary' but are not both sufficient and necessary for the hypothesized relations to hold. Perhaps the classic example of this was the early attempt to formulate a five-stage law of demographic transition (the so-called demographic 'cycle'). To explain further this notion of necessary and sufficient conditions, let us digress briefly.

Assume that we have a distribution such as that represented in Fig. 1. Then we state that at certain stages the curve reaches a maximum *A*, a minimum *B*, and has an inflection point *C* which is neither a maximum nor a minimum. Using differential calculus we can define the conditions for the curve to have an inflection point (i.e., that the slope equals zero or the first derivative of the function's equation equals zero). This condition is a necessary condition for determining the nature of the inflection point but it is not by itself sufficient to tell which of the three situations exist. To determine the sufficient conditions for distinguishing between the points, the slope of the line is examined at an infinitely small distance beyond each inflection point to determine whether the curve rises or falls. Thus the second derivative defines the sufficient condition to determine the nature of the point.

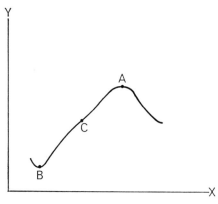

Fig. 1.   Distribution

The reason for this digression is to highlight the nature of imperfection in attempts to formulate dynamic laws. To use a trivial example, the generalization inferred by the statement:

If factories mechanize extensively, then they will increase their profits – (statement 1),

(an implicit dynamic law) does not warrant an inference that:

if factory *A* increases its profits, then it is extensively mechanized . . . (statement 2).

Statement 1 outlines a specific condition for increasing factory profits. Whereas it may highlight a critical condition for profit improvement, the only way statement 2 can justifiably be inferred from statement 1 would be if the prerequisite 'if and only if' preceded the word 'factories' in the first statement. Under such circumstances both necessary and sufficient conditions for profit improvement would be spelled out.

Keeping these limitations in mind, let us examine types of laws incorporating temporal change.

HISTORICAL LAWS

Incorporating historical elements allows the construction of a schema for laws that predict the future not from the present alone but from the present in conjunction with some information about the past. At a particular moment of time our actual knowledge in an area is, was, or will be either partly or predominantly historical. In a sense every dynamic law is historical.

The historical law is the name of a certain type of generalization which is based not so much on the fact that instances on which it is structured lie in the past, but rather that the law conforms to a certain schema with respect to time. For example, given states of a system A, B, and C, then the schema for the historical law is:

If (B now and A earlier), then (C later).

or

If (B now), then (if A earlier then C later).

Examples of studies which may be reinterpreted to fit this framework are numerous in geography. The most obvious examples occur in studies of population growth,[26] and in time-oriented studies between physical features (e.g., slope of land), occupance types (crop types), and cultural and technological attributes of populations.[27] Reconstructions of the changing locations of economic and social activities through time,[28] and some migration studies,[29] have provided a mass of empirical evidence which may contain a number of historical laws.

> In the field of location of industry a typical historical schema might be: If members of industry X were located near phenomena Y at time t_n, and members of the same industry were located adjacent to the same class of phenomena at previous times (t_{n-1}, t_{n-2}, ...), then members of industry X will be located near phenomena Y at time t_{n+1}.

The predictive power of such a schema is often dependent on a subset of ancillary statements that accompany the proposed lawful statement. For example, it may be necessary to add side conditions of constant technology or immobility of labor, or renewable resources, before a locational statement of this type could be regarded as being lawful. Economics has provided a handy phrase 'all other things being equal' which dispenses with a listing of side conditions. Of course, if factors influencing location change drastically from t_n to t_{n+1}, then exceptions to this 'law' will be found and the law would have to be revised to account for the exceptions. This might result in the inclusion of specific side constraints into the originally proposed law or the formulation of other specific laws concerned with the influence of changes in individual location factors. The whole set of such laws could then be used to produce a dynamic theory for the location of industry.[30]

DEVELOPMENTAL LAWS

Often it is difficult to distinguish between historical and developmental laws, since both involve temporal change. The essential difference is found in the schema the laws follow. As opposed to the 'if (B now and A earlier) then (C later)' form of the historical law which may use past and present information to predict the future, a developmental law takes the form:

If (B now) then (A earlier *and* successively C and then D later).

This is a four stage developmental law anchored at the second stage. Freely translated, this states that if a system of a certain kind has at a certain time the character B, then it will under normal conditions at some earlier time have had the character A and at some later time will successively have the characters C and D.

Particular examples of attempts to formulate generalizations of this type can be found where longitudinal development is stressed, as in Rostow's stages of economic growth, the law of demographic transition, and the diffusion process. As an example, the law of demographic transition can be phrased as follows:

> If country Z has at time t^0 the demographic characteristics of high crude birth rates and high crude death rates, then at time t^1 and t^2 respectively it will, under 'normal' conditions, successively have the characteristics of high crude birth rates and low crude death rates, and low crude birth rates and low crude death rates.

Examples of direct and indirect attempts to formulate dynamic laws related to the diffusion process can be found in the works of Hägerstrand, Morrill, Rapoport, and Brown.[31] Earlier studies whose conclusions are expressed in such a way as to lend themselves to reinterpretation as developmental laws include those of Sauer[32] and some of the geomorphological writings of Davis, Penck, Peltier, and Strahler.[33] Although not couched in lawful terms, studies by Borchert and Ward provide interesting material that contains the germs of developmental laws relating to urban evolution.[34]

So far it has been stressed that the laws mentioned have various degrees of imperfection. The discussion of developmental laws, however, made explicit a phrase that was implicitly assumed to be part of previous laws, namely 'under normal conditions.' Essentially this phrase means that the 'law' holds in many cases though not in all, and that it probably holds without exception under conditions as yet either completely unknown or but partially known. The pursuance of this line of thought leads directly into notions of probability and allows us to approach a fifth type of law – statistical law.

STATISTICAL LAWS

A statistical law states that if each member of a class of objects has the character A, then a certain proportion $p(o < p < 1)$ of the objects have the character B. If such a class has N members, then N_1 of its members have the character B, where $N_1 = (p.N)$. For example, given $N = 100$ and $p = .5$, then in a certain number of k classes (in which each member exemplifies A), each of the k-classes should have $N_1 = 50$. Yet we do not expect successive samples from the population to yield N_1's exactly equal to 50. In general we consider the law confirmed if the N_1's cluster or scatter around 50 in a certain fashion.

The antecedents of such laws invariably contain a probability law to the effect that the frequency of a certain character in successive samples from a large population converges toward a certain number in a random manner. However, we do have a degree of imperfection in such laws in that they do not tell us whether or not a single object which has the character A also has the character B.

Attempts to find and/or create statistical lawfulness occur frequently in current geographical research.[35] Attempts to generalize from sample studies or small scale studies to large populations or universes or phenomena are a common part of the statistical procedure. This constitutes an attempt to extend the result of the sample study to a more general level of knowledge. Thus it is possible to build on the results of a local study (such as Häger-strand's studies of intranational migration in Sweden) to produce a more comprehensive statement:[36] if out-migration occurs from town A, then seventy to eighty percent of destinations will be within fifty kilometers of A and from antecedent information about the characteristics of town A and the area in which it lies, we can hypothesize that: if out-migration occurs from a town of size A in an area with characteristics (k, f, and g), then seventy to eighty percent of destinations will be within fifty kilometers of the town of origin.

This hypothesis is now phrased in a manner suitable for testing in a number of different areas and for towns of various sizes. If further testing of migrations from any given size category of towns in similar areas confirms this hypothesis, then a statistical law can be produced by substituting the appropriate size class of towns and specifying the antecedent information concerning characteristics (k, f, and g). Finding and stating instances of this type of lawfulness provides a solid base from which to expand in the search for more complete knowledge in geography.

Every article which has a statistical bias is not, however, an example of an attempt to express a statistical law. Many, such as those of Yeates, Thomas, and Stevens,[37] are either expressions of types of functional relationships that were found in isolated case studies or examples of techniques that can be used to help solve empirical problems. However, by using their results and conduct-ing further empirical investigations, it may later be possible to develop statistical laws relating to the phenomena in which they were interested. On the other hand, a serious attempt at formulating a statistical law for pattern analysis is evident in the work of Dacey,[38] and the search for laws concerning meteorological and climatological regularities can be found in the works of Rayner and Curry.[39] In the field of population studies, the assumptions, postulates and models tested in de Cani's[40] article provide another example of attempts to formulate statistical laws. The recent trend in geography towards the use of probability models may expedite the search for and discovery of useful statistical laws.

COMPOSITION RULES AND PROCESS LAWS

The value of a law depends on its scope, i.e. the number of different systems whose laws can be deduced from the original law. There are certain rules for increasing the scope of a law. Composition rules, for example, state how to make laws out of other laws. Such rules may be discovered, and they may break down at certain levels of complexity. Their place in a schema can be described as follows. If we have the equality that: $N_2 + R = N_3, N_4, \ldots N_{999}$, then $N_3, N_4 \ldots N_{999}$ do not follow from N_2 alone but do follow if the composition rule R is added.

A common (and rather frustrating) experience in geography is to find that apparently good laws break down at a certain level of complexity. For example, N_2 may hold for any state (k) from 1 to 999, but not for any state $k \geqslant 1000$. Under these circumstances, Bergmann argued that the following conclusions can be drawn:[41]

1   there is no law for $k \geqslant 1000$;
2   there is a law for $k \geqslant 1000$, but it involves new variables;
3   there is a law for $k \geqslant 1000$ that involves *no* new variables but that cannot be made by a composition rule out of N_2;
4   there is a law for $k \geqslant 1000$ involving no new variables, and it can be made out of N_2 by means of a composition rule other than R (e.g., R').

Scientists search for and expect to find composition rules. Examples in geography can be seen in the rules which make local, national, and international migrations subject to the same law, or the rules which generalize the functional relationships between numbers of functions and hamlets, villages, towns, and cities.

One of the effects of the discovery of composition rules may be the development of a process law. To know a process law is to know everything there is to be known about its relevant variables. For example, Bergmann argued that[42] fully articulated process laws are differential equations and are the most accomplished dynamic laws or laws of temporal change.

If a process law is known about a system, then any two or more of its states can by means of the process law be inferred from each other. For example, given a system of C with two states, S^{t1} at time t_1 and S^{t2} at time t_2, then any state of the system is a necessary and sufficient condition of any other.

$$e.g., \text{(C and } S^{t1}) \rightarrow S^{t2}$$
$$\text{(C and } S^{t2}) \rightarrow S^{t1}$$
$$C (S^{t1} \leftrightarrow S^{t2}).$$

In this case the logical connective 'if and only if' (as exemplified by the symbol '$\leftrightarrow$') states the necessary and sufficient conditions of the law. As contrasted with the familiar cross-sectional schema 'if A then B' in which A is only a sufficient condition of B, the appropriate schema of the process law is 'A if and only if B.' This implies:

1   If A then B.
2   If not A then not B, and
3   If B then A,
4   If not B then not A.

It does not appear that geography has any laws of the process type at present, but it is possible that, in the long run, as more knowledge is gained about spatial phenomena and this knowledge is formalized into lawful statements, some process laws may be found.

Prediction and explanation

The great advantage of a law or of a generalization is that it allows us to predict from what is known about a part of a class to the whole class, even though the whole class has not been examined. In other words, a law allows an extension of knowledge beyond what has actually been observed. As Bergmann put it:[43]

> One might say, (then), that a law always predicts from what we know at the moment what is as yet unknown, whether it lives in the future or whether we have not inspected it though it lies in the present, or whether we would have to look it up in a record of the past.

Thus to go back to our previous example on predicting and explaining the spatial pattern of climatic types, the laws used in that case are efficient and valid enough to predict the distribution of climatic types either in the distant past or the future as well as in the present. Only the facts change in the various time periods.

Besides prediction, laws or generalizations also provide explanation of either statements of fact or other laws. For example, suppose it is observed that two facts $2X$ and $3Y$ are located side by side at a specific location. Suppose further that a law is known that connects these two facts. Then explanation of either Y or X proceeds deductively, e.g., by a Law of Association: Wherever one finds $2X$, there $3Y$ will also be found;

Observed Fact: $2X$ is at point A (Fact (1))
Conclusion: $3Y$ is at point A (Fact (2))

If some causal relationship was known between *2X* and *3Y*, then it is possible to state that *3Y* obtains at location *A* as a result of the presence of *2X*. In both examples *3Y's* existence in space is explained because the law of association that connects the two facts *2X* and *3Y* is known. This law gives a reason for the existence of *3Y*. Fact (2) (*3Y*'s existence at point *A*) logically follows from the premises because these premises logically imply the conclusion. Furthermore, under deductive reasoning, the conclusion is only making explicit what is implicit in the premises. Hence, if the premises are true and the argument is valid, then the conclusion must be true. The explanation of laws may follow the same deductive reasoning process that was used above in the prediction and explanation of statements of facts. Alternatively, conclusions concerning either new facts or new laws may be derived using one of the other forms of explanation suggested earlier.

For predictive purposes one last question is of importance: What happens when a prediction inferred from a set of premises turns out to be either true or false? If the prediction turns out to be true, then the generalization may be further confirmed. This statement is strictly correct only if the reasoning is logical, the arguments are valid, and the facts involved are true. It is possible to make correct predictions using invalid arguments – which appear superficially to confirm the generalization but which actually do not. Many statements put forward as laws have survived over time on the basis of correct predictions but have ultimately been disproved because of irregularities in their antecedents or in their logical structure. If predictions turn out to be false, then three things are possible: the argument is invalid, the generalization is false, or the statement of fact used in making the prediction is false. In attempting to correct a false prediction we usually modify the argument and/or the generalization; this is because it is commonly accepted that we are less likely to be wrong in what we have observed as fact.

It is possible to argue that as our knowledge expands we will continually find exceptions to existing laws. It becomes relevant, therefore, to raise the question about the number of 'exceptions' that may exist before a law becomes 'no law.' This particular question has no absolute answer unless laws are considered to be strict universals. As has been suggested, however, the idea of laws being strict universals no longer dominates in the interpretation of the concept of 'laws.' In the same way that no one can say how many instances of a relation must exist before that relation is lawful, so there is also no one who can give a strict numerical definition of the transition point between law and no law.

Finally, we should point out that acceptance of the ultimate existence of laws is predicated on accepting the principle of rationality or non-randomness in a universe. Such an assumption is basic to this paper, and for those accepting the principle of rationality (in whatever degree) our argument is that the presentation of research results in an appropriately structured form may lead ultimately to the development of one or another of the types of laws discussed in this paper. The presentation of research results in a form such that they can be tested (given similar antecedent conditions) is a major part of the movement towards the production of laws and adequate theory. Continued reassessment of the validity of laws is, of course, a necessary part of the evolution of knowledge, and in a rapidly developing discipline such as geography it is inevitable that some proposed 'laws' will turn out to be no laws. However, if we evince an aim to develop an adequate framework of knowledge about the discipline, a start must be made somewhere. Presenting research results in a rigorous manner and examining the possibility that a form of lawfulness may have been discovered is as good a way as any to start developing a meaningful and sound philosophical base in the discipline.

References

1　Reprinted from *Annals of the Association of American Geographers,* Vol. 58 (December 1968), pp. 760–74.
2　SCHAEFER, F. K. 'Exceptionalism in Geography: A Methodological Examination,' *Annals of the Association of American Geographers,* Vol. 43 (1953), p. 229.
3　We do not mean to imply that geographers were not interested in explanations, lawful statements, or theories, in the past. Quite the contrary; for example, during the period 1750 to 1850 Carl Ritter and Alexander von Humbolt both viewed geography as a science concerned with the construction of laws about the spatial distribution of phenomena on the surface of the earth. Both scholars consistently stressed the importance of establishing the nature of the relationships between facts. In more recent times specific statements on the advocation of explanation, prediction, laws, theories and, in general, the scientific approach to geography have been written by the following geographers: Schaefer, op. cit., footnote 2; Mc Carty, H. H., Hook, J. C. and Knos, D. S., *The Measurement of Association in Industrial Geography,* Iowa City: University of Iowa, Department of Geography (1956), Report I; Ackerman, E. A., 'Where is a Research Frontier?' *Annals of the Association of American Geographers,* Vol. 53 (1963), pp. 429–40; and Bunge, W., *Theoretical Geography,* Lund, Sweden: Gleerup, *Lund Studies in Geography,* Series C, General and Mathematical Geography (1962). This, of course, does not exhaust the list of geographers who have written on this subject. However, it is not our purpose to argue about whether or not geographers have neglected issues of explanation, lawfulness, and theories; instead, we desire to convey our impression that, as in most social sciences, the present emphasis in geography appears to be shifting so that more weight is now placed on these things than in the past. Ackerman expressed this feeling about the minor role scientific emphasis has received in past geographic research by stating: 'We are, naturally, especially interested in the place that geography occupied in this advancing front of science. There is no reason to avoid frankness. I am sure that all but a few here would agree that our contributions have been modest thus far. We have not been on the forward salients in science nor, until recently, have we been associated closely with those who have.' Ackerman, op. cit., p. 430.
4　NAGEL, E. *The Structure of Science: Problems in the Logic of Scientific Explanation,* New York: Harcourt, Brace and World, Inc. (1961), p. 5.
5　Here 'meaning' is used in a definitional sense only. Where 'meaning' infers 'significance,' the latter word will be used.

6   BERGMANN, G. *Philosophy of Science,* Madison: University of Wisconsin Press (1958), p. 12.

7   For example, 'either it is raining or it is not raining' is a tautology. Truth tables can be used to check whether or not a statement is a tautology. For example, see Korfhage, R. R., *Logic and Algorithms,* New York: John Wiley & Sons, Inc. (1966), pp. 50–9.

8   NAGEL, op. cit., p. 49.

9   POPPER, K. 'Unity of Method in the Natural and Social Sciences,' in Braybrooke, C. (Ed.), *Philosophical Problems in the Social Sciences,* New York: Macmillan (1965), p. 36.

10   POPPER, op. cit., p. 36.

11   NAGEL, op. cit., p. 5.

12   BERGMANN, op. cit., p. 30.

13   Analytic statements can be produced by the deductive process. As such they say nothing new or nothing that is not implied by the original axioms. An analytic statement that is true is a tautology. A false analytic statement is a contradiction. However, as in the case of Euclidean proofs, one must go through the deductive process before any theorems can be produced. Once the theorem is known, and if it is true, it can be stated as a tautology.

14   We make no judgment here about the validity or efficiency of the theories we cite; this is still another argument which involves the central issues of testing and methods of testing. Neither time nor space permit us to entertain any of these issues.

15   BERELSON, B. and STEINER, G. A. *Human Behavior: An Inventory of Scientific Findings,* New York: Harcourt, Brace and World, Inc. (1964), p. 593.

16   NAGEL, op. cit., p. 22.

17   BERGMANN, op. cit., p. 141.

18   It is stressed that this list is by no means exhaustive but it a list which we suggest is useful for geographers. Alternative names given to these laws are probabilistic (statistical), laws of invariable sequences of events (developmental, historical), ultimate (process), and functional dependence (cross-section, equilibrium). Other suggested types of laws are 'experimental,' 'derivative,' and 'fundamental.' For a discussion of these latter types see Nagel, op. cit., pp. 57–80. Examples of the use of Bergmann-type terminology can be found in Kaplan, A., *The Conduct of Inquiry: Methodology for Behavioral Science,* San Francisco: Chandler Publishing Co., (1964); Brodbeck, M., 'Logic and Scientific Method in Research on Teaching,' in Gage, N. L. (Ed.), *Handbook of Research on Teaching,* American Educational Research Association, New York: Rand McNally and Company (1963); Palermo, D. and Lipsitt, E., *Research Readings in Child Psychology,* New York: Holt, Reinhart and Winston, Inc. (1963); Alexander, P., *A Preface to the Logic of Science,* New York: Sheed and Ward (1963); Ritchie, A. D., *Scientific Method: An Inquiry into the Character and Validity of Natural Laws,* Paterson, New Jersey: Littlefield, Adams and Co. (1960); and Schroedinger, E., *Science, Theory and Man,* London: Ruskin House, Allen and Unwin, Ltd. (1957).

19   According to Dewey's philosophy, statements of this type become, by virtue of their form alone, definitions – which are analytic and cannot be classed as laws. For example, using Dewey's system, this scheme reduces to a definition of A in terms of B. Bergmann argued, however, that deletion of the structural connection between $(X \, \& \, A)$ and $(X \, \& \, B)$ reduces the statements to tautologies and under such a system the implications of the original cross-section law are lost in the new format. For a more detailed discussion of this point see Bergmann, op. cit., pp. 102–9.

20   CHRISTALLER, W. *The Central Places of Southern Germany,* translated by Baskin, C., Englewood Cliffs, New Jersey: Prentice-Hall Inc. (1966).

21   Examples include: MAYFIELD, R., 'A Central Place Hierarchy in Northern India,' in Garrison, W. (Ed.), *Quantitative Geography,* Evanston: North-western University Studies in Geography (1965); Snyder, D. E., 'Commercial Passenger Linkages and the Metropolitan Nodality of Montevideo,' *Economic Geography,* Vol. 39 (1962), pp. 95–112; Thomas, E. N., 'The Stability of Distance – Population Size Relationships for Iowa Towns from 1900–1950,' in Norberg, K. (Ed.), *Proceedings of the International Geophysical Union Symposium in Urban Geography,* Lund, Sweden: C. W. K. Gleerup (1962), pp. 13–30; King, L. J., 'The Functional Role of Small Towns in Canterbury,' *Proceedings of the 3rd New Zealand Geography Conference,* New Plymouth, New Zealand: Avery Press Ltd. (1961), pp. 139–49; Murdie, R. A., 'Cultural Differences in Consumer Travel,' *Economic Geography,* Vol. 41 (1965), pp. 211–33.

22   DACEY, M. F. 'The Geometry of Central Place Theory,' *Geografiska Annaler,* Vol. 47B (1965), pp. 111–24.

23   BERRY, B. J. L. *Geography of Market Centers and Retail Distribution,* Engle-wood Cliffs, New Jersey: Prentice-Hall Inc., Foundation of Economic Geography Series (1967), p. 78.

24   FETTER, F. A. 'The Economic Law of Market Areas,' *Quarterly Journal of Economics,* Vol. 38 (1924), pp. 520–9.

25   For example: GREEN, F. W. H. 'Urban Hinterlands in England and Wales: An Analysis of Bus Services,' *The Geographical Journal,* Vol. 116 (1950), pp. 64–88; Duncan, O. D., 'Service Industries and the Urban Hierarchy,' *Papers and Proceedings, Regional Science Association,* Vol. 5 (1959), pp. 105–20; Kiji, S., 'Retail Trade Areas of Kyoto,' *Human Geography,* Vol. 6 (1954), pp. 23–7; Huff, D. L., 'A Probabilistic Analysis of Shopping Center Trade Areas,' *Land Economics,* Vol. 39 (1963), pp. 81–9; Carroll, J. D., 'Spatial Interaction and the Urban Metropolitan Regional Description,' *Papers and Proceedings of the Regional Science Association,* Vol. 1 (1955), pp. D1–D14; Stewart, J. P., 'Potential of Population and its Relationship to Marketing,' in Cox, R. and Alderson, W. (Eds.), *Theory in Marketing,* Chicago: R. D. Irwin, Inc. (1950), pp. 19–40; Golledge, R. G., 'Concep-tualizing the Market Decision Process,' *Journal of Regional Science,* Vol. 7, No. 2 (supplement) (1967), pp. 239–58.

26   Some generalizations, references, and empirical evidences can be found in: Hauser, O. M., *Population Perspectives,* New Brunswick, New Jersey: Rutgers University Press (1960); Hauser, P. M. and Duncan, O. S. (Eds.), *The Study of Population,* Chicago: University of Chicago Press (1959).

27   Many examples of studies of this type have emerged in geography. Examples of studies which may provide relevant information for the development of this type of historical law include: Sauer, C. O., 'The Morphology of Land-scape,' *University of California Publications in Geography,* Vol. 2 (1925), pp. 19–53; Morgan, W. B. and Moss, R. P., 'Geography and Ecology: The Concept of the Community and its Relationship to Environment,' *Annals of the Association of American Geographers,* Vol. 55 (1965), pp. 339–50; Moss, R. P., 'Soils, Slopes and Land Use in a Part of S.W. Nigeria,' *Transactions of the Institute of British Geographers,* No. 32 (1963), pp. 143–68.

28   PRED, A. *The Spatial Dynamics of United States Urban-Industrial Growth 1800–1914,* Regional Science Studies Series, Cambridge, Massachusetts: M.I.T. Press (1966).

29   Sources of reference for relevant articles here include: Banks, V. J., *Migration and Farm Population: An Annotated Bibliography,* Washington, D.C.: Miscellaneous Publication #954, Economic and Statistical Analysis Division, Economics Research Series, U.S. Department of Agriculture (1963); and Olsson, G., *Distance and Human Interaction: A Review and Bibliography,*

Philadelphia: Regional Science Research Institute, Bibliographic Series No. 2, (1965).

30   Evidence of a trend toward accomplishing this can be found in PRED, op. cit., footnote 28.

31   HÄGERSTRAND, T. 'Migration and Area,' *Lund Studies in Geography,* Series B., Human Geography, No. 13, Lund, Sweden: C. W. K. Gleerup (1957); ibid., 'A Monte Carlo Approach to Diffusion,' *European Journal of Sociology,* Vol. 6 (1965), pp. 43–67; Morrill, R., *Simulation of Central Place Patterns Over Time,* Lund, Sweden: C. W. K. Gleerup, Lund Studies in Geography, Series B. Human Geography No. 24, (1962), pp. 109–20; Rapoport, A., 'The Diffusion Problem in Mass Behavior,' in Von Bertalanffy, L. (Ed.), *General Systems Yearbook* (1956), pp. 48–55; Brown, L., 'Towards a General Theory of Spatial Diffusion', Iowa City: Department of Geography, University of Iowa, Xerox (1967); and *Diffusion Dynamics: A Review and Revision of the Quantitative Theory of the Spatial Diffusion of Innovation,* Lund, Sweden: C. W. K. Gleerup, Lund Studies in Geography (1967).

32   SAUER, C. 'American Agricultural Origins: A Consideration of Nature and Culture,' in *Essays in Anthropology Presented to A. L. Kroeber,* Berkeley: University of California Press (1936), pp. 279–97; 'Theme of Plant and Animal Destruction in Economic History,' *Journal of Farm Economics,* Vol. 20 (1938), pp. 765–75; *Agriculture Origins and Dispersal,* Bowman Memorial Lectures, Series 2, New York: American Geographical Society (1952).

33   DAVIS, W. M. *Geographical Essays,* Boston: Ginn and Company (1909); Penck, W., *Morphological Analysis of Land Forms: A Contribution to Physical Geology,* translated by Czech, H. and Boswell, K. C., New York: St. Martin's Press (1953); Strahler, A. W., 'Equilibrium Theory of Erosional Slopes Approached by Frequency Distribution Analysis,' *American Journal of Science,* Vol. 248 (1950), pp. 673–96 and 800–14; Peltier, L. C., 'The Geographic Cycle in Periglacial Regions as it is Related to Climatic Geomorphology,' *Annals of the Association of American Geographers,* Vol. 40 (1950), pp. 214–36.

34   BORCHERT, J. 'American Metropolitan Evolution,' *Geographical Review,* Vol. 57 (1967), pp. 301–32; Ward, D., 'A Comparative Historical Geography of Streetcar Suburbs in Boston, Massachusetts and Leeds, England 1850–1920,' *Annals of the Association of American Geographers,* Vol. 54 (1964), pp. 477–89.

35   Some examples include: DACEY, M. F., 'A Probability Model for Central Place Locations,' *Annals of the Association of American Geographers,* Vol. 56 (1966), pp. 550–69; Curry, L., 'The Random Spatial Economy: an Exploration in Settlement Theory,' *Annals of the Association of American Geographers,* Vol. 54 (1964), pp. 138–46; Berry, B. J. L. and Garrison, W. L., 'Alternate Explanations of Urban Rank Size Relationships,' *Annals of the Association of American Geographers,* Vol. 48 (1958), pp. 83–99; Kulldorf, G., *Migration Probabilities,* Lund, Sweden: C. W. K. Gleerup, Lund Studies in Geography, Series B, Human Geography, No. 14 (1955); Melton, M. A., 'Intravalley Variation in Slope Angles Related to Micro-Climate and Erosional Environment,' *Geological Society of America Bulletin,* Vol. 71 (1960), pp. 133–44; Stewart, C. T. Jr., 'The Size and Spacing of Cities,' *Geographical Review,* Vol. 48 (1958), pp. 222–45.

36   HÄGERSTRAND, op. cit., p. 60.

37   YEATES, M. 'Some Factors Effecting the Spatial Distribution of Land Values 1910–1960,' *Economic Geography,* Vol. 41 (1965), pp. 57–70; Thomas, E. N., 'Areal Association Between Population Growth and Selected Factors in the Chicago Urbanized Area,' *Economic Geography,* Vol. 36 (1960), pp. 158–79; Stevens, B. H., 'An Application of Game Theory to a Problem in Location Strategy,' *Papers and Proceedings Regional Science Association,* Vol. 7 (1961), pp. 143–57.

38   DACEY, M. F. 'Modified Poisson Probability Law for Point Patterns More Regular than Random,' *Annals of the Association of American Geographers,* Vol. 54 (1964), pp. 559–65.

39   RAYNER, J. N. *An Approach to the Dynamic Climatology of New Zealand,* Christchurch, New Zealand: University of Canterbury, Ph.D. Thesis (1965). Other examples appear in Berry, B. J. L. and Marble, D. F. (Eds.), *Spatial Analysis: A Reader in Statistical Geography,* Englewood Cliffs, New Jersey: Prentice-Hall Inc. (1967), specific examples in this book include Curry, L., 'Climatic Change as a Random Series,' pp. 184–94 and Sabbagh, M. E. and Bryson, R. A., 'Aspects of the Precipitation Climatology of Canada Investigated by the Method of Harmonic Analysis,' pp. 250–68.

40   DE CANI, J. S. 'On the Construction of Stochastic Models of Population Growth and Migration,' *Journal of Regional Science,* Vol. 3, No. 2 (1961), pp. 1–13.

41   BERGMANN, op. cit., p. 140.

42   Ibid., p. 127.

43   Ibid., p. 88.

9 Geography and analogue theory[1]

Richard J. Chorley
University of Cambridge, U.K.

One of the most striking characteristics of geographical analysis which this subject has in common with the other natural and social sciences is the high degree of ambiguity presented by its subject matter and the attendantly large 'elbow room' which the researcher has for the manner in which this material may be organized and interpreted. This characteristic is a necessary result of the relatively small amount of available information which has been extracted in a very partial manner from a large and multivariate reality, and leads not only to radically conflicting 'explanations' of geographical phenomena but to differing opinions regarding the significant aspects of geographical reality which are worth exploring. Even within a circumscribed body of information there is no universally appropriate manner of treatment, and such treatment is often conditioned either by the general systematic framework which one (often subconsciously) adopts as an appropriate setting for the information[2] or by the type of question which one is prepared to ask about the 'real world'. The change in character between the geographical methodologies of the nineteenth and twentieth centuries (and, for that matter, between the methodologies of botany or social anthropology), for example, lies very largely in the abandonment of attempts at causal explanation in favour of functional studies.

Where such ambiguity exists, scholars commonly handle the associated information either by means of *classifications* or *models*. Much has been written on the subject of geographical classifications, which usually result from the accumulation of a backlog of information which is then dissected and categorized in some convenient manner. Model-building, which sometimes may even precede the collection of a great deal of data, involves the association of supposedly significant aspects of reality into a system which seems to possess some special properties of intellectual stimulation. This is not to imply that classifications and models do not share some common ground, and, indeed, the 'genetic classifications' of national economic stages[3] and of shorelines[4] form something of a link between them. However, in their extreme forms, classifications and models are sharply differentiated, and it is with the employment of model-building, or *analogue theory*, in geography that this paper is concerned.

186

The use of models

The use of analogy has long been recognized as a powerful tool both in the reasoning process and in throwing a new light on reality. The concept of analogy was first applied by the Greeks to expose a similarity in geometrical proportion between two objects, but was later extended to include a much wider range of similarities, including qualitative ones.[5] Reasoning by analogy involves the assumption of a resemblance of relations or attributes between some phenomenon or aspect of the real world in which one is interested and an analogue or model. The basic assumption involved is that 'two analogues are more likely to have further properties in common than if no resemblance existed at all, and that additional knowledge concerning one consequently provides some basis for a prediction of the existence of similar properties in the other.'[6] A model, or analogue, must belong therefore to a more familiar realm than the system to which it is applied.[7] In other words, the problem is translated into more familiar or convenient terms such that a useful model involves a more simplified accessible, observable, controllable, rapidly developing, or easily formulated phenomenon from which conclusions can be deduced, which, in turn, can be reapplied to the original system or real world.

'In general, any two things, whether events, situations, creatures, or objects, can be said to be analogues if they resemble each other to some extent in their properties, behaviour or mode of functioning, so that, in practice, the term analogue is used rather loosely to cover a wide range of degrees of resemblance.'[8] This wide range of resemblance may lead the model-builder to employ as varied analogues as a past situation or happening in the real world, a mathematical simplification, or a construction built of string and wire.

Obviously the use of analogues in any process of reasoning involves very apparent difficulties and dangers, some of which will be treated in later sections. However, their use can be extremely productive providing there is no reason to believe that the points of similarity between the analogue or model and the real world situation are irrelevant to the matter under investigation, or that there is 'no *a priori* knowledge that the two analogues differ in such a way as to prohibit any similarity in the particular property which is the subject of the prediction.'[9] In spite of the dangers of analogy, physical scientists have used it freely in the past because the conclusions can usually be checked by experiment or further observation;[10] but, largely deprived of these checks, social scientists have been more cautious in their use. One of the more obvious exceptions to this has been the development of 'social physics,'[11] in which physical principles have been used as models for the behavior of man in society, such that, for example, the size and spacing of towns have been examined from the standpoint of Newton's gravitational law, the equalization of social classes and races have been viewed in the context of a thermodynamic model as a tendency towards maximum 'social entropy,' and, in somewhat the same manner, the political state and human society have been treated by the social Darwinists in the light of the organic model provided by the theory of evolution. Although many of these attempts have failed, for reasons which will be treated later, some have proved important vehicles for analysis and research. As Bunge has pointed out, it is one of the features of geographical methodology that the idea of uniqueness of geographical phenomena (the 'idiographic' notion) has retarded the overt application of analogue theory to the subject, although the latter is very commonly employed as a means of explanation and analysis in less obvious ways.[12]

Finally, in the field of definition, it is important to differentiate between models and theories. Analogy is often a fruitful source of suggestion for hypotheses for further inductive investigation, but alone it cannot 'prove' anything. A model becomes a theory about the real world only when a segment of the real world has been successfully mapped into it, both by avoiding the discarding of too much information in the stage of abstraction (see next section) and by carrying out a rigorous *interpretation* of the model results into real world terms. 'As a theory, (a model) . . . can be accepted or rejected on the basis of how well it works. As a model, it can only be right or wrong on logical grounds. A model must satisfy only internal criteria; a theory must satisfy external criteria as well.'[13] In short, reasoning by analogy is a step toward the building of theories, such that 'a promising model is one with implications rich enough to suggest novel hypotheses and speculations in the primary field of investigation.'[14]

A model for models

Following the efforts of Coombs, Raiffa, and Thrall[15] and of Aronow[16] to formalize broad sectors of the analogy reasoning processes, an attempt has been made in Fig. 2 to provide a more detailed analysis of the way in which models may be employed both in scholarship as a whole and in geography in particular. It is important to stress at the outset that this diagram differs from a board designed for a child's dice game in that success is seldom measured in terms of the model-builder's ability to negotiate all the hurdles on the course (although the most important model-builders have done so), the only proviso being that the course must be completed in some manner. The diagram is simply composed of a series of *steps* (A_{1-6}) each of which contains some aspect of the real world, model, observation, or conclusion; these are connected, sometimes in a very loose and varied manner, by *transformations* (T_{1-6}) whereby the reasoning process is advanced or checked upon. Each of these transformations introduces (in the language of the information theorist) the possibility of 'noise', insofar as the processes of simplification and translation involve, on the one hand, the discarding of information some of which may eventually be proved to have been useful information (popularly, though rather inaccurately, termed 'throwing out the baby with the bath water'), and, on the other, the possibility of the introduction of new information which is irrelevant (commonly termed 'clouding the issue' or 'introducing

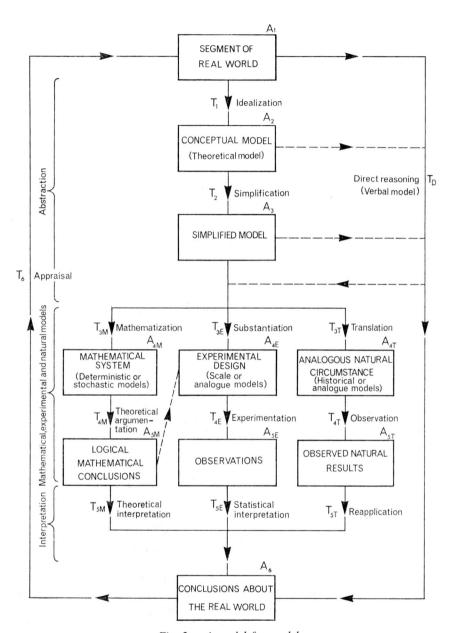

Fig. 2.   A model for models.

a red herring'). The most useful employment of analogues is that which succeeds in causing the least noise.

1   ABSTRACTION

This sequence of steps represents the most difficult type of model-building because, in developing a simplified but appropriate model for a given object system or segment of the real world (A_1), huge amounts of available information are being discarded (T_1 and T_2), and therefore much noise is being potentially introduced. The transformation of *idealization* (T_1) is especially difficult and its successful negotiation is very much a function of the intuition, luck, knowledge, experience, and genius of the model-builder. It involves extracting from the mass of information about the real world those aspects which are held to be especially significant, in that they seem to fit together into some sort of pattern,[17] and it is readily apparent that the appropriateness and significance of those parts of reality so abstracted depend very much on the creative ability and vision of the model-builder.[18] One of the major sources of noise in this transformation is introduced by the phenomenon of *feedback*. This occurs because our belief as to what 'facts' about the real world are important, appropriate, significant, or even existent is to a certain extent conditioned by the framework of thought which we have consciously or unconsciously adopted.[19] Idealization largely involves reaching some conclusion about the relationships in the real world which really matter[20] so that the irrelevant 'dead wood' can be pruned away to expose the significant relationships then susceptible to further examination. An example of the operation of feedback is provided by geophysics in which the selection and interpretation of the 'facts' of seismograph recordings largely depend on the previous notions of the operator regarding the internal structure of the earth.

By such means a *conceptual model* (A_2) is derived which, having some basis of observed facts and regularities, contains a mental image of the significant 'web of reality' that may have come quite unambiguously from the simplification of previous empirical knowledge or may seem to have sprung largely from intuition or imagination.[21] Many attempts at model-building depart from formality at this point and pass by *direct reasoning* (T_D) to a *hypothesis* or some *conclusions about the real world* (A_6) which, if successfully *appraised* (T_6) against the real world, may form the basis of a *theory*. Usually, however, the conceptual model is still too complex to handle, and it is subjected to further *simplification* (T_2) by the discarding of still more extraneous information to lay bare to the bone what are considered to be the simplest and most significant aspects of the basic matter at issue. This transformation is somewhat less noisy than T_1 because both ends of it are more completely understood[22] and in practice it may represent a whole series of simplifications the aim of which is to retain the simplest and most significant aspects of the problem, while removing any irrelevant material which might obscure the fundamental relationships and prevent a satisfactory solution. The final product of abstraction is often a *simplified model* (A_3) which, if at all successful, has been reduced, by the laying aside of minor variables and the discarding of irrelevant information, to a condition where the fundamental symmetries and relationships lie exposed for precise definition and further treatment.

Usually, however, the simplified model is to a greater or lesser extent unsatisfactory, and this is because the whole process of abstraction forces the model-builder to strike a balance between his desire to distill the problem down to its essence and to eliminate the possible sources of noise which excessive simplification may introduce. There are consequently few examples of largely successful simplified models. Newton's model of gravitation is the most outstanding. In geography most attempts at model-building by abstraction have met with minimal success, but it is notable that those which are currently judged as most successful are the conceptual models in which excessive simplification has been avoided. These contrast with the overly simplified geographical models of a deterministic character so popular in the nineteenth century. Thus the simplified 'place, work, people' concept of Le Play[23] seems to us to be a less real and appropriate model than the 'regional synthesis' of Vidal de la Blache,[24] in which too much information has been retained but very little noise introduced. So too the simplified economic models of Von Thunen[25] and Weber,[26] with their assumption of fixed markets and sources of raw materials, seem less attractive than the freer and more complex models of Christaller[27] and Lösch.[28] It is natural that the excessive pruning of simplified models should recommend them to certain scholars, and it is very interesting that Wooldridge has written of the 'superbly "clean" and intellectually attractive quality' of W. M. Davis's cyclical concept of landform development.[29] To these scholars the less simplified model of landform development adopted by G. K. Gilbert has little to commend it by comparison.[30] In short, one often has to choose between the clean but noisy model of Davis and the less clean but relatively quiet model of Gilbert.

Simplified models abound in geography and related fields, exhibiting various degrees of failure to measure up to reality within an oversimplified framework. Ritter's notions regarding the development of peoples under differing geographical conditions,[31] Mackinder's 'Heartland' theory,[32] Malthus' population growth model,[33] Frederick Jackson Turner's 'frontier hypothesis' of national development,[34] and Suess' eustatic theory[35] are examples of such simplified models in which too much truth has been sacrificed for too much simplicity. It must not be assumed, however, that in all instances (except under the hand of genius) increasing simplification produces detrimental results. Some simplification may be significant or successful in that it transforms a segment of the real world into a new dimension which is neither 'better nor worse' but merely different and interesting. One has only to compare a realistic landscape painting from the Dutch School, with an impressionist landscape painting from the French Impressionist School and an abstract modern 'landscape' painting to be impressed both by this transformation and by the unity of knowledge which is fostered by attention to analogue theory. A Cezanne painting is cleaner but more noisy than one by Van Ruysdael!

2   MATHEMATICAL, EXPERIMENTAL, AND NATURAL MODELS

The derivation of some form of simplified model enables structures and relationships to appear which are capable of further exploitation, commonly in such a manner that prediction can be attempted. This exploitation can be pursued by means of mathematical, experimental, or natural models.

a.   Mathematical models   Mathematical models include those which have been most spectacularly employed by the model-builders of genius, and involve the adoption of 'a number of idealizations of the various details of the phenomena studied and in ascribing to the various entities involved some strictly defined properties.'[36] The essential features of the phenomena are then 'analogous to the relationship between certain abstract symbols which we can write down. The observed phenomena resemble closely something extremely simple, with very few attributes. The resemblance is so close that the equations are a kind of working model, from which we can predict features of the real thing which we have never observed.'[37] The first task in the construction of a mathematical model is the language transformation (T_{3M}) from the words of the simplified model to mathematical symbols (*mathematization*) so as to produce a *mathematical system* (A_{4M}). These systems can be of two broad kinds, deterministic or stochastic.[38]

Deterministic models are based on the classic mathematical notion of direct cause and effect, and consist of a set of mathematical assertions from which consequences can be derived by logical mathematical argument.[39] Commonly the reasoning exploits the assumed simple or multiple relationships between a number of interlocked factors which have been identified in the simplified model.[40] The best example of the application of this type of reasoning to geomorphology is provided by Jeffreys, who developed a mathematical model for the denudation of the land surface by runoff, and by this means deduced theoretically the form of the resulting peneplain.[41] The most popular geomorphic application of deterministic mathematical models has been in the deduction of the forms accompanying slope recession, summarized by Scheidegger,[42] and a similar method has been employed by Miller and Zeigler[43] to predict the expected patterns of coastal sediment size and sorting. Most mathematical models of the atmospheric circulation are of the deterministic sort, as are those involved in numerical forecasting.[44]

Deterministic mathematical models having relevance to human geography commonly possess a strongly experimental flavor, demonstrating the lack of reality in any rigid separation of mathematical, experimental, and natural models, which inevitably grade into one another. Thus, Beckmann has attempted to minimize the costs of interlocal commodity flows by the theoretical use of the hydrodynamics 'equation of continuity';[45] Lighthill and Whitham used the principle of kinematic waves, suggested by theories of flow around supersonic projectiles and of flood movement in rivers, to investigate the flow and concentration of traffic on crowded arterial roads;[46] and Richards similarly investigated traffic flow on the basis of the mathematics of fluid flow.[47]

Stochastic models, based upon probability instead of mathematical certainty, are obviously of more promise in investigations relating to human geography rather than to the physical branches of the discipline. However, Leopold and Langbein have used the thermodynamic engine model to illustrate the steady-state concept applied to the longitudinal profile of a river, as well as the mathematical random-walk model to rationalize the stream network geometry of a drainage basin.[48] In the human field Isard has constructed a spatial economic model of a statistical character,[49] Hägerstrand's 'innovation wave' model is similarly based,[50] as is the mathematical theory of population clusters by Neyman and Scott, which was inspired by the kinetic gas theory.[51] More recently, Garrison has suggested the employment of an electronic computer to develop a stochastic mathematical model for city growth.[52]

Both the deterministic and stochastic mathematical models must be susceptible to logical *mathematical argumentation* (T_{4M}), a transformation which is virtually noise-free and involves the solution of the equations forming the basis of the mathematical system to provide *logical mathematical conclusions* (A_{5M}). Although these conclusions are susceptible to *theoretical interpretation* (T_{5M}) into *conclusions about the real world* (A_6), it is important to recognize that, of themselves, mathematical models do not provide explanations of the real world, but merely allow conclusions to be drawn from the original mathematical assumptions.[53] Some authors even take the extreme position that 'The mathematician, as such, has no responsibility whatever for the degree to which idealizations may represent a real situation. His work consists in discovering, developing, and/or applying methods of abstract logic dealing with form, structure, quantity, etc., and his responsibility is discharged completely and with honour if he avoids *internal* error.'[54] A more balanced view of the need to anchor mathematical models to reality in the earth sciences has been given by Jeffreys, who wrote 'Such treatment is . . . highly desirable; for a mathematical investigation enables us to specify accurately the causes we are taking into account, and the correspondence or divergence between the effects it predicts and the actual phenomena indicated the extent to which we have succeeded in tracing the more important causes. The differences revealed may then lead to the discovery of further causes, and thus observed facts may gradually become understood in greater completeness and detail.'[55]

b.   Experimental models   A second manner in which a simplified model can be further treated, in order to examine certain phrases of its operation or to attempt predictions about it, is by *substantiation* (T_{3E}). In this process the concepts of the simplified model are reproduced as tangible structures (i.e., the usual concept of a 'model') by a transformation which is inherently very noisy. Two types of experimental models may be conveniently differentiated, the scale model and the analogue model.

Scale models are closely imitative of a segment of the real world, which they resemble in some very obvious respects (i.e., being composed mostly of the same type of materials), and the resemblance may sometimes be so close that the scale model be considered as merely a suitably controlled portion of the real world.[56] Thus, scale models have been employed with much success in physical geography. Friedkin describing the use of scale models to investigate the phenomenon of meandering,[57] and Rouse the use of model techniques in meteorological research.[58] The most obvious advantages of the use of scale models are the high degree of control which can be achieved over the simplified experimental conditions and the manner in which time can be compressed. However, the fundamental problem attending their construction is that changes of scale affect the relationships between certain properties of the model and of the real world in different ways. Problems of scaling and of dimensional similarity have been well treated by Langhaar[59] and Duncan,[60] while Strahler[61] and Stahl[62] have treated the application of dimensional analysis to the natural sciences. This inherent lack of ability to produce faithful scale models is most strikingly exemplified by the discrepancy between geometric and kinematic model ratios. Such discrepancies can be circumvented in any of three interrelated ways. First, a distortion of one important attribute can (usually by rule of thumb) be reduced or eliminated by the distortion of another attribute – for example, a distortion of the vertical scale of river models enables the effects due to turbulence to be more or less faithfully reproduced. The second, and most important, way in which analogous model ratios can be produced is by dimensionless combinations of attributes. Thus a combination of density, velocity, depth, and viscosity (combined in the Reynold's number) enables viscous effects to be accurately reproduced; and a combination of velocity, length, and the acceleration of gravity (e.g., the Froude number) is important where gravity effects need to be accurately scaled in the model. Third, one or more of the media can be changed in the model to assist the true scaling of other effects, and early German experiments with wave tanks even involved the substitution of lower density brandy for the more conventional water! Such considerations naturally lead into the second type of experimental models, the analogue model.

Analogue models involve a radical change in the medium of which the model is constructed. They have a much more limited aim than scale models in that they are intended to reproduce only some aspects of the structure or web of relationships recognized in the simplified model of the real world segment.[63] Such a transformation is obviously rather difficult and a great potential source of noise. Although such analogue models can, at best, merely furnish the basis for plausible hypotheses, rather than proofs, their use can be extremely valuable. An electronic computer may be considered as an experimental analogue model of the human brain. Other such models of especial interest to the physical geographer include the construction of a hydraulic analogue model to simulate freezing and thawing of soil layers,[64] Lewis and Miller's use of a kaolin mixture to simulate some features of the deformation and crevassing of a valley glacier,[65] Reiner's rheological models for the simulation of different types of deformation;[66] and Starr's 'dishpan' analogue in which the circulation of water in a heated and rotating pan was made to exhibit many of the gross features of the atmospheric circulation in a hemisphere.[67] It is, however, in the fields abutting on human geography that analogue experimental models have been most strikingly applied. Hotelling early used a non-constructed heat flow analogue to present a theory of human migration.[68] But it is the spatial economists who have been especially interested in the use of these models. Enke employed electric circuits to attack a problem of spatial price equilibrium, in which voltage was identified with price, and current with real goods;[69] Brink and de Cani similarly employed an electrical analogue machine in determining the location of service points to serve a number of customers of known demand, so as to minimize transport costs;[70] and Stringer and Haley constructed a pulley and weight 'transport optimizing machine.'[71] Needless to say, most such machines have been displaced in favor of the electronic analogue computer.

Whether a scale or analogue experimental model is used, their construction is followed by *experimentation* (T_{4E}), leading to a set of experimental *observations* (A_{5E}) which remain to be interpreted in terms of the real world in a manner similar to the theoretical interpretation of the conclusions from a mathematical model. Very occasionally these results from a mathematical model are transformed (T_s) into experimental terms (a very noisy process) as a further step in testing their agreement with the real world.[72]

c.   *Natural models*   The third manner in which simplified models can be exploited and used as a basis for further analysis and prediction is by their *translation* (T_{3T}) into some *analogous natural circumstance* (A_{4T}) believed to be simpler, better known, or more readily observable. Such translations are inherently of two kinds – historical and analogue.

The employment of historical natural models involves a translation of the simplified model into a different time and/or place, on the assumption that what has happened before will happen again, or that what happens here will happen there.[73] Such a translation is an obvious source of noise because the multi-component character of most models means that, even where a reasonable analogue can be found, 'it is to be expected that the importance of each of these factors will be different in the two cases, and frequently the dominant influence in each case is exerted by a different factor'.[74] However, some historians have naturally been attracted to the historical analogue concept (e.g., Toynbee). For example, assumptions have been made that the present demography of India has some features in common with that of pre-industrial revolution Europe, or that similarities exist between the characteristics of feudalism in seventeenth to eighteenth century Russia and Medieval Europe. The most obvious geographical analogue which is applied to a different time (apart from some of the broad historical geographical analogues used in the last century, e.g., by Ritter and Ratzel) is presented by the analogue forecasting methods at present being developed in meteorology.[75] Reasoning from 'experience' of analogous situations elsewhere is a common basis for current geographical planning, but an obvious general example of such a spatial analogy translation is the fashion model who is considered by the prospective purchaser of a dress to share some analogous features with her!

Analogue natural models imply the translation of the simplified model into a different natural medium – a process of the most noisy kind. Problems of social geography are being attacked on the basis of physical analogues. Bunge[76] has compared the shift of highways with the shift of rivers, and has applied the mathematics of crystallography to central place theory in considering the growth and shrinkage of market areas as similar to the patterns associated with two-dimensional crystals; and Garrison has developed an 'ice-cap analogy' of city growth.[77] The social physicists obviously utilize such models; and in the field of physical geography, Chorley has attempted some mechanical explanation for the variation in drumlin shape on the basis of some of the factors controlling the shape of birds' eggs.[78] The introduction of noise into such a transformation stems from the very real danger of comparing the original phenomenon with another which is equally, or even more, complicated and equally unknown.[79] Comparisons with living organisms are thus especially dangerous, as has been shown by failure of organic theories of the state popular among geopoliticians. Even the use of the 'loaded' terms of 'youth', 'maturity', and 'old age' by Davis has introduced into modern geomorphology a great deal of quite extraneous evolutionary noise.[80]

In any event, an appropriate analogous natural circumstance, if subjected to appropriate *observation* (T_{4T}), may yield *observed natural results* (A_{5T}) which, through *reapplication* (T_{5T}), may lead to some testable conclusions about the real world situation in which one is interested.

3   INTERPRETATION AND APPRAISAL

Little further remains to be said regarding the problems of interpreting the conclusions derived from mathematical, experimental, or natural models (T_5) to reach conclusions which are appropriate to the real world situation (A_6), except to stress the potential noise of this transformation. It is interesting that many lines of reasoning in geography reach these conclusions by a process of more or less *direct reasoning* (T_D) – sometimes with a minimum of model building, by what has been graphically termed the 'eyeball' method. The value of this method is obviously very much determined by the degree of shortsightedness of the operator and the tint of his spectacles!

Successful *appraisal* (T_6) involves the checking of conclusions derived from model-building with the real world, so that a hypothesis can be developed. If sufficient checks of this kind lead to similar conclusions, a *theory* may be developed. This appraisal is an indispensable step, no matter what the simplicity or sophistication of the previous model-building, and on this the success or failure of the whole reasoning process hinges.

Conclusion

The use of analogous models, in the wide sense here employed, is an important, and it might be argued instinctive, adjunct to the reasoning process in the natural and social sciences. In these disciplines, where so much ambiguity exists in the manner in which phenomena may be interpreted, it seems characteristic that much of the most satisfying scholarship includes the utilization of some sort of model, the successful construction of which involves the highest levels of intellectual attainment. It is important, therefore, that the geographic applications of analogue theory should be placed on as rigorous a footing as possible, so that we may recognize in the work of ourselves and others the possibilities and dangers attendant upon reasoning by analogy. These dangers have been stressed throughout the present analysis, and involve questions of the following character: Is the analogue model truly analogous in the sense in which it has been applied? Has an excessive discarding of information introduced so much noise that the simplified model is effectively divorced from reality? Is reality of too multivariate a character to be susceptible to reasoning by analogy? Are leaps into different domains justifiable? Above all, does the use of models introduce too great a detour into the reasoning process, and the repeated 'decanting and filtering' destroy the reality which must necessarily be preserved in scientific analysis? It would appear that, where analogy is used in a controlled and careful manner, the answer to all these questions must be a satisfactory one. Just as no analogue model is completely successful, and even the most significant ones (e.g., that of Newton) are eventually superseded, so, few models are without some value. Model theories are like torches of varying size and intensity, shining in different directions and each illuminating some novel aspect or relationship of reality. With all their defects and distortions, they are often the most convenient vehicles for analyzing, interpreting and expressing our concepts of the real world. Any better understanding of analogue theory cannot but help us to reach towards a deeper appreciation of the methodological basis, essential character, and limitations which geographical scholarship shares with other natural and social sciences.

References

1   Reprinted from *Annals of the Association of American Geographers,* Vol. 54, No. 2 (March 1964), Washington, D.C.

2   CHORLEY, R. J. 'Geomorphology and General Systems Theory,' *U.S. Geological Survey, Professional Paper* 500–B (1962).

3   ROSTOW, W. W. *The Stages of Economic Growth,* Cambridge: Cambridge University Press (1960).

4   JOHNSON, D. W. *Shore Processes and Shoreline Development,* New York: John Wiley and Sons, Inc. (1919).

5   DANIEL, V. 'The Uses and Abuses of Analogy,' *Operations Research Quarterly,* Vol. 6 (1955), pp. 32–46.

6   ROSS, A. W. 'Approximate Methods in Operational Research,' *Proceedings of the First International Conference on Operational Research,* London: English Universities Press (1958), Ed. by Davies, M., Eddison, R. T., and Page, T., p. 58.

7   BLACK, M. *Models and Metaphors,* Ithaca: Cornell University Press (1962), Chapter 13.

8   ROSS, op. cit., p. 58.

9   Ibid., p. 59.

10   DANIEL, op. cit.

11   See, for example, ZIPF, G. K. *Human Behavior and the Principle of Least Effort,* Addison-Wesley, Massachusetts (1949).

12   BUNGE, W. 'Theoretical Geography,' *Lund Studies in Geography,* Series C, No. 1 (1962), pp. 7–13.

13   COOMBS, C. H., RAIFFA, H. and THRALL, R. M. 'Some Views on Mathematical Models and Measurement Theory,' in *Decision Processes,* by Thrall, R. M., *et al.,* New York: John Wiley and Sons, Inc. (1954), pp. 25, 26.

14   BLACK, op. cit.

15   COOMBS, RAIFFA, and THRALL, op. cit.

16   ARONOW, S. 'A Theory of Analogs,' *Proceedings of the First National Biophysics Conference,* Ed. by Quastler, H. and Morowitz, H. J., New Haven: Yale University Press (1959), pp. 27–34.

17   Ibid.

18   COOMBS, RAIFFA, and THRALL, op. cit.

19   CHORLEY, op. cit.

20   DANIEL, op. cit.

21   ANDREWARTHA, H. G. 'The Use of Conceptual Models in Population Ecology,' *Cold Spring Harbor Symposia on Quantitative Biology,* Vol. 22 (1957), p. 226.

22   ARONOW, op. cit.

23   LE PLAY, F. *Les Ouvriers Européens,* 1st ed. Paris (1855).

24   VIDAL DE LA BLACHE, P. *Le France de L'Est,* Paris (1917).

25   VON THÜNEN, J. H. *Der isolierte Staat in Beziehung auf Landwirtschaft und Nationalökonomie,* Hamburg (1826).

26   WEBER, A. *Ueber den Standort der Industrien,* Pt. 1, Tübingen (1909).

27   CHRISTALLER, W. *Die zentralen Orte in Süddeutschland,* Jena: G. Fischer (1935).

28   LÖSCH, A. *The Economics of Location,* 2nd ed., trans. by Woglom, W. G., New Haven: Yale University Press (1954).

29   WOOLDRIDGE, S. W. 'The Study of Geomorphology: A Review of "Geographical Essays," by W. M. Davis,' *Geographical Journal,* Vol. 121 (1955), p. 90.

30   HACK, J. T. 'Interpretation of Erosional Topography in Humid Temperate Regions,' *American Journal of Science,* Bradley Volume, Vol. 258–A (1960), pp. 80–97; Chorley, op. cit.

31   RITTER, K. *Die Erdkunde im Verhältnis zur Natur und zur Geschichte des Menschen,* Berlin (1817–18), 2 vols.

32   MACKINDER, H. J. *Democratic Ideals and Reality: A Study in the Politics of Reconstruction,* London: Constable and Co. Ltd. (1919).

33   MALTHUS, T. R. *An Essay on the Principle of Population as it Affects the Future Improvement of Society,* London (1798), 1st ed.

34   TURNER, F. J. 'The Significance of the Frontier in American History,' *Proceedings of the State Historical Society of Wisconsin,* Vol. 41 (1894), pp. 79–112.

35   SUESS, E. *The Face of the Earth,* Vol. II, Oxford University Press (1906), translated by H. B. C. and W. J. Sollas.

36   NEYMAN, J. and SCOTT, E. L. 'On a Mathematical Theory of Populations Conceived as a Conglomeration of Clusters,' *Cold Spring Harbor Symposia on Quantitative Biology,* Vol. 22 (1957), p. 109.

37   DANIEL, op. cit., p. 34.

38   ANDREWARTHA, op. cit.

39   COOMBS, RAIFFA, and THRALL, op. cit.

40 HAYNEL, J. 'Mathematical Models in Demography,' *Cold Spring Harbor Symposia on Quantitative Biology*, Vol. XXII (1957), pp. 97–102.

41 JEFFREYS, H. 'Problems of Denudation,' *Philosophical Magazine*, 6th Ser., Vol. 36 (1918), pp. 179–90.

42 SCHEIDEGGER, A. E. 'Mathematical Models of Slope Development,' *Bulletin of the Geological Society of America*, Vol. 72 (1961), pp. 37–49.

43 MILLER, R. L. and ZEIGLER, J. M. 'A Model Relating Dynamics and Sediment Pattern in Equilibrium in the Region of Shoaling Waves, Breaker Zone and Foreshore,' *Journal of Geology*, Vol. 66 (1958), pp. 417–41.

44 SUTTON, O. G. *Understanding Weather*, Penguin Books Ltd (1960), pp. 134–56.

45 BECKMANN, M. 'A Continuous Model of Transportation,' *Econometrica*, Vol. 20 (1952), pp. 643–60.

46 LIGHTHILL, M. J. and WHITHAM, G. B. 'On Kinetic Waves: II, A Theory of Traffic Flow on Long Crowded Roads,' *Proceedings of the Royal Society*, London, Series A, Vol. 229, No. 1178 (1955), pp. 317–45.

47 RICHARDS, P. I. 'Shock Waves on the Highway,' *Journal of the Operations Research Society of America*, Vol. 4 (1956), pp. 42–51.

48 LEOPOLD, L. B. and LANGBEIN, W. B. 'The Concept of Entropy in Landscape Evolution,' *U.S. Geological Survey*, Professional Paper 500–A (1962), 20 pp. 20 pp.

49 ISARD, W. *Location and Space Economy*, New York: John Wiley and Sons, Inc. (1956).

50 HÄGERSTRAND, T. 'The Propagation of Innovation Waves,' *Lund Studies in Geography*, Series B, No. 4 (1952).

51 NEYMAN and SCOTT, op. cit., pp. 109–20.

52 GARRISON, W. L. 'Notes on the Simulation of Urban Growth and Development,' *Department of Geography, University of Washington, Discussion Paper* 34 (1960).

53 BLACK, op. cit.

54 CAMP, G. D. 'Models as Approximations,' *Proceedings of the Second International Conference on Operational Research* (Aix-en-Provence, 1960), edited by Banbury, J. and Maitland, J. (1961), p. 22.

55 JEFFREYS, op. cit., p. 179.

56 ARONOW, op. cit.; BLACK, op. cit.

57 FRIEDKIN, J. F. 'A Laboratory Study of the Meandering of Alluvial Rivers,' U.S. Waterways Experiment Station, Vicksburg, Mississippi (1945).

58 ROUSE, H. 'Model Techniques in Meteorological Research,' in *Compendium of Meteorology*, edited by Byers, H. R., *et al.*, American Meteorological Society, Boston (1951), pp. 1249–54.

59 LANGHAAR, H. L. *Dimensional Analysis and Theory of Models*, New York: John Wiley and Sons, Inc. (1951),

60 DUNCAN, W. J. *Physical Similarity and Dimensional Analysis*, London: Edward Arnold and Co. (1943).

61 STRAHLER, A. N. 'Dimensional Analysis Applied to Fluvially Eroded Landforms,' *Bulletin of the Geological Society of America*, Vol. 69 (1958), pp. 279–300.

62 STAHL, W. R. 'Similarity and Dimensional Methods in Biology,' *Science*, Vol. 137 (1962), pp. 205–12.

63 BLACK, op. cit.

64 MASSACHUSETTS INSTITUTE OF TECHNOLOGY, 'Design and Operation of an Hydraulic Analog Computer for Studies of Freezing and Thawing of Soils,' *Soil Engineering Division, Department of Civil Engineering, M.I.T. Technical Report, No. 62, Corps of Engineers, U.S. Army* (May 1956).

65 LEWIS, W. V. and MILLER, M. M. 'Kaolin Model Glaciers,' *Journal of Glaciology*, Vol. 2 (1955), pp. 535–38.

66　REINER, M. 'The Flow of Matter,' *The Scientific American,* Vol. 201, No. 6 (1959), pp. 122–38.

67　STARR, V. P. 'The General Circulation of the Atmosphere,' *The Scientific American,* Vol. 195, No. 6 (1956), pp. 40–5.

68　HOTELLING, H. 'A Mathematical Theory of Migration,' unpublished Master's thesis, University of Washington (1921).

69　ENKE, S. 'Equilibrium Among Spatially Separated Markets: Solution by Electric Analogue,' *Econometrica,* Vol. 19 (1951), pp. 40–7.

70　BRINK, E. L. and DE CANI, J. S. 'An Analogue Solution of the Generalized Transportation Problem, with Specific Application to Marketing Location,' *Proceedings of the First International Conference on Operational Research,* London: English Universities Press (1957), edited by Davies, M., Eddison, R. T., and Page, T. (1958), pp. 123–37.

71　STRINGER, S. and HALEY, K. B. 'The Application of Linear Programming to a Large-scale Transportation Problem,' *Proceedings of the First International Conference on Operational Research,* London: English Universities Press (1957), edited by Davies, M., Eddison, R. T., and Page, T. pp. 109–22.

72　ARONOW, op. cit.

73　VON BERTALANFFY, L. 'General System Theory – A Critical Review,' *General Systems,* Vol. 7 (1962), p. 15.

74　ROSS, op. cit., p. 59.

75　SUTTON, op. cit., pp. 170–2.

76　BUNGE, op. cit., pp. 27–31.

77　GARRISON, W. L. (1962), lecture given in the 'Regional Science Seminar,' held at the University of California, Berkeley, in August 1962.

78　CHORLEY, R. J. 'The Shape of Drumlins,' *Journal of Glaciology,* Vol. 3 (1959) pp. 339–44.

79　DANIEL, op. cit.

80　DAVIS, W. M. 'The Geographical Cycle,' *Geographical Journal,* Vol. 14 (1899), pp. 481–504.

10   The logic of regional systems[1]

David Grigg
The University of Sheffield, U.K.

ABSTRACT   It is argued in this paper that regionalization is a similar process to classification. The terminology and procedures of the two are compared and it is concluded that there is a close similarity except that there is no direct analogy in regionalization with the individual of classification. It is suggested that operationally defined individuals partially overcome this problem. The histories of ideas about the nature and purpose of regional systems and classification systems are briefly compared and again found to be similar. Ten principles of classification derived from the work of logicians and taxonomists are then stated and the methods of constructing regional systems are examined in the light of these principles. It is shown that most of the points revealed by such an examination have been previously discussed by geographers, although they are arguing from different premises. Some of the principles, however, do raise problems which have hitherto received little attention.

It has often been noticed that whereas many regional systems have been devised by geographers there have been relatively few attempts to suggest any principles of regional division.[2] By principles we mean procedures which should be followed in constructing any system of regions, and which would be as applicable to a division of the whole world as for a county, and for agricultural regions or climatic regions as much as, say, religious regions. Presumably one person could, if sufficiently industrious, patient, and perceptive infer from existing systems a number of common principles. Yet this would assume that regionalization is a procedure which has no counterpart in any other science. The argument here, which has been implicit in the work of many other geographers,[3] is that regionalization is similar to classification.

If this is so then the principles of classification can be applied to the construction of regional systems. The principles of classification which are discussed later in this work are derived from two sources. First, since the time of Aristotle philosophers have discussed the logical procedures which should be observed when grouping objects into classes. Such rules of formal logic are designed to prevent internal inconsistencies arising within a classification scheme.[4] Second, the theory of classification and division has its most obvious application in taxonomy. In the last quarter of a century the established classifications in botany, zoology, pedology, and other sciences have been subject to increasing criticism, as advances in knowledge have undermined the basis of classifications established in the nineteenth century. Three trends have particular relevance to the problems of regionalization: First, the attempts to apply logical principles to existing botanical and zoological classifications;[5] second, the rise in zoology of numerical taxonomy, which is an attempt to quantify the description and classification of animals;[6] and third, the continuing reappraisal by pedologists of the purpose and methods of soil classification.[7]

Although it must be admitted that there are important differences between the classification of objects and the classification of areas, we believe that the application of the principles of classification to the construction of regional systems is a worthwhile approach to the problems of regionalization. The purpose of this paper is, first, to outline the procedures of classification; second, to examine the extent to which regionalization and classification are analogous procedures; and third, to consider the construction of regional systems in the light of the principles of classification. We hope that this paper will illustrate how similar many of the problems and confusions of classification and regionalization have been over the last 150 years.

The procedures of classification

Classification is the grouping of objects into classes on the basis of properties or relationships they have in common. Such a grouping can be reached by two distinct methods, classification and division.

In classification objects are grouped on the basis of *properties* they have in common. The objects which are to be classed are called *individuals* and the total number of individuals considered in any given classification system, the *universe* or *population*. Thus, for example, if the individuals being considered are human beings then they have properties such as skin color, height, weight, nationality, and religious affiliation. In the first stage of classification one property which is possessed in some degree by all the individuals is selected as the basis of grouping: the *differentiating characteristic*. Thus, for example, human beings could be grouped, on the basis of skin color, into three groups: those with yellow skins, those with black, and those with white skins. Now if the differentiating characteristic is carefully chosen, then other properties of the individuals will be found to change as the differentiating characteristic changes. Such a property is called an *accessory characteristic*. Thus in the example chosen, type of hair may be found to change as skin color changes. Or, to take another example, if soils are classified on the basis of soil texture, then as texture changes, so soil moisture content will change. When two properties change in such a manner they are often said to display *covariance*.

The nineteenth century taxonomists were concerned primarily with properties which were inherent in the objects classified. Modern biologists have, however, found a need to classify plants and animals on the basis of *relationships*. Simpson has made a useful distinction between these two approaches.

Following the terminology of a school of psychologists, Simpson recognized that grouping into classes can be made either on the basis of similarity between objects or on the basis of a relationship between connected and different objects. The former case he called *association by similarity*; this is the usual procedure in classification. The latter he called *association by contiguity*, and he illustrated his meaning as follows:[8]

> Association by contiguity (for our purposes) is a structural and functional relationship amongst things that, in a different psychological terminology, enter into a single *Gestalt*. The things involved may be quite dissimilar, or in any event their similarity is irrelevant. Such for instance, is the relationship between a plant and the soil in which it grows, between a rabbit and the fox that pursues it, between the separate organs that compose an organism, amongst all the trees of a forest. . . . Things in this relationship to each other belong both structurally and functionally to what may be defined in a broad but technical sense as a single system.

Individuals may then be grouped into classes on the basis of either similarity or relationship. The first stage in classification is to select a differentiating characteristic and group all the individuals in the universe into classes. This first set of classes is sometimes called a *category*. But the selection of one property and the formation of only one category of classes may give insufficient insight into the things being classified. The classes of the first category may then be grouped on the basis of a second differentiating characteristic into a second category. The classes of the first category then are included within and subclasses of the classes of the second category. Clearly this procedure can be carried further until all the classes are included within one superclass or class of all classes. This is illustrated in Fig. 3. A hierarchy of classes is thus established; the sets of classes or categories are sometimes referred to as classes of the first order, classes of the second order, and so forth. Perhaps the most celebrated example of a hierarchy is the orders of the Linnean system. It is perhaps significant to remember that Linné, when illustrating his hierarchy, took as an analogy the political map;[9] certainly this illustrates the hierarchial system extremely well. Thus, for example, in the United Kingdom the classes of the first order are parishes; every twenty or so parishes are grouped into hundreds. A number of hundreds grouped together constitutes a county whereas counties are themselves grouped together to make four countries, England, Ireland, Wales, and Scotland, which are all subordinate to the highest class, the United Kingdom.

A procedure closely allied to classification is that of *logical division*.[10] But whereas in classification individuals are grouped into classes and classes then included within superclasses, in division an initial class is taken as the

1.CLASSIFICATION: each rank of classes is a SET or CATEGORY

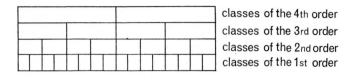

classes of the 4th order
classes of the 3rd order
classes of the 2nd order
classes of the 1st order

2. LOGICAL DIVISION

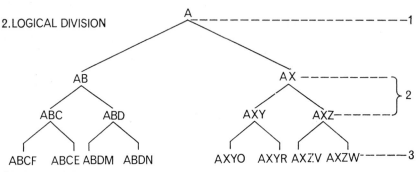

1. summum genum
2. subaltern genum or constituent species
3. infima species

Fig. 3. Diagrammatic representation of classification and logical division.

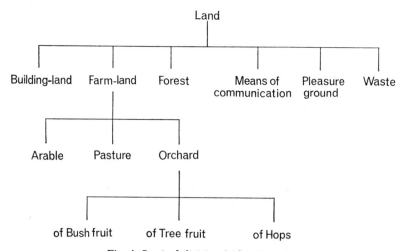

Fig. 4. Logical division (After Joseph).

universe and this class divided into subclasses on the basis of some principle. This is illustrated in Figs. 3 and 4. The class to be divided is called the *genus*. In Fig. 4 the genus is Land, and the principle upon which it is divided is the use of land. The genus is divided into its *constituent species*, and each species can be further subdivided on the same principle, or as it is sometimes called, *fundamentis divisionis*. In Fig. 4 only farmland is further subdivided although clearly each species could be further subdivided as they have been in Fig. 3.

Logical division and classification are distinct but allied processes and they produce the same result, a classification system with a hierarchy of orders. Although logicians commonly separate the two proceses for discussion, as has been done here, many point out that in practical classification the scientist will use both procedures to establish his system.[11] Further, the principles of division, which are discussed later, apply equally well to classifi-

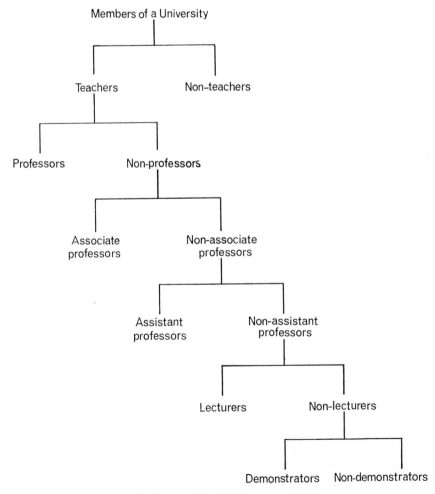

Fig. 5. An example of dichotomous division.

cation. However, there is one form of division which is quite different from classification and that is *dichotomous division*. The purpose of dichotomous division is not, as in logical division, to distinguish all the species within a genus, but simply to isolate a single species. The genus is divided upon the principle that a class must either possess a given characteristic or not possess it. An example is given in Fig. 5. Such a procedure normally is of little value in the sciences because it does not allow the formation of categories and there is no possibility of comparison and thus of useful generalizations. However, as will be seen later, there appears to be a closely analogous procedure in regionalization.

THE NEED FOR CLASSIFICATION

As Simpson put it:[12]

> If each of the many things in the world were taken as distinct, unique, a thing in itself unrelated to any other thing, perception of the world would disintegrate into complete meaninglessness.

The purpose of classification, then, is to give order to the objects studied. Without classification it would be impossible to:

1   Give names to things,
2   To transmit information,
3   To make inductive generalizations.

Classification is a necessary preliminary in most sciences; it is often argued that the state of classification is a measure of the maturity of a science. Thus Stebbing writes:[13]

> The earliest stage of a science is the classificatory stage: it is not long since botany passed beyond this stage and sociology has hardly done so yet.

Let us briefly consider the relevance of these three aims to the purposes of regionalization.

1   Names are given to parts of the earth's surface in a number of ways. The inhabitants of a particular area may themselves give it a name; alternatively an influential visitor to an area may give it a name which passes into common usage. Such names may have wide acceptance but rarely define the limits of an area with any precision. Thus, for example, names such as the Weald, the Lake District, or the Fens are used in every-day speech in England but have no precise connotation. A more rigorous areal classification is the administrative division which most countries possess. Such an areal classification is an essential prerequisite for effective government, for without it taxes cannot be raised, mail delivered, nor any of the every-day matters of organized living be pursued. It was dissatisfaction with this form of areal classification which led in the eighteenth century to the rise of the idea of the region. But

no regional system devised by geographers has ever had such widespread acceptance that administrative and political classifications have received. Indeed confusion has arisen in geography because of the use of the same name for different areas in different regional systems. Hence the varying usage of names such as 'the Middle East,' 'the Far East,' or 'the Midwest.' One might here draw an analogy with nomenclature in botany. Most common plants receive local names; but the same plant will have different names in different areas. Thus the necessity for a uniform classification system.

Whereas the exact naming of parts of the earth's surface is important, classification is the basis of a more important form of geographical nomenclature. Geographers require names or terms for similar features of the earth's surface which occur repeatedly. Again every-day language provides much of this terminology; e.g., mountains, plains, and valleys. But if understanding of the earth's surface is to advance, a more precise terminology is needed, and a necessary preliminary is a widely accepted areal classification of these features. Much of the terminology used by geographers is in fact derived from allied systematic sciences; but within the subject areal classification has produced a terminology. Thus, for example, although many geomorphologists are now skeptical of much of W. M. Davis's work, at the time when he wrote, his terminology, based on a genetic classification of landforms, was invaluable.

2   If a science is to progress, knowledge must be easily exchanged and easily passed on to new students of the subject. Now obviously no one person can know all facts about the character of all parts of the earth's surface; but if similar parts of the earth's surface are grouped together, then statements can be made about that areal class which are applicable to all the smaller parts of the class. This form of generalization is a form of intellectual shorthand which facilitates the transmission of knowledge. Thus, for example, no one person can know all the climatic characteristics of all the world's weather stations. But if the similar stations are grouped into climatic regions then the student can obtain an understanding of climatic differences.

3   But perhaps the most important purpose of classification systems is to permit inductive generalizations to be made about the objects studied. Areal classification, or regionalization, serves a similar purpose. Many generalizations in geography arise from the comparison of two different regional systems; thus, for example, if in any one country an areal classification on the basis of soil type is compared with a classification according to land use, then a number of generalizations about the relationships between the two may be inferred.

SOME MISCONCEPTIONS ABOUT CLASSIFICATION

It is generally agreed by logicians that there can be many valid classifications of a given universe of individuals.[14] Thus human beings could be classified according to their religion, their weight, or the color of their skin. The property chosen as the differentiating characteristic depends primarily upon the purpose of the classifications. Such would be a modern view: but the history of classification has been bedevilled by the distinction made by some logicians between *natural* classifications and *artificial* classifications. The definition of a natural classification varies. According to Jones it is that classification which correlates the greatest amount of knowledge about the objects under study.[15] To put it another way, a natural classification is one where the differentiating characteristic has the maximum number of accessory characteristics. An artificial classification is one which is designed for much more limited purposes and the differentiating characteristic has few if any accessory characteristics. Unfortunately, the interpretation put upon this distinction has led to confusion. Latta and Macbeath,[16] for example, defined a natural classification as one which attempts to express the actual order or system of things classified. Some taxonomists assume that if a differentiating characteristic with a large number of accessory characteristics is found then such a natural classification will be the correct classification and will serve all purposes equally well. Hence the art of the classifier is to find the one correct classification.

Indeed so much confusion has arisen from the use of the term 'natural classification' that it would seem wise to follow the suggestion of Gilmour and Walters[17] and use instead general-purpose and special-purpose classifications instead of natural and artificial.

Modern logicians and taxonomists would agree that there can be no natural classification in the sense that there is one and only one classification which will serve all purposes. Geographers will see an obvious echo of this controversy in the history of ideas about regions. Hartshorne[18] noted that, at the beginning of the nineteenth century and again in this century, there have been attempts to establish regions as real entities existing in nature. If regions were real entities then there could be a correct regional system. Although this view has been examined and rejected by geographers since Bucher,[19] it is a belief that still persists among some Western geographers; in the Soviet Union the majority believe that there can be a correct system of regions.[20]

It could be added here that even among those who recognize that regional systems are not a classification of entities that exist in nature there is still a tendency to forget that lines on a map are rarely real and that any given classification or regional system is but one way of looking at the world. Well-established regional systems tend to become, in the mind of some readers, the reality rather than simply a device for representing sections of reality. Such attitudes lead to curious consequences. It was, for example, once suggested that Köppen's classification of climate should be made the standard climatic classification.[21] In England Herbertson's system of natural regions has received similar veneration. It seems certain that neither Köppen nor Herbertson hoped that their classifications would be viewed so uncritically by later generations.

Although the belief that there can be a 'correct' classification has had unfortunate consequences in both the biological sciences and in geography, the misinterpretation of the word 'natural' has led to particular difficulties in the study of regions. The use of this word has been admirably surveyed by Hartshorne.[22] In England the term 'natural regions' was introduced by Herbertson,[23] who clearly initially confined its use to regions based solely on features of the physical environment, although in his later work he apparently hoped to extend the term to include human activity.[24] Unstead[25] used the term *geographical region* to describe regions where there was a high degree of homogeneity in both physical environment and human activity. The belief that an area can be logically divided into a system of such geographical regions is one that had some currency in England until recently. It is analogous to the belief in the biological sciences that all aspects of the objects studied can be incorporated into one natural classification.

Given the belief that there can be one 'correct' natural classification it is not difficult to envisage a situation where the construction of classifications becomes an end in itself. But a prime aim of classification is to enable inductive generalizations to be made about the objects studied,[26] not simply to arrange objects in classes. As Jevons pointed out:[27]

> There can be no use in placing an object in a class unless something more than the fact of being in the class is implied.

Some taxonomists have fallen into this trap. Classification becomes no more than a game of identification; once the object has been satisfactorily classed it is forgotten. In soil science some workers have been misled to the point that one recent critic of existing soil classifications could write that few workers are clear 'about what they are doing or why they are doing it.'[28]

A similar situation has arisen in geography. Many regional systems seem to be devised as an end in themselves; the definition of a region or regions becomes important in itself rather than as a means to understand the area in question. Here we may profitably return to the distinction between general-purpose classifications and special-purpose classifications. The natural system of the zoologist has, since Darwin, been based on the assumption that the classification explained the differences between the objects as well as arranging them in classes. In the last two decades some zoologists have pointed out that the properties chosen to classify animals have been weighted, for taxonomists have selected those properties which appear to demonstrate phylogenetic relationships.[29] It has been questioned whether in fact such a relationship is a sufficient explanation of differences. Further it has been pointed out that if differentiating characteristics are sought with such a purpose in mind, then the classification will not serve the purpose of many modern zoologists. One consequence of this criticism has been the rise of numerical taxonomy. This is an attempt to replace the genetic natural classification of animals by purely descriptive classification. Instead of those properties which are presumed to indicate phylogenetic relationships being selected, an attempt is made to use a much wider range of properties as the basis of classification, and further, to quantify the properties.[30]

A similar development may be seen in the history of ideas about regions. Systems of geographical regions were an attempt to embrace the totality of an area within a single classification; they are thus natural classifications in the technical sense. Further, they also tried to explain regional variations in the totality of things seen on the earth's surface. It was assumed that the physical environment and human activity covaried spatially because human activity was controlled primarily by the physical environment. Just as the concept of phylogenetic relationships has seemed to many biologists to be an unsatisfactory basis for classification, so geographical determinism has lost currency and validity among geographers as a satisfactory basis for classification.[31]

Thus it can be seen that classifications based upon such an assumption are unlikely to produce any fruitful generalizations about the objects under study. But this does not mean that general-purpose classifications have no value; they may still be used to transmit information although they may not offer any explanations about the objects classified. One might add that some would claim that a natural classification of areas that was also a satisfactory explanation of differences between those areas is still theoretically possible. But this seems a chimera upon whose pursuit geographers have already spent too much time.

Are regions areal classes?

Many geographers have used the terminology of classification theory in discussing regional systems and some have recognized that many of the principles of classification are applicable to regionalization.[32] But the closest comparison between the two modes of thought has been made but recently by William Bunge.[33] Bunge argued that the individual is represented in regionalization by 'place'; the properties of places are commonly called *elements* whereas differentiating characteristics are usually termed *criteria*. Places can be grouped into regions or areal classes. Bunge found the similarity between the two procedures to be remarkably close; indeed he went so far as to state that[34]

> ... geographers have independently rediscovered the entire logic of classification systems. ...

Bunge's argument seems well founded, and it can be further extended. We must begin by considering the types of regions commonly recognized by geographers. Berry and Hankin, following Hartshorne, recognized three types of region:[35]

1   The 'region' in the general sense in which the region is given *a priori*.
2   A homogeneous or uniform region: this is defined as an area . . . . within which the variations and co-variations of one or more selected characteristics fall within some specified range of variability around a norm, in contrast with areas that fall outside the range.
      Such a region, unlike that previously described, but like the functional region, is a result of the process of regionalization and is not given *a priori*.
3   A region of 'coherent organization' or a 'functional' region. This region

is defined as one in which one or more selected phenomena of movement connect the localities within it into a functionally organized whole.

Although this latter definition could perhaps have been more happily expressed, the distinction between functional and formal regions is one which most geographers would agree is valid and valuable; yet before the 1930's most regional systems were made up of uniform regions. But as Hempel[36] has pointed out, modern science deals increasingly not with the properties of objects but with their function and with the relationships between such objects. The rise of the idea of the functional region which deals essentially with interconnections between objects rather than with the similarities between those objects, seems to have stemmed from studies of the influence of urban centers over the surrounding areas, although the concept has since become much more sophisticated.[37] But it is of course not only geographers who have felt a need to classify objects on the basis of relationship between objects rather than simply on the basis of the properties of those objects. It was noticed earlier that Simpson[38] has distinguished between association by similarity and association by contiguity, a distinction which he himself took from the terminology of psychology and applied to zoology. Hence it seems that the development of functional and formal regions parallels a development in taxonomy; but it is not apparently a distinction commonly made in logic.

A perhaps more fundamental distinction between methods of arriving at regional systems is between the *synthetic* and *analytical* approach. Synthetic regionalization is a procedure frequently used by geographers, but the term was coined and is rightly associated with Unstead.[39] Unstead argued that the smallest unit area of the earth's surface is the *feature*, and that features (or individuals) can be grouped on resemblance to give first order regions called *stows*; such classes can be successively grouped into higher order categories called *tracts*, *subregions*, *minor regions*, and *major regions* in a hierarchy with five orders or categories. This would seem to be a process very similar to that of classification.

In contrast to this procedure, Herbertson's regional system was obtained by beginning with the world and dividing it into subclasses on the basis of a number of principles.[40] This is a quite different process from synthetic regionalization and has been called analytical regionalization.

This fundamental distinction between methods of arriving at regional systems has been recognized by Hartshorne and Whittlesey.[41] Hartshorne concluded that the two processes are complementary and not mutually exclusive. This is, of course, a statement that most formal logicians would agree with, for the two methods of regionalization appear to correspond to the two methods of grouping objects, classification (synthetic regionalization) and logical division (analytical regionalization). A similar divergence of approach characterizes other sciences, particularly plant ecology and pedology. Whittaker, in an excellent review of the problems of classifying plant communities, distinguished between the methods of Braun-Blanquet in Europe and Clements in the United States. The Braun-Blanquet school has emphasized the importance of floristic composition, detailed analysis, and careful sampling:[42]

> Classification might naturally proceed upward from small-scale units derived from local study to higher vegetation units, forming a hierarchy based on floristic relations.

In Clements' work on the other hand,

> ... regional vegetation types defined by physiognomy and dominance provided an initial general picture of major community types of a continent. More detailed classification proceeded downward from these geographic units.

As Whittaker points out, these two approaches reflect the different ecology of the two areas studied and the number of botanists available to undertake the work. Nonetheless they also represent the two main methods of grouping objects, either by classification or division. A similar divergence of approach has been noticed by G. Manil[43] in his review of methods of classifying soils. He distinguished between the analytical and descending method, which starts from general facts and principles and goes down to detailed categories, and the synthetic and ascending method, which begins with detailed categories (individuals) and works upward to higher categories.

Thus the distinction between classification and division is reflected not only in geography but also in plant ecology and in soil science. To some extent the distinction is a function of the scale of the classification. Any area may be divided; but to establish a classification for the world, when we need to know the measurable properties of very small unit areas, is extremely difficult. And although it may be theoretically possible to obtain a consistent hierarchy of co-ordinate classes, we may share Manil's skepticism about the practical possibilities when the individual we begin with is as small as a soil profile.

But there are perhaps more fundamental implications of the distinction between classification and division. In classification individuals are grouped together on the basis of some observable and measurable properties. No assumptions are made, or need to be made, about the cause of the differences and similarities between the individuals grouped in the same class. But in division the *genus* is divided on some principle and thus a division presupposes that there is some understanding of the system being constructed.[44] Analytical regionalization then often implies knowledge of the causes of similarities and differences between the objects studied. Hence it can be argued that a genetic classification is only valid if our understanding of the causes of these differences is full and complete.[45]

It is this point which has led to considerable controversy in the biological sciences. It is argued by many taxonomists[46] that a genetic classification of animals is misleading, partly because a classification, even if its causality is fully understood, does not serve all the purposes of modern zoology; and partly because they consider the traditional understanding of phylogenetic relationships to be incomplete. A similar situation has arisen in pedology.

The Russian school of soil classification was based on the assumption that climate is the major determinant of soil differences. This view is now held to be only partially true and thus genetic classifications based on this assumption are of limited value.[47] In geography similar instances may be cited. The belief, for example, that agriculture is mainly differentiated because of differences in climate has given rise to systems of agricultural regions which are based in the first place upon a climatic division of the earth's surface.[48]

DICHOTOMOUS DIVISION AND THE REGION

So far we have examined the similarities between regional systems and classification systems. Yet geographers have often delimited and discussed the content of a single area or region. To what extent is there a parallel between the methods of delimiting such a region and the procedures of classification?

Consider the procedure by which a geographer arrives at such a region, which for the purposes of this argument we will assume to be a uniform or formal region called the English Fenland. We have visited this area, examined it, read about it, and are sensibly persuaded that it is markedly distinct from adjacent areas. The properties which are considered to distinguish the area are then mapped and an approximate accordance found (Fig. 6). We may then feel justified in describing this area as 'a region' and may then proceed to analyze it in greater detail.

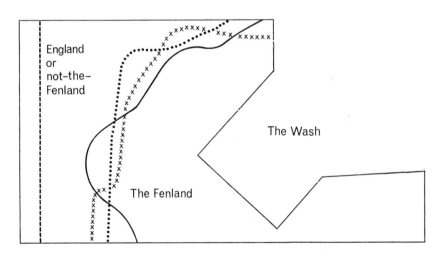

••••••••••••••• 15 ft Contour

xxxxxxxxxx Landward limit of peat and silt soils

—————————— Limit of area where arable land exceeds 80% of total farmland

------------- Boundary of Eastern England

Fig. 6. The delimitation of the English Fenland.

Now there seems to be no parallel in either classification or logical division for such a procedure, for only one class is formed. But the procedure seems to be very similar to that of dichotomous division. The purpose of dichotomous division is to isolate a single species within a genus, not to create coordinate classes or species.[49] The procedure is described by Stebbing.[50]

Any given class can be subdivided into two mutually exclusive and collectively exhaustive subclasses on the basis of a given characteristic which is possessed by every member of one class and is not possessed by any member of the other class.

Figure 7 illustrates how such a procedure is applied to the example of the English Fenland. The class which is subdivided is England; it is initially divided into Eastern England and not-Eastern England, and the successive principles of subdivision are similarly applied. We end by isolating two classes or regions, the Fenland and not-the-Fenland. Such a procedure seems to parallel the process of dichotomous division, and furthermore, is equivalent to the first type of region recognized by Berry and Hankin, those regions arrived at *a priori*.

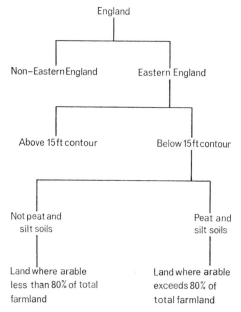

Fig. 7. The English Fenland by dichotomous division.

THE PROBLEM OF THE GEOGRAPHICAL INDIVIDUAL

So far the reader may agree that there is a considerable similarity between the procedures of classification and regionalization. But a number of objections can be raised: they all relate to the difficulty of finding a logical equivalent in regionalization to the *individual* of classification theory.

1.   It can be argued that all parts of the earth's surface are unique; classification and regionalization obscure this fundamental fact.

2.   Even if this objection is overcome it can be argued that parts of the earth's surface all have the property of location; all locations are unique by definition, hence this is a property that cannot be used as a differentiating characteristic. But a geographical classification which neglects location is of limited value.

3.   In both taxonomy and logic individuals are assumed to be discrete and separate entities; but the earth's surface, considered in its totality, or even in more limited terms, does not consist of a mosaic of entities. It is a continuum.

4.   Lastly, even supposing that we can define a geographical individual, a problem arises. In classification location is ignored in grouping similar individuals into classes; but in regionalization, when we group individuals into regions, it is assumed that all the individuals which are similar are also contiguous. If they are not, then a region is not formed.

There have been a number of discussions of 'uniqueness' but generally in a wider context than is necessary here.[51] Although it is certainly true that all parts of the earth's surface are unique, this does not seem an insuperable obstacle. First, it is also true that the objects grouped in taxonomy are unique. The uniqueness of parts of the earth's surface is neither more nor less unique than the objects studied in taxonomy. Second, experience shows that some properties and some combinations of properties repeat themselves over the earth's surface. Such spatial repetition invites explanation, and classification is a preliminary to explanation.

But if we consider location to be an essential property of places, which are our individuals, we are presented with a considerable dilemma. All locations are unique by definition; hence, whereas location can be a property, it cannot be a differentiating characteristic, for a differentiating characteristic must be a property possessed in some degree by all members of the universe. Both Bunge and Hartshorne recognized this problem and pointed out that the map is essential to areal classification and is indeed the only way of adequately expressing location in a classification system.[52]

At a higher level of abstraction geographers have traditionally distinguished between *generic* and *specific* regional systems. Generic regions are[53]

> ... those which fall into types and may therefore be said to be of a generic character, all the representatives of a particular type resembling each other in certain essential respects, according to the criteria selected – e.g., climate, character of vegetation, or human use.

Such regions may occur in different parts of the world but location is not a property used in their classification. A specific region is, however, a region.

> ... whose character is determined not only by the intrinsic conditions of the area in question but by its location and geographic orientation.

This distinction seems justified. In generic regional systems we consider an area independent of its relationship to adjacent areas. Clearly much that is discussed in geography deals with the relationship of areas to other areas, and a system of regions which excludes such relationships may be thought to be of limited value. But we have already noted the difficulty of incorporating the property of location into a classification system. If this is so, then systems of specific regions cannot be said to be analogous to classification systems. We may ask then if a system of specific regions can serve the purposes that classification systems serve. In particular can such systems be productive of inductive generalizations?

When we turn to our third problem we are on less debatable ground, but one nonetheless strewn with difficulties. In taxonomy there is no doubt as to what is being classified. In regionalization the issue is less clear.

We must begin by noticing that the problem of the individual does not arise in logical division (or analytical regionalization). A class is divided and subdivided but not reduced to the parts or individuals which make up the lowest set of classes (or the infima species).[54] It is true, as Hartshorne[55] pointed out, that if subdivision of the earth's surface is continued indefinitely there comes a point where we must inquire if there is an irreducible part of the earth's surface. But in practical division this rarely arises. In classification, in contrast, we begin with the individuals and proceed to group them into classes on the basis of similarity; thus the individual must be defined. Now many of the problems of the individual diminish if we assume that there cannot be a rigorous classification of the earth's surface based on all the properties it possesses; or to put it another way, based on the totality of the earth's surface. If we attempt such a natural classification it is clearly difficult to find individuals suitable for all the properties which geographers normally contend make up the totality of the earth's surface, such as climate, soils, vegetation, economic activity, and so forth.

But earlier we have decided that such classifications are impossible; let us now consider the problem of the individual in more limited classifications. The problem of defining an individual arises most acutely in the classifications of elements of physical geography such as climate, vegetation cover, soil types, and landforms. Discussion of what exactly the individual is in such studies has received less attention from geographers than it has perhaps merited.[56] Unstead[57] recognized the problem and called the individual 'the feature,' but gave little attention to the recognition or definition of such individuals. Linton[58] attempted a morphological classification: he argued that landforms are ultimately reducible to *flats* and *slopes*, and that these could be grouped together to form classes. But the recognition of these individuals seems to have been highly subjective. Bunge[59] has discussed the problem of the individual and considers 'place' to be the geographical equivalent. Although this is true in a theoretical analogy, it gives little guide to the nature of the individual in a practical classification of properties of the earth's surface.

But the problem of the individual is by no means unique to geography.

It arises in any subject where the things studied occur in a continuum or have areal expression, and we may profitably turn to the problem as it arises in other sciences. Two subjects where the individual is a matter of concern are phytosociology (phytocenology or plant ecology) and pedology. In phytosociology, plant communities are studied. It is obvious that a particular combination or plants of differing species occurs over the earth's surface; the classification of such communities is essentially areal as well as floristic, so that some individual is necessary. Much the same problems arise in classifying plant communities as arise in geographical classifications, and unfortunately there appears to be as great a confusion of method and terminology as in geography. These problems have been admirably discussed in a review article by Whittaker[60] who points out that phytosociologists are essentially seeking an equivalent, for plant communities, to the species in plant taxonomy. But botanists are split initially into two schools; those who consider that there is some particular natural unit inherent in vegetation and that the problem of the plant ecologist is to find the correct solution; and those who think natural communities are so complex that different properties may be with equal justification used as the basis of classification. This division of opinion is clearly analogous to the difference in geographical classification between those who believe the region to be a real entity and those who regard regions as simply 'intellectual constructs.' But whatever the approach adopted some 'fundamental units,' or individuals, must be grouped to form classes. In practice sampling has to be practised and the *stand* forms the individual.

Some ecologists have adopted a more holistic approach and have attempted to classify not only the vegetation of an area but the soils and the surface morphology as well. Here the problem of the ecologist becomes even closer to that of the geographer; but for the most part the problems of classification remain unresolved. A frequently quoted example is the work of Bourne who attempted to classify landscapes. His fundamental unit he called a *site* defined as:[61]

> an area which appears, for all practical purposes, to provide throughout its extent similar local conditions as to climate, physiography, geology, soil and edaphic factors in general.

Sites are unique, but sites of the same type occur again and again; thus sites may then be grouped into regions. Bourne's ideas appear to have influenced Unstead and certainly Linton, but they were worked out only for Southeastern England. The recognition of *sites* is a highly subjective process; and again such sites would not form useful individuals for classification of the phenomena dealt with by human geographers.

Classification has recently become a major issue in soil science and a number of writers have questioned the aims and assumptions of existing systems. The most significant result has been the revision of the United States soil classification. But even this interesting work has not resolved the methodological problems of soil classification. Soils change imperceptibly and one soil type merges into another. Thus the population, as in geography and ecology, is a continuum. Pedology is further hampered by the fact that hitherto the soil profile has been used as the basis of classification. This has two consequences. Short of covering the surface of the earth with soil pits, soil classifications must be based on sampling. Second, the soil profile is two-dimensional, whereas the classifications based upon them are three-dimensional. These problems have been partly overcome by basing classification on the principle of the modal profile. This has been explained lucidly by Cline:[62]

> One may visualise a class as a group of individuals tied by bonds of varying strength to a central nucleus. At the centre is the modal individual in which the modal properties of the class are typified. In the immediate vicinity are many individuals held by bonds of similarity so strong that no doubt can exist as to their relationship. At the margins of the group, however, are many individuals less strongly held by similarity to this modal individual than to that of any other class.

This interpretation of soil classification will be recognized by many geographers as a process that they themselves adopt in practical classification. But it is a method which has been subject to considerable criticism.

In a recent article Jones made two searching criticisms of existing methods of soil classification. In the first place he pointed out that the soil profile, hitherto used as the basis of classification, is a 'vertical section' and hence two-dimensioned. On the other hand, the soil is a continuum which is three-dimensioned, and a soil classification can hardly be based on a two-dimensioned entity. Jones also criticized the principle of the modal profile:[63]

> It can easily be appreciated that the complete range of any modal quality at each level of the classification will decrease as one approaches the lower orders of the classification. It must then become extremely difficult, if not impossible, to recognize either the extremes of the modal quality or even the modal expression itself. No soil property has a discrete or circumscribed existence ... Instead it moves immediately and imperceptibly either in the direction of greater intensity or towards complete absence. This state of affairs enforces upon our thinking a kind of calculus outlook in which we aim to build the structure of our classification from an infinite number of finite units ... The interval between each finite unit must be regarded as sufficiently small that in the ultimate summation the final structure will be endowed with a quality of continuity. In this manner the discrete quality of the soil profile is overcome and the three dimensional quality of the soil mass as a whole is recognized.

Jones doubted if the modal principle offers a logical basis for soil classification unless an astronomical number of profile pits are used. He concluded that it is fundamentally unsound to accept the soil profile as an:

> expression of soil cover, since soil is essentially part of landscape and a soil classification must contain within itself an inherent reflection of three dimensions.

He suggested that an entirely new approach to soil classification is necessary.

It is no part of this paper to discuss the methodological problems of ecology and pedology. All that needs to be noted from the preceding paragraphs is that there is no agreement in either of these sciences on the nature of the *individual*; or at least there is no agreement at the existence of a 'natural' and indisputable individual. Certainly in pedology quite fundamental disputes have arisen simply because the individual has not been clearly defined. Now this confusion may perhaps console geographers; but we cannot, however, hope to decide the nature of the geographical individual by analogy from allied sciences. The problem remains.

In fact what most geographers do in practice is to use an 'operationally defined' individual;[64] here they follow the practice of most sciences which deal not with objects with a separate discrete existence but with a continuum. Such an individual does not occur in the landscape. It is dictated by the availability of statistics and clarity of definition. Above all it is a function of the purpose of the classification. Say, for example, that we are interested in the agriculture of an English county. There appear to be no 'natural' individuals to base our classification upon; and indeed upon reflection it will be clear that the individual in practice will depend upon both the purpose of the classification and the availability of statistics. Let us suppose we wish to divide the county into land-use regions. Although such differences can be seen, nonetheless we are likely to work from a land-use map of the county. The individual is thus dictated by the specification of the map, and the smallest clearly defined unit on a land-use map is the field. Hence for a land-use classification of the county the 'field' is the individual. But a land-use classification is but a beginning. We wish to know more about the farming systems of the area and for this a classification of crops and livestock is necessary. Now agricultural statistics for English counties are available only for parishes, and each parish may contain several farms; thus here the parish must be the individual. If we wish to press our investigation further, say to discuss regional differences in farm income, then each farmer must be questioned; hence the farmer becomes the individual.

Similar principles apply to any type of areal classification. The natural individual does not exist, and must be chosen with regard to the availability of data and the purpose of the classification. The individual may in fact be a quite arbitrarily chosen unit. Thus, for example, Hammond,[65] who wished to devise an areal classification of the landforms of the United States, took simply the squares of a grid system. In establishing climatic regions the basis of classification is not a unit of the atmosphere but the readings taken at weather stations. There are clearly hazards in adopting this type of individual. But there seems no alternative.

The last problem which is concerned with the individual also deals with location. In classification objects may be grouped into classes on the basis of some similarity. Such objects need not be near to each other, indeed they are considered in most cases quite independently of location. But we have argued that regions are areal classes. Having decided what our individual is, we select some property of the individuals and use this as a differentiating characteristic to group individuals into classes or regions. But a region must consist of contiguous individuals, otherwise a region will not be formed. Clearly such an assumption is not necessary in taxonomy. But one must immediately ask why similar individuals should occur next to each other? Now in dealing with the phenomena of the physical world, soil profiles, 'stands,' and so forth, a metaphysical assumption is made that there is order in the world.[66] Hence in taking a number of soil profiles within a small area, we may assume that they represent soil types which are not greatly different in kind. The assumption of order assumes that distributions can be logically explained and are not merely random. To some extent such an assumption implies that we must know the causes of differences in soils before a classification can be undertaken. Now we may be entitled to make this assumption in making classifications of such phenomena as vegetation types or soil types. Is it equally valid to make such an assumption when dealing with the cultural features of the earth's surface? Thus, for example, if we took every farm in England and grouped these farms into classes on the basis of some similarity, does it necessarily follow that similar farms will occur next to each other if they are then plotted upon a map?[67] This is a problem which is not easily resolved and worth more attention than is given here.

Many readers may reasonably conclude that the analogy between classification and regionalization founders upon the problem of the geographical individual. This is not a conclusion drawn by plant ecologists or soil scientists although they would surely admit that the individual and its interpretation in classifications is a considerable obstacle to progress. We will proceed on the assumption that whereas classification and regionalization are not directly analogous procedures, they are sufficiently similar for the geographer to study the principles of classification with some profit.

PRINCIPLES OF CLASSIFICATION AND DIVISION

Although this section is called 'Principles of classification and division' it should not be thought that these are inflexible rules which must be followed inexorably in practical classification. In the following list the first three principles relate to the strategy, rather than to the tactics, of classification and are general statements about the nature and purpose of classification systems which most logicians and taxonomists agree are useful guides; but they do not have the status of logical principles. Principles five, six, and seven are the principles of division established by logicians; four, eight, and nine are simply rules which have been found by many logicians and taxonomists to be useful in constructing classification systems. The last rule follows from the preceding rules. Two cautions are necessary here. In the first place most logicians have warned of the difficulties in applying the principles of logic to practical classification.[68] The principles are counsels of perfection which are not always possible to observe. So we may claim no more than that it is rewarding to examine regional systems in the light of these principles.

Nor do we contend that regional systems have not been examined in this light before. Indeed most of the conclusions which can be drawn from a study of regional systems in this manner are conclusions which have been made before by geographers, but arguing from different premises. Such a concordance of views is doubly satisfying. It suggests first, that the principles of classification have a relevance to practical classification; and secondly, that much, but perhaps not all, existing regional theory has a secure foundation.

The principles are listed below and the relevance of each to regional systems is discussed separately later.

1   Classifications should be designed for a specific purpose; they rarely serve two different purposes equally well.[69]
2   There exist differences in kind between objects; objects which differ in kind will not easily fit into the same classification system.[70]
3   Classifications are not absolute; they must be changed as more knowledge is gained about the objects under study. Jevons put it thus:[71]

> . . . almost every classification which is proposed in the early stages of a science will be found to break down as deeper similarities of the objects come to be detected.

4   The classification of any group of objects should be based upon properties which are properties of those objects; it follows then that differentiating characteristics should be properties of the objects classed.[72]
5   In logical division the division should be exhaustive.[73]
6   In logical division and classification the species or classes should exclude each other.
7   In division, the division should proceed at every stage, and as far as possible throughout the division, upon one principle.
8   The differentiating characteristic, or the principle of division, must be important for the purpose of the classification.[74]
9   Properties which are used to divide or classify in the higher categories must be more important for the purpose of the division than those used in the lower categories.[75]
10   The use of more than one differentiating characteristic or principle of division produces a hierarchy of classes; the logical consistency of the hierarchy will only be maintained if rules five, six, seven, eight, and nine are observed.

THE PURPOSE OF REGIONAL SYSTEMS

Logicians and taxonomists have all stressed the importance of considering the purpose of a classification system in both its construction and use. Geographers have also emphasized the importance of purpose in the use of regional systems.[76] The implications of this merit discussion.

It has already been pointed out that a division of the earth's surface based upon the totality of its properties is unattainable, and that such a system, which in the terminology of classification theory is a natural classification, has its expression in geography in the search for geographical regions. Such a system seems logically impossible, for it is surely implausible that all the properties of the individuals will display covariance.

But the distinction between natural and artificial classifications (or general-purpose and special-purpose classifications) is not one confined solely to the distinction between classifications that attempt to embrace the totality of the earth's surface and those that confine themselves to a single property such as vegetation cover or type of farming. Even when considering such single property classifications of area the purpose of the system determines the construction and use. Kuchler[77] has shown, for example, that there may be many different areal classifications of vegetation, for not all the properties of vegetation may be encompassed within a single system or map. Chisholm[78] has demonstrated the importance of considering purpose when constructing type-of-farming regions. He argues that a number of workers in this field have attempted to find 'objective' classifications; but the properties of the objects studied (in this case the individuals are farms), are many and no classification is likely to find covariance between such diverse properties as farm income, type of tenure, crop combination, or methods of management. Hence a classification of agriculture must, if it is to be productive of useful generalizations, be quite clear about its purpose. This does not mean that a general-purpose classification of agriculture has no value: such classifications are useful in teaching and may have some value as the basis of planning if used with caution.[79]

Clearly then the purpose of a classification must be borne in mind before attempting its construction; classifications which attempt to cover too broad a field of properties, and are often based on the assumption that there are genetic links between a diverse range of properties, lose their precision as instruments of analysis, for as Preston James has written:[80]

> ... the attempt to recognize combinations of several elements often has the effect of obscuring the realities rather than throwing additional light on them.

But if attention must be paid to purpose in the construction of classifications equal care must be taken in their use, for a classification constructed for one specified purpose cannot necessarily be used for an allied but different purpose. An example of this may be taken from the field of geomorphology. The classification of landforms has until recently been on the basis of origin; hence landforms have been grouped together into classes on the basis of similar genesis. Now this is perfectly valid but it is not always suitable for a study of the human geography of an area. This of course is a point now widely accepted and in recent years there have been several attempts to classify and map landforms on the basis of their observable and measurable properties, not upon their presumed origin.[81]

Once it has been realized that the purpose of a system is of paramount importance in construction, many disputes about the respective merits of different systems could be resolved. Arguments about whether one system is better than another often become meaningless when we realize that the respective classifications have different purposes. This point may be illustrated by considering the classification of climate. If we wish to classify climates with regard to their influence upon potential agricultural development, different criteria will be selected than if we are concerned with the impact of climate upon human physiology, the routes of aircraft, or the origin of climate itself.[82] It is too much to hope that one system could be devized which would serve all these purposes.

DIFFERENCES IN KIND

Mill,[83] although believing in the possibility of a natural classification, as did many of his contemporaries, also pointed out that there were, in some cases, such great differences in kind between objects that they should not be included within the same classification system. One would recognize such a difference, for example, between plants and stones.

What relevance has this principle to regionalization? Bunge[84] pointed out that in areal classification land and sea are normally assumed to be of different kingdoms; that is they are so different that they are but rarely included within the same areal class. But the principle also adds to the argument against the feasibility of geographical regions; geographical regions assume that properties of the earth's surface, unspecified in number but including properties of both the physical and the social environment, will co-vary spatially. Here clearly objects very different in kind are included within the same classification system. James has pointed out the dangers of such a system:[85]

> ... an attempt to define regions based on phenomena produced by a variety of different processes is dangerous and could lead to serious errors of interpretation. We may find ourselves trying to add things like cabbages and kings.

CLASSIFICATIONS AND CHANGE

It would be generally agreed that the classification of any universe of objects must be changed as knowledge about these objects becomes more complete; often as a science progresses the assumptions on which the established classifications are based become increasingly subject to criticism and the need for change becomes apparent. But a considerable problem arises here, for, as Simpson has pointed out,[86] in those sciences where classification is the basis of further investigation, some compromise between stability and change is essential. The great danger is that a system will become accepted and uncriticized from sheer inertia. Kellog has written that:[87]

> ... no system of classification should ever become so sacred or so classical that the system becomes an end in itself.

This point has already been briefly discussed. But the issue of changing classifications has special problems for the geographer. First, the human geographer has a double problem, for not only does his knowledge of the objects he is studying change, but the objects themselves change.[88] This is a difficulty not faced by the biologist, although it could be perhaps argued that vegetation does change sufficiently in the short run for classifications of relatively small areas to merit revision. The problem is far greater in the classification of aspects of human activity such as manufacturing, industry, towns, or types of farming. Few classifications of agriculture at the turn of the century would have much relevance to the present world. Chisholm[89] discussed the difficulties, indeed the impossibilities, of establishing an objective (by which he means *natural*) classification of types of farming in England. He points out that few of the existing classifications take into account the changes going on in British farming. He concludes that the construction of such classification involving Herculean labors may be of doubtful value and suggests that geographers might turn to different forms of research. Although we need not necessarily accept this conclusion, his discussion does illustrate some of the difficulties of classifying the human features of the earth's surface.

But the properties commonly classified by physical geographers are also in need of constant review, on two grounds. First, classifications of features of physical geography have often been devised by workers in the allied systematic field; the classifications in these fields change as we learn more about the phenomena studied. But geographers are not always cognizant of these changes. Nowhere is this more true than in the complex field of study which attempts to interrelate climate, vegetation, and soil. Maps of types of vegetation and soil form an important part of elementary geographies, but they are not always judiciously chosen nor are they invariably the result of the most recent work in the field. A classic example of this is noted by Thornthwaite.[90] Köppen's 'climatic regions' were an attempt to find meteorological limits to vegetation formations; yet the vegetation formations on which he based his work were to some extent discredited even when he wrote. Nonetheless, Köppen's work, and modified forms of Köppen's work, still loom large in geographical studies.

Second, it is by no means sure that the classifications borrowed from other fields are always really suited to the geographer's purpose. Examples of this have already been discussed and we may give one final instance. The classification of soils has been traditionally based upon genesis, and many have argued that soil classification should be based upon soil-forming factors.[91] Whether this is valid is not at issue here; more relevant is whether such a classification is useful to the geographer. Geographers are interested in soils partly because of their influence on agriculture and hence upon areal variations in economic activity. But it is by no means certain that the existing world classifications of soils are most suited to this purpose, for the properties upon which the classification is based are not in every case those which determine the suitability of a soil for various farming activities. It is a common criticism of the as yet incomplete soil classification of Britain that it is of limited value to the farmer and, incidentally, to the geographer.

THE DIFFERENTIATING CHARACTERISTICS OF CLASSIFICATION

In classification objects are grouped together into classes on the basis of a property they have in common called the differentiating characteristic. It might seem self-evident then that a differentiating characteristic must be a property of the objects grouped. Yet few rules are so frequently infringed.

Soil science offers an example which has been touched upon before. Much modern pedology derives from the work done in Russia in the later nineteenth century. Pedologists were impressed by the striking differences between one natural region and another, and attributed these differences ultimately to differences in climate. Hence the major classes of soil types were not based upon properties of the soils themselves but upon climatic properties of the areas in which those soils occurred. This method of approach, which confuses climatic characteristics with empirical characteristics of the soil profile, has led to a great deal of confusion in soil classification.[92] A further example has already been cited. Whereas it is now widely accepted that classifications of agriculture should be based upon properties of agricultural systems and not on the factors that influence regional differences in such systems, agricultural regions were for long based upon differences in either climate or soil type, and some writers may still hold this view.[93]

Here we must return to a point discussed earlier. What part should genesis play in the construction of regional systems? Climate was made the basis of both soil and agricultural classifications because it was held to be the preponderant cause of regional differences in soil type and farming. Similar assumptions have been made in other classification systems. Now it can be argued that there is one overriding factor affecting the properties of objects. Where there is an obvious classification of the objects which is also a natural classification, in that the differentiating characteristic has the maximum possible number of accessory characteristics, Gilmour[94] argued that the existence of such a powerful factor makes a natural classification possible; but that this is not the same as saying that the classification is based upon that powerful factor. Thus it might be possible to classify types of farming of the world upon the basis of the predominant crop grown. Such a property might have a large number of accessory characteristics such as the relative importance of livestock, and so forth. Such a classification might be said to be natural because it maximizes the number of accessory characteristics; the reason for this is that climate is the preponderant factor determining the distribution of crops over the earth. Hence a natural classification is possible because of the overriding importance of climate; but the classification is not based upon climate.

But having made this point it still seems clear that differentiating characteristics should be properties of the objects classed. Genetic classifications, where the differentiating characteristic will, to use regional terminology, be a factor and not an element, will in the long run be misleading.

ARE THERE NON-REGIONAL AREAS?

Our fifth rule of classification is in fact the first principle of division in traditional logic and it simply states that any division must be exhaustive; that is, if a *genus* is to be divided upon some principle then all the constituent species must be recognized. Mourant[95] points out that in practical classification this may be impossible. Thus if roses are divided into species we can never be sure that all the roses in the world are known and hence one can never be sure that all the constituent species have been recognized. In terms of regionalization the principle can be stated as follows: if any area is to be divided or classified into regions, then the whole area must be divided or classified. This principle may seem self-evident, but there are difficulties. We may conveniently recognize two aspects of this problem: first, the rule as applied to geographical regions, and second, the rule applied to generic regional systems.

If any area is to be divided exhaustively into geographical regions then a problem soon arises that has been aptly put by Barbour:[96]

> . . . a study of the *pays* of the Sudan would certainly have a number of intermediate areas that could not readily be assigned to one *pays* or another; and yet had no obvious name or distinctive character of their own.

Any attempt to divide an area will eventually face this dilemma; must we conclude then that there are areas which are non-regional, as Crowe[97] pertinently inquired? Some geographers have at least implied this. Jones and Bryan,[98] in the preface to their book on North America, argued that although regions exist it by no means follows that a whole continent can be divided up into well-marked areas. Now it will be apparent from previous discussions of the geographical region that an attempt to divide an area into such regions presupposes a spatial correlation between a number of properties, some of which must be properties of the physical and some properties of the human environment. We have doubted whether this is logically possible. The implausibility of a set of geographical regions becomes more evident when we apply the rule of exhaustiveness to the idea, for it seems unlikely that human and physical properties of the earth's surface will co-vary continuously over any large area. We will always find areas where the relationship breaks down. The possible exceptions to this may be in relatively small areas where the dominant form of economy is agrarian, where farming techniques are backward, and where there is a close adjustment between soil type and land use. Wrigley[99] pointed out that to a large extent the great success of French regional geographers of the nineteenth century was attributed to their dealing with a society; nonetheless it would probably be difficult to devise an acceptable exhaustive set of geographical regions even for nineteenth-century France, even though the delimitation and discussion of single regions carried some conviction.

The fifth principle of classification, then, is yet another means of demonstrating the logical inconsistencies of the geographical region. But when we turn to consider the rule of exhaustiveness in relation to generic regional systems important new issues arise. It will be profitable to discuss these together with the sixth principle of classification. This states that classes should be mutually exclusive as well as exhaustive. Thus if we divide Mankind into four classes, Catholic, Protestant, Muslim, and Negroid, the classes are not exhaustive because there are men who fall into none of these classes. Nor are the classes exclusive because a man can be a Catholic and also a Negro. In division care must be taken to see that classes do not overlap, whereas in regionalization regions should be mutually exclusive; to put it another way, an individual must be in one region and only one region in any given regional system.

So far the limits of classes have not been considered. When objects are grouped on the basis of some property the classification can be undertaken in two ways. In the first case objects are placed into only two classes according to whether they possess a property or not. In the second case they are grouped into a number of classes according to the degree of expression of the property. Consider an area which is classified on the basis of mean annual precipitation. The area could be classified into only two regions, one which receives no annual precipitation at all, and one which receives some precipitation. But of course a much more common procedure is to base the classes on *degrees of expression* of the property. Hence individuals could be placed in one of five classes: less than ten inches, ten inches to nineteen inches, twenty inches to twenty-nine inches, thirty inches to thirty-nine inches, and forty inches and over.

The problem is how such limits, which are also the boundaries of areal classes, should be determined. In some cases quite arbitrary classes may be chosen, as in the example above. But an attempt to find natural breaks in a statistical series should presumably be attempted. Thus if an area was being classified according to annual precipitation, of the one hundred readings available, ranging from one inch to one hundred inches, suppose twenty readings fell between five and fifteen inches, thirty between thirty and forty inches, and forty between fifty-five and sixty-five inches. Such a clustering would suggest obvious class limits. But in most cases the series would be more continuous and the problem remains. Fortunately in recent years geographers have turned increasingly to statistical methods as an aid to the determination of class or regional limits,[100] and it is to be hoped from the present debate some recognized and generally acceptable procedures will emerge.

Where properties used as differentiating characteristics can be measured the problems of deciding class limits are minimized and classes are necessarily mutually exclusive. But in classifications of geographical phenomena where qualitatively assessed characteristics are used, the possibility of overlap is considerable. This is particularly true of classifications of world agriculture; it is impossible to base a classification on measurable properties simply because the data do not exist. Such classifications usually proceed by defining a given number of major types of agriculture and then mapping the extent of each type. A general criticism is that insufficient attention has been paid to the dangers of class exclusion in defining the types, and thus an examination of the maps and the properties of the types will so reveal parts of the earth's surface which could be in more than one class. It must, however, be admitted that such instances are often difficult to avoid given the present state of knowledge about the world.

THE USE OF THE PRINCIPLE OF DIVISION

One seventh principle, which is the third principle of traditional logic, requires first, that the division should proceed at every stage upon the same principle; and second, that as far as possible the division should proceed throughout upon one principle. Some logicians[101] have doubted whether this principle, and particularly the second part, can be applied to practical classification; yet their consideration raises interesting problems in regionalization.

The first part of the principle is surely relevant and is indeed a corollary of previous statements. If we divide an area into regions of the same order, the basis of each region should be some degree of expression of the same property.[102] Yet in many regional systems regions of the same order are delimited on the basis of a number of different principles. To take an imaginary example we may find that a country has been divided into four major regions: the West, East, South, and North. We find that the West is defined on the basis of it being an area of young folded mountains, the East because it has a low annual rainfall, the South because it has a distinctive type of land tenure, and the North because it has the majority of the country's manufacturing industry. Now it will be readily seen that such systems are likely to have classes that overlap and, further, it will often be impossible to make the division exhaustive. It must be admitted that there is no reason why such a framework should not be the basis of a description of the geography of a country; but the system itself is unlikely to be productive of any valid generalizations.

The second part of the principle is more difficult to apply rigorously; and it also requires some explanation. Stebbing expressed the rule in the following terms: the successive steps of the division must proceed by gradual stages. She illustrates this as follows:[103]

> If, for example, we were able to divide university students first into science and arts students, and were then to subdivide science students into polite and impolite, and arts students into dark, fair and medium-complexioned, the division would serve no useful purpose.

Now in analytical regionalization it is customary to divide the world upon some initial principle and then further subdivide upon some different principle. Thus an area may be initially divided upon the principle of landforms, then upon precipitation, and finally upon vegetation type. Such procedure establishes a hierarchy of classes and each set or category of classes is arrived at by the use of one principle. But does the division proceed from category to category by gradual stages? Are we in fact entitled to use at each stage principles which are different in kind? Crowe expressed doubts about such a procedure:[104]

> There is not the same type of connection (that is, genetic) between climate and physiography or between either of these and vegetation. We cannot build up a rational chain of climate–landform–vegetation, because of the magnitude of the missing links. The mental agility of geographers in trying to skip these gulfs has been truly remarkable, but it has usually been an acrobatic display that has just failed to come off. Regional synthesis based primarily on these three factors must ignore their incongruity and invalidate itself thereby. The popular divide of making broad divisions on one basis and then sub-dividing them according to another is no solution; it is not synthesis but arbitrary classification.

This criticism was aimed primarily at the systems of natural and geographical regions established by British geographers before the Second World War. But even if we concern ourselves with more limited classifications, dealing with, for example, climate, or vegetation, or agriculture, rather than attempts to classify the total character of the earth's surface, we are presented with similar problems. First, how do we select properties to be differentiating characteristics? Second, in division, what determines the order in which the principles selected are applied?

THE ORDER OF PRIORITIES

Any group of objects is likely to possess a number of properties which could be chosen as differentiating characteristics. Which of these are to be selected? Mill argued that they should be those properties which are important for the purpose of the classification. But how is *important* to be judged?

An analogy from the history of biological classification may make this point clearer. Post-Linnean classifications sought to construct classes which had the maximum correlation of properties and were based on the real essence of the things classified. With the advent of Darwin the idea of phylogenetic relationships came to dominate classification so that differentiating characteristics were selected with this in mind. The selection of properties was then weighted or influenced by a value judgment about which properties were most important. The recent rise of numerical taxonomy has attempted to avoid this preselection of properties based on assumptions about origins. The theory of numerical taxonomy is in fact a return to the principles of classification put forward by Michael Adanson before Linné. He assumed that all characters are of equal weight and that affinity, the basis of grouping individuals into the same class, is a function of the proportion of features (properties or characters or elements) in common. There is thus no assumption made about origin.[105]

The recent development of statistical methods in regionalization presents a parallel to the rise of numerical taxonomy in zoology. But as Gilmour[106] pointed out, even the use of computers cannot eliminate value judgments from the selection of differentiating characteristics. We may be able to determine statistically significant properties from a given list: but these must be initially selected on the basis of subjective judgment.

Neither logic nor taxonomy can give us any positive guide, then, to the means of selecting differentiating characteristics other than our eighth principle of classification which states quite simply that a differentiating characteristic should be important for the purpose of the classification. An example may show the relevance of this rule. Let us suppose we are studying the farms of a given area and we are concerned with the relationship between size and other characteristics of the area. Now farms, as operational units, have a great number of properties; they could be classified according to total area, number of workers, amount of fixed capital, or total annual revenue. Which of these is chosen as the differentiating characteristic will depend primarily, assuming all the requisite data is available, upon the purpose of the classification. Thus a classification based upon the total acreage of each farm would not be very helpful if our concern was to trace relationships between, say, farming intensity and crop combinations. A classification of farm size based on the number of employees would not necessarily bear any relationship to acreage, cropping, or capital inputs. The system of classification would have to be devised with the purpose in mind.

In most classifications more than one differentiating characteristic is used and hence at least two orders, or classes, or regions, are formed. In such a hierarchy the order in which the differentiating characteristics are selected is of significance. This may be most easily discussed in terms of division. If an area is to be divided into a number of regions, the whole area is first divided into a number of classes or regions upon the basis of one principle; each of these subclasses is divided on the basis of a second principle, whereas the resultant classes may be further divided. What decides the order in which these principles are applied?

Hettner[107] considered this problem in his discussion of the possibilities of a natural classification of the earth's surface. He pointed out that the selection of the principles from a great number of properties is essentially subjective; he went on to indicate that a logical order of priorities could only be obtained if genetically linked properties were used as principles of division. Hartshorne, in discussing this contention, concluded:[108]

> ... that there can be no fixed order of importance of the different elements that is applicable to all parts of the world. ...

Hettner discussed a classification of the total character of the earth's surface; in such terms it is indeed difficult to see how there can be a logical and fixed order of priorities for such diverse properties as landforms, climate, vegetation, and soil type. But the problem of the order of priorities remains even if we confine ourselves to the classification of but one property of the earth's surface. The only guide that we can derive from the principles of classification is our ninth rule: that the principle used to divide in the higher categories should be more important than those in lower categories. Cline[109] gave an instance of this in soil science. In Shaw's soil classification parent material is a differentiating characteristic in a higher category and thus separates similar profiles in the lower categories. Clearly this rule again emphasizes the importance of purpose in classification. Consider a world classification of agriculture. If the highest category is divided into classes on the basis of the degree of commercialization we would find that, in the lower orders, farms which are identical in terms of crop combinations, methods, and systems of land tenure are found in different classes. Conversely if the highest category is initially divided on the basis of methods of management then in the lower orders we may find farms with identical crop combinations, systems of land tenure, and degree of commercialization separated into quite different classes. Hence the purpose of the particular classification must be borne in mind when deciding the order of priorities.

One important implication of the order of priorities remains to be discussed. If classes in any given category are determined on the basis of one property alone, then the number of properties and their degree of expression determines the number of classes. The number of classes may soon become very large. Helburn,[110] for example, cited eleven properties of farms which he considers should be incorporated into a classification of types of agriculture. But as he went on to point out, if each property is considered to have only three degrees of expression then there would be some 59,000 possible classes. This problem is often avoided in practical classification in two ways. First, a category of classes may be formed on the basis of not a single property but of a number of properties. Second, many classifications are uncompleted; that is either the higher or the lower orders are omitted. Thus Whittlesey's[111] classification of agriculture claims only to deal with first-order regions, whereas, as Manil[112] pointed out, some soil classifications consist only of either the higher or the lower categories.

THE HIERARCHY OF REGIONS

If objects are grouped into classes and more than one differentiating characteristic is used; or if a genus is subdivided and subdivided again on some second principle, then a hierarchy of classes of different orders is formed into regions, and then these regions grouped together on the basis of some second principle, then a hierarchy of regions is formed. Such an idea has a long tradition in the history of regionalization. Hettner, Herbertson, Unstead, and others have all discussed the possibility of forming regional hierarchies.[113] But many geographers have attempted to construct such hierarchies based on the total character of the earth's surface; opinion of most geographers now is that this is logically impossible. Certainly we must agree with Hartshorne[114] that there is no way of forming a hierarchy of regions based upon total character in any way analogous to the orders of biological classifications.

But this does not mean that regional hierarchies themselves must be rejected; the great value of a hierarchy of classes or regions, is as Lazarsfeld[115] pointed out, that generalizations may be made about the same objects of study at different levels of abstraction, thus saving time in organizing and reorganizing similar material. The need for regional hierarchies is widely recognized, not only by geographers but also by workers in allied fields such as soil science and plant ecology. Among geographers Russian workers have particularly stressed the need for hierarchial approach and a uniform nomenclature.[116] Yet by no means are all regional systems organized upon a hierarchial basis and some regard this as a principal defect in existing regional systems. Thus Whittlesey wrote:[117]

> The general neglect of the meaning of differences of scale or degree of generalization is a lacuna in geographic thinking which . . . should be filled as soon as possible.

Whittlesey went on to suggest a standard nomenclature for regions of different orders: locality, district, province, and realm. Other geographers have suggested standard names; thus Herbertson recognized four orders: locality, district, region, and continent, whereas Unstead recognized five orders: stow, tract, subregion, minor region, and major region.[118] Russian geographers have also discussed a variety of different possible terms for regions of different orders.[119]

The need for an accepted hierarchy and nomenclature is often discussed; the thinking behind this approach is well expressed by Lebedev when discussing geomorphic regionalization:[120]

> Unfortunately they are still being compiled on the basis of different approaches and through the use of various systems of regional units, which make it impossible to compare materials covering different areas.

Presumably it is assumed that if a standardized hierarchy of regional geomorphic units could be established, the generalizations for a given order of a hierarchy for one part of the world would be applicable to another part of the world where the standard system had also been applied. One may share Bunge's[121] skepticism of the validity of this approach. A hierarchy depends not upon any features inherent in the landscape but upon the number of differentiating characteristics chosen for the system. Whittaker, in reviewing the similar problem in plant ecology commented:[122]

> Hierarchies are not inherent in landscapes; a given landscape offers material which can be fashioned into hierarchies in innumerable ways.

But in any given classification of a given area on the basis of certain properties, a hierarchy is inevitably established. If such a hierarchy is to be a useful basis for the making of generalizations about the objects classified then the rules cited at the beginning of this section must be borne in mind, particularly principles five to nine. If such principles are grossly infringed then the system ceases to be internally consistent and the possibilities of the system producing any valid generalizations are greatly reduced. Two examples of this may be given. One of the most familiar regional systems found in textbooks of geography is where a country is divided into higher order regions on the basis of landforms, then subdivided perhaps on the basis of climatic differences, whereas the lower order regions are established on the basis of some feature of human occupance. It is these lower order regions which are used for the description of the country. Yet it will be readily apparent that the lower order regions, because the higher order regions are formed upon a principle greatly different in kind from the lower order regions, often place in different regions very similar areas of human occupance. Now admittedly such regional systems are intended not as a means of analysis but simply as a framework of description. But not only are they valueless for the former purpose but of very limited significance for the latter aim.

The failure to maintain class exclusion is another common fault. Whittlesey's world system of agricultural regions gives instances of this.[123] Whittlesey based his classification of agricultural regions upon five properties: each of these properties may be assumed to have at least two degrees of expression. This would give at least thirty-two lower order classes in a consistent system; yet he recognized only fourteen lower order regions. Clearly there must be cases of overlapping.

But it must be admitted that there is a great gulf between the demands of logic and the expediencies of practical classification. Suppose we wish to establish a world classification of types of farming; we take as individuals, farms, and let us further suppose that statistical data on all these farms are available. Theoretically we could group these individuals into classes according to some differentiating characteristic, then group those classes upon some second characteristic, and so on. Alternatively, we may begin with the world and subdivide. In both cases a system of hierarchy of classes could theoretically be obtained; but there would clearly come a point where the system and the reality it represents would radically differ. Manil, in discussing this problem with reference to soils, doubted whether such a classification which embraced the world as the highest category and a soil profile as the individual is feasible.[124] Even if logically possible it is still a formidable task.

Conclusions

We hope this article has adequately demonstrated the similarity between the procedures of classification and those of regionalization. Once attention has been drawn to the differing terminologies, we can see that taxonomists and geographers are grappling with fundamentally the same methodological problems. This is most aptly shown by comparing the history of classification and that of regionalization; there is a remarkable similarity in the development of ideas on the natural system and the geographical region.

If we accept that the two procedures are similar, and it must be admitted that the problem of the geographical individual is a difficulty, then the principles of classification can be applied to the construction of regional systems, and in particular to generic regional systems. An examination of the principles confirms most of the conclusions reached independently by geographers. To a large extent these principles give negative guides to procedure; that is, they tell us what should be avoided rather than what should be done. Nonetheless we hope they will prove useful to those who are constructing systems or assessing the value of existing systems.

References

1   It will be apparent that I owe a great deal to reading the work of J. S. L. Gilmour, M. G. Cline, G. G. Simpson, R. Hartshorne and W. Bunge. I hope that I have properly understood and correctly interpreted their views. I have also received a great deal of helpful criticism on earlier drafts of this paper from many colleagues and I would like to thank D. W. Harvey, P. Hagget, R. J. Chorley, B. Garner, C. Board, B. H. Farmer, D. Anderson, R. L. Wright and R. A. G. Savigear and in particular G. M. Lewis and R. Hartshorne. Needless to say they do not necessarily share the opinions expressed here.
         Reprinted from *Annals of the Association of American Geographers,* Vol. 55, No. 3 (September 1965), Washington, D.C.
2   Noted, for example, by HARTSHORNE, R. *The Nature of Geography,* Lancaster, Pennsylvania: Association of American Geographers (1939), p. 362. Hartshorne surveys some of the previous attempts, particularly those by Hettner, A. (pp. 288, 290, 291, 294, 298, 306); later work is reviewed by Hartshorne in *Perspective on the Nature of Geography,* Chicago: Rand McNally (1959), p. 128 and references.
3   That regionalization is a form of classification appears to have been implied by Hettner as reported in HARTSHORNE, op. cit. (1939), the idea has recently been put forward by Bunge, W., *Theoretical Geography,* Lund: Lund Studies in Geography. Series C. General and Mathematical Geography, No. 1 (1962), pp. 14–23.
4   Most textbooks of logic discuss the principles of classification and division. Among the most useful are: Jevons, W. S. *The Principles of Science,* London: Macmillan (1887), Chapter 30; Mill, J. S., *A System of Logic,* London: Longmans (1959), pp. 465–74; Joseph, H. W. B., *An Introduction to Logic,* Oxford: Clarendon Press (1946); Stebbing, L. S., *A Modern Elementary Logic,* London: Methuen (1963).
5   GILMOUR, J. S. L. 'Taxonomy and Philosophy,' *The New Systematics,* Huxley, Julian (Ed.), London: Oxford University Press (1952), pp. 461–74; 'The Development of Taxonomic Theory since 1851,' *Nature,* Vol. 168 (1951), pp. 400–2; Gilmour, J. S. L. and Walters, S. M., 'Philosophy and Classification,' *Vistas in Botany,* Turril, W. B. (Ed.), Vol. IV, London: Pergamon Press (1964), pp. 1–22; Gilmour, J. S. L., 'Taxonomy,' in *Contemporary Biological Thought,* McLeod, A. M. and Cobley, L. S. (Eds.), Edinburgh: Oliver and Boyd (1961); Simpson, G. G., *Principles of Animal Taxonomy,* New York: Columbia University Press (1961).

6   SOKAL, R. R. and SNEATH, P. H. A. *Principles of Numerical Taxonomy,* San
     Francisco: Freeman (1963); Sneath, P. and Sokal, R. R., 'Numerical Taxo-
     nomy,' *Nature,* Vol. 193 (1962), pp. 855–60.

7   CLINE, M. G. 'Basic Principles of Soil Classification,' *Soil Science,* Vol. 67
     (1949), pp. 81–91; Manil, G., 'General Considerations on the Problem of
     Soil Classification,' *The Journal of Soil Science,* Vol. 10 (1959), pp. 5–13;
     Bidwell, O. W. and Hole, F. D., 'Numerical Taxonomy and Soil Classifica-
     tion,' *Soil Science,* Vol. 97 (1964), pp. 58–62; Leeper, G. W., 'The Classifica-
     tion of Soils,' *The Journal of Soil Science,* Vol. 9 (1958), pp. 9–19; Jones,
     T. A., 'Soil Classification – a Destructive Criticism,' *Journal of Soil Science,*
     Vol. 10 (1959), pp. 196–200.

8   SIMPSON, op. cit., p. 3.

9   LORCH, J. 'The Natural System in Biology,' *Philosophy of Science,* Vol. 28
     (1961), p. 283.

10   This section follows JOSEPH, op. cit., pp. 115–21.

11   This section follows STEBBING, L. S. *A Modern Introduction to Logic,* London:
     Methuen (1930), pp. 435–6.

12   SIMPSON, op. cit., p. 2.

13   STEBBING, op. cit., footnote 4, p. 108.

14   CHAPMAN, F. M. and HENLE, P. *The Fundamentals of Logic,* New York:
     Charles Scribner (1933), p. 283; Jones, A. L., *Logic Deductive and Inductive,*
     New York: Henry Holt (1909), p. 32; Gilmour and Walters, op. cit., p. 3;
     Jevons, op. cit., p. 677.

15   JONES, op. cit., footnote 14, p. 34.

16   LATTA, R. and MACBEATH, A. *The Elements of Logic,* London: Macmillan
     (1929), p. 54.

17   GILMOUR and WALTERS, op. cit., p. 5.

18   HARTSHORNE, op. cit., (1939), footnote 2, p. 250; op. cit. (1959), footnote 2,
     p. 31.

19   HARTSHORNE, op. cit., (1939), footnote 2, p. 46.

20   NAMPIYEV, P. M. 'The Objective Basis of Economic Regionalization and its
     Long-Range Prospects,' *Soviet Geography,* Vol. 11, No. 8 (1961), pp. 64–74;
     Saushkin, Y. G., 'On the Objective and Subjective Character of Economic
     Regionalization, *Soviet Geography,* Vol. 11, No. 8 (1961), pp. 75–81.

21   Noted by THORNTHWAITE, C. W., 'Problems in the Classification of Climates,'
     Geographical Review, Vol. XXXIII (1943), p. 253.

22   HARTSHORNE, op. cit. (1939), footnote 2, pp. 296–300.

23   HERBERTSON, A. J. 'The Major Natural Regions: an Essay in Systematic
     Geography,' *Geographical Journal,* Vol. XXV (1905), p. 300–10; *Natural
     Regions: Abstract of Remarks Opening a Discussion* (British Association for
     the Advancement of Science, Birmingham, 1913); 'The Higher Units,'
     Scientia, Vol. XIV (1913), pp. 199–212.

24   HERBERTSON, A. J. 'Regional Environment, Heredity and Consciousness,'
     The Geographical Teacher, Vol. 8 (1915–1916), pp. 147–53.

25   UNSTEAD, J. F. 'A Synthetic Method of Determining Geographical Regions,'
     Geographical Journal, Vol. XLVIII (1916), pp. 230–49; 'Geographical
     Regions Illustrated by Reference to the Iberian Peninsula,' *Scottish Geographi-
     cal Magazine,* Vol. XLII, No. 111 (1926), pp. 159–70; 'The Lötschental: a
     Regional Study,' *Geographical Journal,* Vol. LXXIX (1938), pp. 298–317;
     'A System of Regional Geography,' *Geography,* Vol. XVIII, Part 3 (1933),
     pp. 175–87.

26   BROWN, R. *Explanations in the Social Sciences,* London: Routledge and Kegan
     Paul (1963), p. 171; Jevons, op. cit., p. 675.

27   JEVONS, op. cit., p. 675.

28   LEEPER, op. cit., p. 59.

29   SNEATH and SOKAL, op. cit. (1962), footnote 6, pp. 855–60; Bidwell and Hole, op. cit., pp. 58–62; Ehrlich, P. R., 'Problems of Higher Classification,' *Systematic Zoology*, Vol. 7 (1958), pp. 180–4.

30   SOKAL and SNEATH, op. cit. (1963), footnote 6, pp. 20–30.

31   WRIGLEY, E. A. *The Dilemma of Vidal de la Blache,* unpublished mimeographed manuscript (1964).

32   The work of Unstead and Herbertson shows an awareness of the terminology of classification but perhaps not always of the logical procedures. On the other hand, Hettner's work shows a close understanding of the principles of formal logic and their application to the problems of regionalization. For examples see discussions of Hettner's views in HARTSHORNE, op. cit. (1939), footnote 2, pp. 290, 294, 298, 306.

33   BUNGE, op. cit., pp. 14–22; Bunge follows the terminology of regions used by Whittlesey, D., 'The Regional Concept and the Regional Method,' in *American Geography: Inventory and Prospect*, James, P. E. and Jones, C. F. (Eds.), Syracuse, New York: Syracuse University Press (1954), pp. 19–67.

34   BUNGE, op. cit., p. 23.

35   BERRY, B. J. L. and HANKINS, T. D. 'A Bibliographic Guide to the Economic Regions of the United States,' Chicago: University of Chicago Press, *Department of Geography Research Paper*, No. 87 (1963), p. X.

36   HEMPEL, C. G. 'Fundamentals of Concept Formation in Empirical Science,' *International Encyclopedia of United Science*, Vol. II, No. 7, Chicago: University of Chicago Press (1952), pp. 5–6.

37   ROBINSON, G. W. S. 'The Geographical Region; Form and Function,' *Scottish Geographical Magazine*, Vol. 69 (1953), pp. 49–58.

38   SIMPSON, op. cit., p. 3.

39   UNSTEAD, op. cit. (1916), footnote 25, pp. 232, 235; op. cit. (1926), footnote 25, p. 159.

40   HERBERTSON, op. cit. (1905, 1913), footnote 23.

41   HARTSHORNE, op. cit. (1939), footnote 2, pp. 291–2; Whittlesey, op. cit. (1954), footnote 33, p. 35.

42   WHITTAKER, R. A. 'The Classification of Natural Communities,' *The Botanical Review*, Vol. XXVIII (1962), pp. 1–160; quotations from pp. 77–8, for both items from the two authors.

43   MANIL, op. cit., pp. 8–9.

44   LATTA and MACBEATH, op. cit., footnote 16, p. 151.

45   Hettner's views of genetic classifications were discussed by HARTSHORNE, op. cit., footnote 2, pp. 307–11. He believed that classifications should be genetic, a view that reflected the beliefs of the majority of both the logicians and taxonomists of his day.

46   SOKAL and SNEATH, op. cit., (1963), footnote 6, pp. 5–30.

47   BASINSKI, J. 'The Russian Approach to Soil Classification,' *Journal of Soil Science*, Vol. 10 (1959), pp. 14–26.

48   Such systems are critically examined by WHITTLESEY, D. 'Major Agricultural Regions of the Earth,' *Annals of the Association of American Geographers*, Vol. 26 (1936), pp. 199–213.

49   JOSEPH, op. cit., pp. 121–35.

50   STEBBING, op. cit. (1963), footnote 4, p. 109.

51   HARTSHORNE, op. cit. (1939), footnote 2, pp. 379–84; op. cit. (1959), footnote 2, pp. 146–53; Bunge, op. cit., pp. 8–13; Schaefer, F. K., 'Exceptionalism in Geography: a Methodological Examination,' *Annals of the Association of American Geographers*, Vol. 43 (1953), pp. 226–49.

52   BUNGE, op. cit., p. 16; Hartshorne, op. cit. (1939), footnote 2, pp. 304–5.

53 'Classification of Regions of the World,' *Geography,* Vol. 22 (1937), pp. 253–82; see also Hartshorne, op. cit. (1939), footnote 2, p. 305.

54 MOURANT, J. A. *Formal Logic: An Introductory Textbook,* New York: Macmillan (1963), p. 64.

55 HARTSHORNE, op. cit. (1939) footnote 2, p. 285.

56 For German contributions to this topic see HARTSHORNE, op. cit. (1939), footnote 2, pp. 264–6; see also Robinson, op. cit., pp. 49–55.

57 UNSTEAD, op. cit. (1933), footnote 25.

58 LINTON, D. L. 'The Delimitation of Morphological Regions,' in *London Essays in Geography,* Stamp, L. D. and Wooldridge, S. W. (Eds.), London: Longmans, Green (1951), pp. 199–217.

59 BUNGE, op. cit., p. 16.

60 WHITTAKER, op. cit., p. 14.

61 BOURNE, R. 'Regional Survey and its Relation to Stock-Taking of the Agricultural and Forest Resources of the British Empire,' *Oxford Forestry Memoirs,* 13, Oxford: Clarendon Press (1931), pp. 16–17.

62 CLINE, op. cit., p. 82.

63 JONES, op. cit., footnote 7, pp. 196–200.

64 A term now widely used in the social sciences but derived from BRIDGMAN, P. *The Logic of Modern Physics,* New York: Macmillan (1927), pp. 3–32.

65 HAMMOND, E. H. 'On the Place, Nature and Methods of Description in the Geography of Land Forms', *Technical Report No. 1, Contract 1202 (01), Procedures in the Descriptive Analysis of Terrain, Geographical Branch, Office of Naval Research, Wisconsin (1957); Procedures in the Descriptive Analysis of Terrain,* Madison, Wisconsin: University of Wisconsin (1958).

66 CLINE speaks of geographic order, op. cit. (1949).

67 CHISHOLM, M. 'Problems in the Classification and Use of Farming-type Regions,' *Transactions of the Institute of British Geographers,* No. 35 (1964), p. 102.

68 JEVONS, op. cit., p. 689; Cohen, M. R. and Nagel, E., *An Introduction to Logic and Scientific Method,* London: Routledge (1949), p. 242; Chapman and Henle, op. cit. (1933), p. 287, Eaton, R. M., *General Logic: an Introductory Survey,* New York: Scribners (1931), p. 283.

69 CLINE, op. cit. (1949), p. 81; Simpson, op. cit. (1961), footnote 5, p. 25; Gilmour, op. cit. (1951), footnote 5, p. 401; Stebbing, op. cit. (1963), footnote 4, pp. 107–8; Latta and Macbeath, op. cit., p. 153; Mace, C. A., *The Principles of Logic, an Introductory Survey,* London: Longmans, Green (1933), p. 197.

70 MILL, op. cit., pp. 470–1; Cline, op. cit., p. 87.

71 SIMPSON, op. cit., p. 111; Jevons, op. cit., p. 691.

72 CLINE, op. cit., p. 86.

73 Principles five, six, and seven are all from the same reference; see JOSEPH, op. cit., pp. 117–19; Mourant, op. cit., p. 65; Cohen and Nagel, op. cit., pp. 241–2; Stebbing. op. cit. (1963), footnote 4, p. 109. The three rules of division and their application to classification in the social sciences are discussed by Lazarsfeld, P. F. and Barton, A. H., 'Some General Principles of Questionnaire Classification,' *The Language of Social Research,* Lazarsfeld, P. W. and Rosenberg, M. (Eds.), Glencoe, Illinois: The Free Press (1957), pp. 84–5; also by Lazarsfeld in 'Qualitative Measurement in the Social Sciences; Classification, Typologies and Indices,' in *The Policy Sciences,* Lerner, D. and Lasswell, H. D. (Eds.), Stanford, California: Stanford University Press (1951), pp. 115–92.

74 MILL, op. cit., p. 468; Cline, op. cit., p. 88.

75 LATTA and MACBEATH, op. cit., p. 155; Cline, op. cit., p. 88.

76   HARTSHORNE, op. cit., (1939), footnote 2, p. 290; (1959), footnote 2, pp. 77–8, 137–44; James, P. E. 'Toward a Further Understanding of the Regional Concept,' *Annals of the Association of American Geographers,* Vol. 42 (1952), p. 207; Ullman, E., 'Human Geography and Area Research,' *Annals of the Association of American Geographers,* Vol. 43 (1953), p. 58.

77   KUCHLER, A. W. 'Classification and Purpose in Vegetation Maps,' *Geographical Review,* Vol. 46 (1956), pp. 155–6.

78   CHISHOLM, op. cit. (1964), pp. 91–7.

79   JAMES, op. cit., footnote 76, p. 109; Hartshorne, op. cit. (1959), footnote 2, pp. 77–8.

80   JAMES, P. E. 'A Regional Division of Brazil,' *Geographical Review,* Vol. 32 (1942), pp. 493–5.

81   HAMMOND, op. cit. (1957), footnote 65; Van Lopik, J. R. and Kolb, C. R., 'A Technique for Preparing Desert Terrain Analogs,' Vicksburg: Technical Report No. 3–506, U.S. Army Corps of Engineers (1959).

82   A good review of some of the problems of climatic classification is by HARE, F. K., 'Climatic Classification,' in *London Essays in Geography,* Stamp, L. D. and Wooldridge, S. W. (Eds.), London: Longmans, Green (1951), pp. 111–34.

83   MILL, op. cit., pp. 407–71.

84   BUNGE, op. cit. (1952), footnote 3, p. 20.

85   JAMES, op. cit. (1952), footnote 76, p. 204.

86   SIMPSON, op. cit., pp. 111–12.

87   KELLOG, C. E. 'Why a New System of Soil Classification?' *Soil Science,* Vol. 96 (1963), p. 5.

88   A point recognized by UNSTEAD, op. cit. (1916), footnote 25, p. 241.

89   CHISHOLM, op. cit., pp. 91–103.

90   THORNTHWAITE, op. cit., p. 242.

91   KELLOG, op. cit., p. 4; Smith, G. D., 'Objectives and Basic Assumptions of the New Soil Classification System,' *Soil Science,* Vol. 96 (1963), pp. 6–16; Kubiena, op. cit., Leeper, op. cit.

92   BASINSKI, op. cit., p. 16; Manil, op. cit., p. 9.

93   This point was discussed by WHITTLESEY, op. cit. (1936), footnote 48, and op. cit. (1954), footnote 33, p. 38. For a recent version of the older view see Lewis, A. B., *Land Classification for Agricultural Development,* Rome: F.A.O. Development Paper No. 18 (1952), pp. 31–3.

94   GILMOUR, op. cit. (1961), footnote 5, pp. 32–4.

95   MOURANT, op. cit., p. 65.

96   BARBOUR, K. M. *The Republic of the Sudan, a Regional Geography,* London: University of London Press (1961), p. 130.

97   CROWE, P. R. 'On Progress in Geography,' *Scottish Geographical Magazine,* Vol. 54 (1938), p. 9.

98   JONES, L. R. and BRYAN, P. M. *North America,* London: Methuen (1950), p. vi; cf. Vogel, quoted in Hartshorne, op. cit. (1939), footnote 2, p. 274.

99   WRIGLEY, op. cit., footnote 31; See also Kimble G., 'The Inadequacy of the Regional Concept,' in *London Essays in Geography,* op. cit., footnote 82, pp. 164–9.

100 There is now a considerable literature on this topic: BERRY, B. J. L. 'A Note Concerning Methods of Classification,' *Annals of the Association of American Geographers,* Vol. 48 (1958), pp. 300–3; Zobler, L., 'Statistical Testing of Regional Boundaries,' *Annals of the Association of American Geographers,* Vol. 47 (1957), pp. 83–95; Duncan, O. D., Cuzzort, R. P., and Duncan, B., *Statistical Geography, Problems in Analyzing Areal Data,* Glencoe, Illinois: The Free Press (1961).

101 CHAPMAN and HENLE, op. cit., pp. 287–8.

102 HETTNER, quoted in Hartshorne, op. cit. (1939), footnote 2, p. 306.

103 STEBBING, op. cit., p. 109.
104 CROWE, op. cit., p. 8.
105 BIDWELL and HOLE, op. cit., footnote 7, p. 58–62.
106 GILMOUR, op. cit., p. 43.
107 HETTNER, quoted by HARTSHORNE, op. cit. (1939), footnote 2, p. 307.
108 HARTSHORNE, op. cit. (1939), footnote 2, p. 306.
109 CLINE, op. cit. (1949), footnote 7, p. 88.
110 HELBURN, N. 'The Bases for a Classification of World Agriculture,' *Professional Geographer,* Vol. IX (1957), p. 6.
111 WHITTLESEY, op. cit. (1936), footnote 48.
112 MANIL, op. cit., p. 9.
113 See also PHILBRICK, A. K. 'Principles of Areal Functional Organization in Regional Human Geography,' *Economic Geography,* Vol. 33 (1957), pp. 299–336.
114 HARTSHORNE, op. cit. (1959), footnote 2, p. 128.
115 LAZARSFELD and BARTON, op. cit., pp. 84–5.
116 PERELMAN, A. 'Geochemical Principles of Landscape Classification,' *Soviet Geography,* Vol. II, No. 3 (1961), pp. 62–3; Lebedev, G. V. 'Principles of Geomorphic Regionalization,' *Soviet Geography,* Vol. II, No. 8 (1961), pp. 59–60; Rikhter, G. D., 'Natural Regionalization,' in *Soviet Geography, Accomplishments and Tasks,* New York: American Geographical Society (1962), pp. 205–9; Grigor'yev, A. A., 'Geographical Zonality,' ibid., pp. 182–7.
117 WHITTLESEY, op. cit. (1954), footnote 33, p. 47.
118 HERBERTSON, op. cit. (1913), footnote 23; Unstead, op. cit., (1933), footnote 25.
119 See footnote 116.
120 LEBEDEV, op. cit, p. 59.
121 BUNGE, op. cit., pp. 21, 96.
122 WHITTAKER, op. cit., p. 122.
123 WHITTLESEY, op. cit. (1936), footnote 48.
124 MANIL, op. cit., p. 9.

11 Approaches to regional analysis: a synthesis[1]

Brian J. L. Berry
University of Chicago, U.S.A.

In my dictionary I find a synthesis defined as 'a complex whole made up of a number of parts united.'[2] The suggestions of complexity and unity are bothersome, however, because the synthesis of approaches to regional analysis presented in this paper is simplistic at best, and we have all found that the parts hardly seem united at times. There is perhaps only one advantage to be gained from the simplification – that poorly developed or new approaches to studying the geography of an area[3] may be identified more readily.

The paper begins with certain assertions concerning geography's role among the sciences. A synthesis of apparently dichotomous approaches to geographic understanding is then proposed,[4] and the concluding remarks are directed to the question of new approaches. The route towards such new approaches begins with analysis of the inadequacies of the proposed synthesis, and continues with discussion of possible solutions to the inadequacies via generalizations produced in General Systems Theory.[5]

Geography among the sciences

James Conant describes science as an interconnected series of concepts and conceptual schemes that have developed as a result of experimentation and observation and are fruitful of further experimentation and observation as man explores his universe. He characterizes the methods of exploration – scientific method – as comprising speculative general ideas, deductive reasoning, and experimentation. Like all brief statements on any subject, these are ambiguous and incomplete outside of the expanded context given them by the author. They do provide a useful setting for the first thesis of this paper, however, that: *Geographers are, like any other scientists, identified not so much by the phenomena they study, as by the integrating concepts and processes that they stress.*[6] James Blaut expresses the point nicely, saying that the objects dealt with by science are not natural entities, ultimate objects, but are rather sets of interlocking propositions about systems.[7]

Systems may be viewed in a variety of ways, and hence the variety of propositions that may be developed concerning them. The particular set of propositions stressed by any science depends upon its point of view, the perspective in looking at systems that it instills into its members as they progress from novices to accepted membership in that select professional core that serves as guardian and proponent of the viewpoint. As Kenneth Boulding has said, subjects 'carve out for themselves certain elements of the experience of man and develop theories and patterns of research activity which yield satisfaction in understanding, and which are appropriate to their special segments.'[8] Within this context, our second and third theses are thus that: *The geographic point of view is spatial* and that *The integrating concepts and processes of the geographer relate to spatial arrangements and distributions, to spatial integration, to spatial interactions and organization, and to spatial processes.*[9]

But the experience of man encompasses many systems, and the geographer does not apply his spatial perspective to all. The second and third theses define the way of viewing, but not that which is viewed. Which system is examined by geographers? Hartshorne properly describes it as comprising 'the earth as the home of man.' A geographer is so trained and inclined that he assumes a spatial perspective in his analysis. But this perspective is not his sole perquisite, for other scientists take such a viewpoint. His contribution is that it is he who provides the spatial perspective so important to any understanding of the system comprising the earth as the home of man. This definition logically excludes from geography studies of other systems from a spatial viewpoint. We are well aware, for example, that when certain physical systems covering the earth are studied apart from their relevance to man, even from a spatial point of view, the job is done by people in other disciplines – geologists, meteorologists, and oceanographers, among others. Similarly, bubble chamber work proceeds from a spatial viewpoint at the microlevel, and is undertaken by physicists.

What is this system comprising the earth as the home of man? It can be described as the complex worldwide man-earth ecosystem.[10] An ecosystem logically comprises populations of living organisms and a complex of environmental factors, in which the organisms interact among themselves in many ways, and in which there are reciprocal effects between the environments and the populations.[11] Biologists, botanists, and ecologists study such ecosystems from a spatial point of view, of course, but the geographer is the person who concentrates upon the spatial analysis of that worldwide ecosystem of which man is a part. The earth as the home of man is a gigantic ecosystem in which man, with culture, has become the ecological dominant. His earthly environments are thus not simply – and less and less – the physical and biological, but also the cultural of his own creating. The fourth thesis thus becomes: *Geography's integrating concepts and processes concern the worldwide ecosystem of which man is the dominant part.*

There is a further problem which emerges at this point. Definition of the system which geography studies from a spatial point of view is perfectly adequate to differentiate geography's role from that of the physical and biological sciences. Many social sciences study the man-made environments, however: political, economic, social, cultural, psychological, and the like, studied by political scientists, economists, sociologists, anthropologists, and psychologists. We resort to our second thesis. None of these sciences examines the man-made environments from a spatial point of view, whether it be to examine spatial distributions or associations of elements, the organization of phenomena over space, or the integration of diverse phenomena in place. Other distributional and organizational themes are stronger and more central to the other social sciences. Thus, whereas it is the system which is studied which differentiates geography from the physical and biological sciences, in studies of man and his works it is the spatial perspective that differentiates. Within the worldwide ecosystem of which man is the dominant part, man creates for himself many environments. These environments are not studied in their totality by geographers, only in their spatial facets.

Dichotomies within geography

Debate about approaches to geographic understanding has traditionally run to dichotomies: natural as opposed to human; topical or systematic versus regional; historical or developmental as contrasted with functional and organizational; qualitative versus quantitative; perks versus pokes. Richard Hartshorne has gone to great lengths to show that many of these dichotomies are either meaningless or useless,[12] but the fact that dichotomies have emerged at all suggests that *the spatial viewpoint has several facets*. In his seminal paper 'Geography as Spatial Interaction' Edward Ullman has gone so far as to argue that the essential intellectual contributions of human geography can be summarized in terms of a dichotomy, the dual concepts of *site* and *situation*.[13] Site is vertical, referring to local man–land relations, to form and morphology. Situation is horizontal and functional, referring to regional interdependencies and the connections between places, or to what Ullman calls spatial interaction.

Existence of several facets poses problems, even if we agree that, as dichotomies, they are of little utility. Boulding argues that the most significant 'crisis in science today arises because of the increasing difficulty of profitable talk among scientists as a whole.' Very descriptively, he says that 'Specialization has outrun Trade, communication ... becomes increasingly difficult, and the Republic of Learning is breaking up into isolated subcultures with only tenuous lines of communication between them. ... One wonders sometimes if science will not grind to a stop in an assemblage of walled-in hermits, each mumbling to himself in a private language that only he can understand. ...' Is this to be our fate within geography, with analytically minded economic urbanists off building their fragile models, anthropologically oriented cultural ecologists sequestered in some primitive backwoods contemplating their navels, and the like? As Boulding continues, 'the spread of specialized deafness means that someone who ought to know something that someone else knows isn't able to find it out for lack of generalized ears.' His solution is 'General Systems Theory to develop those generalized ears . . . to enable one specialist to catch relevant communications from others.'[14]

A system is an entity consisting of specialized interdependent parts. Most systems can be subdivided into subsystems by searching for modules with high degrees of internal connectivity, and lower degrees of intermodule interaction. If larger modules can be partitioned into smaller modules, it is possible to talk of a hierarchy of systems and subsystems.[15]

What we will try to do here is to construct a simple system that depicts the variety of approaches to regional analysis. The traditional dichotomies will be included either as parts of the frame of reference which specifies how the system is separated from the rest of science (the balance of science can be termed the 'environment' of the system) or as modules of the system. It is this system that constitutes the synthesis of approaches to regional analysis. The fact that a system has been created emphasizes the unity of the spatial viewpoint. The many facets are not dichotomous or polychotomous but interdependent; each feeds into and draws upon the others. Moreover, by treating the system so created as one would any other system within the framework of General Systems Theory, poorly developed or new approaches to the geography of large areas may be identified and elaborated. In this way the gift of the 'generalized ears' can be used to catch communications from scientists who have forged ahead of us in the development of their particular sets of propositions about the systems they see and study.

A geographic matrix[16]

Reflect for a moment on the nature of a single observation recorded from the spatial point of view. Such an observation refers to a single characteristic at a single place or location, and may be termed a 'geographic fact.' This geographic fact usually will be one of a set of observations, either of the same characteristic at a series of places, or of a series of characteristics at the same place. The two series need to be examined more closely. If the characteristic recorded at the series of places varies from place to place, it is common to refer to its spatial variations. These variations may be mapped, for just as the statistician's series are arranged in frequency distributions, geographers like to arrange theirs in spatial distributions. Study of the resulting spatial patterns displayed in the map is one of the essentials of geography. As for the series of characteristics recorded at the same place, they are the stuff of locational inventories and the geography of particular places. With such inventories it is the geographer's common practice to study the integration of phenomena in place.

Now assume a whole series of characteristics has been recorded for a whole series of places. Perhaps we can imagine that complete 'geographic data files' are available (whether such a dream may really be a nightmare is another topic). An efficient way to arrange the resulting body of data is in a rectangular array, or matrix. What does this 'geographic matrix' look like? Each characteristic accounts for a row, and each place for a column, as in Figure 8. The intersection of any row and column defines a cell, and each cell is filled by a geographic fact, the characteristic identified in the row, and the place in the column.

At this juncture one might object and say that there is surely an infinity of characteristics and therefore an infinity of possible rows, and at the limit also an infinity of infinitesimal locations on the earth's surface providing an infinite number of possible columns. This is true; all converges to infinity in the long run. However, to quote Keynes' well-worn maxim, in the long run we shall all be dead. In practice, for any particular problem in any particular context there is some specification of rows (characteristics) and columns (places) that is meaningful and useful. The present discussion is phrased so as to be applicable whenever there is such a problem, whatever the problem and consequent specification of the rows and columns may be, just so long as the viewpoint is spatial.

Given a geographic matrix as described above, how many approaches to regional analysis are possible? One can examine:

a.   the arrangement of cells within a row or part of a row; *or*
b.   the arrangement of cells within a column or part of a column.

The former leads to the study of spatial distributions and maps, the latter to the study of localized associations of variables in place, and to locational inventories. Surely we would agree that the two approaches are the bases of all geography.

Next steps might be:

c.   comparison of pairs or of whole series of rows; *and*
d.   comparison of pairs of columns or of whole series of columns.

The former involves studies of spatial covariations, or spatial association. If the columns are complete, running across all characteristics outlined in Figure 8, the latter implies the study of areal differentiation in its holistic sense.[17]

A fifth possibility is:

e.   the study of a 'box' or submatrix (see Fig. 8).

It is evident that this kind of study could involve some or all of steps *a–d* above, but with something additional – the ability to use findings, say, from studies of spatial association to enrich an understanding of areal differentiation in the partitive sense of the box, or of areal differentiation to explain cases which deviate from some generally expected pattern of spatial association between variables. Each approach could indeed feed into and enrich the other.

A THIRD DIMENSION

The definition of a geographic fact presented to this point is deficient in one respect, since a single characteristic observed at a single location must neces-

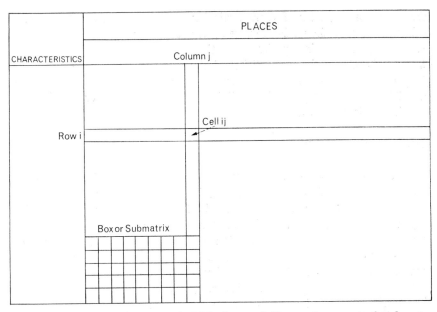

Fig. 8. THE GEOGRAPHIC MATRIX. A row of this matrix presents the place-to-place variation of some characteristic, or a spatial pattern of the variable which can thus be mapped. Each column contains the locational inventory of the many characteristics of some place. Every cell therefore contains a 'geographic fact': the value assumed by some characteristic at some place. Comparison of complete columns is the study of areal differentiation in its holistic sense, and leads to regional geography. Comparison of rows implies the study of spatial covariations and associations, and leads to topical or systematic geography.

sarily also be observed *at a particular point in time.* At any other time it would be different; variation is temporal as well as spatial. Time, too, may be subdivided infinitely, but it is useful to think of the geographic matrix with a third dimension arranged as in Figure 9 in a series of cross sections or 'slices' taken through time in the same manner as rows were drawn through the infinity of characteristics and columns through the infinity of places. Each slice thus summarizes or captures the variations of characteristics from place to place at a certain period of time. Our historical geographers follow this pragmatic procedure. Andrew Clark, for example, noted that 'the cross sections which geography cuts through the dimension of time . . . must have a certain thickness or duration, to provide a representative picture of existing situations.'[18]

    It will be obvious that, for any time period, each of the five possible approaches to geographic analysis previously outlined may be taken. 'Geographies of the past' can be studied in this way. Yet there are additional possibilities introduced by the temporal dimension:

f.   comparison of a row or part of a row through time, the study of changing spatial distributions;

g.   comparison of a column or part of a column through time, the study of the changing character of some particular area through a series of stages, otherwise termed the study of subsequent occupance;

h.   study of changing spatial associations;

i.   study of changing areal differentiation; *and*

j.   comparison of a submatrix through time, a process that could involve all of the preceding approaches individually, but more properly undertaken requires their interplay.

THE TEN APPROACHES

It is thus possible to conceive of ten modes of geographical analysis which may be applied to further an understanding of geographic data files such as are depicted in Figure 9. These ten modes fall into three series. The first (*a*, *c*, *f*, and *h*) includes studies of the nature of single spatial distributions, of the covariance of different distributions at the same period of time or of the distribution of the same phenomenon at different periods of time, and of the covariance of different distributions through time. A similar series of three levels characterizes the second series (*b*, *d*, *g*, and *i*), which spans locational inventories, studies of areal differentiation and of sequent occupance, and

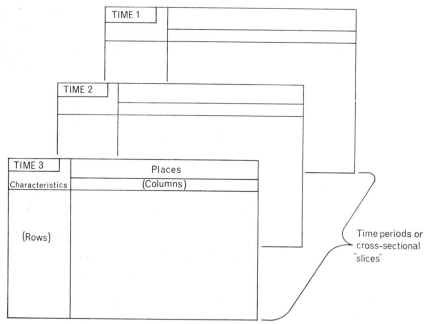

Fig. 9. A THIRD DIMENSION. The third dimension, time, may be introduced by arraying a whole series of geographic matrices such as were presented in Figure 8 in their correct temporal sequence. Each time period thus forms a 'slice' of the three-dimensional cake, and every slice has all the features described in Figure 8. It will be obvious that such an arrangement makes possible examination of rows through time, of columns through time, and of boxes through time.

investigations of changing areal differentiation. The third series (*e* and *j*) involves, at its simplest, the cross-sectional interplay of studies of spatial distributions and associations, locational inventories and areal differentation, and at its more elaborate level the interplay of all nine of the earlier analytic modes.

Traditional groupings of rows and columns

Figure 10 shows the ways in which geographers have traditionally grouped rows and columns of the matrix, and also the conventional ways of grouping the cross-sectional slices, for which we are indebted to historians.[19]

The most common categorization of variables is into one of geography's classic dichotomies, human and physical. Within the human it is conventional to differentiate between variables dealing with collections of people and their numerical and biological characteristics, and those dealing with culture, here used in the holistic sense of the set of man-made variables intervening between man and the earth's surface. These intervening variables may be classified into urban, settlement, transportation, political, economic, and the like. Each of these can be, and has at times been, further subdivided to create further systematic 'fields.' Economic, for example, is often subdivided into: resources, agricultural, manufacturing, and commercial. These in turn involve further subdivisions, until very limited groups of associated characteristics may be said to define 'topical fields.' Such is always the pressure of increasing specialization, and, at the extreme, overspecialization.

Clearly, row-wise groupings of variables of interest correspond with the topical or systematic branches of geography. The essence of this kind of geography is thus the first of the three series of modes of geographical analysis. By the same token, groups of columns form regions (most conventionally, such groupings have been based upon countries and continents, or upon physiographic or climatic criteria). Analysis of such groups of columns is regional geography, with its basis the second series of modes of geographical analysis, emphasizing locational inventories and areal differentiation. *If the object of systematic geography is to find those fundamental patterns and associations characterizing a limited range of functionally interrelated variables over a wide range of places, the object of regional geography is to find the essential characteristics of a particular region – its 'regional character' based upon the localized associations of variables in place – by examining a wide range of variables over a limited number of places.*

Yet neither a topical speciality nor study of a particular region can be sufficient unto itself. More profound understanding of spatial associations can only come from 'comparative systematics' cutting across several topical fields, from an understanding of local variabilities, and from appreciation of the development of patterns through time. Indeed, geography's first, unlamented, theories about man's distribution on the surface of the earth, those of environmental determinism and their wishy-washy derivatives possibilism and probabilism, postulated particular patterns whereby

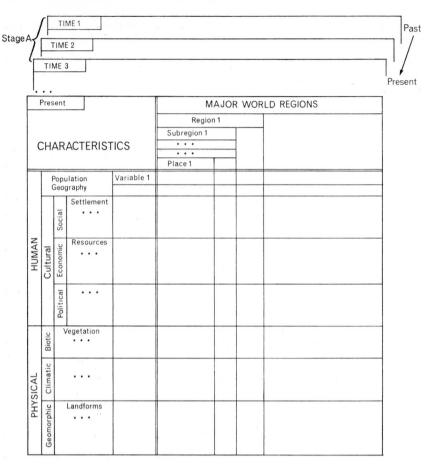

Fig. 10. TRADITIONAL GROUPING OF DIMENSIONS. Geographers have traditionally grouped variables into an ascending hierarchy of rows, the topical sub-fields. The broadest distinction is between human and physical geography. Within the former it is possible to isolate that part concerned with culture in its holistic sense, and within culture, the social, economic, and political. Economic is further subdivided into resources, industrial, etc. Industrial itself has been further subdivided, and so forth. Hartshorne also speaks of the study of areal differentiation as leading towards the identification of a hierarchy of world regions, formed by successive grouping of places and smaller regions into larger more general regions. This is to be seen in the arrangement of the columns. Finally, arrangement of the successive slices into 'stages' is the work of the historian. Given this reference framework, it is possible to locate such things as 'Changing industrial structure of the English Midlands and the Ruhr during the industrial revolution' with ease, and to ascertain their immediate relevance to other undertakings in geography.

arrangements of characteristics from place to place in the 'human rows' of Figure 10 were determined by arrangements of physical characteristics in the physical rows with, in many respects, the former as a reflected image of the latter. The whole idea of study of man–land relationships is the idea of comparative systematics.

· Similarly, 'regional character' can only be evaluated in its integrative sense by proper comparative study of regions, the study of areal differentiation. But here we must pause. What is the basis of regional character? Is it the repetitive appearance of a common theme or themes throughout the entire set of variables recorded for the places within the region, which theme or themes differs from those of other regions? If it is, and there is every reason to believe so, then the understanding of regional character presumes an analysis of spatial associations, simplified because it is undertaken for a relatively small number of places, but complicated because it must be defined for many variables. Only by such study can underlying and repetitive themes be identified. Much the same point can be made for topical studies as well. They are regional because they involve the study of a certain number of variables within the confines of a certain set of places. Whether we call a study topical or regional, then, is basically a function of the relative *length* and *breadth* of the portion of the geographic matrix which is studied. Likewise, whether we classify a study as historical geography or not depends upon the *depth* of the portion of the matrix studied relative to its length or breadth, or else the distance of the slice studied from the present.

To extend the argument further, selection of the columns to be studied is not entirely independent of the rows under investigation in American geography today. If a person is studying things in the economic, urban, and transportation rows, it is likely that his studies will also be confined to those columns encompassing 'modern' urban–industrial societies. Similarly, if the rows under study involve culture in its partitive sense of cultures, settlement forms, language, religion, ecology, and man–land relations, then it is quite probable that the columns embracing the study will be restricted to preliterate and/or 'nonwestern' or 'preindustrial' societies. Although there are different modes of analysis, on no account, therefore, can it be said that the several series are undertaken independently of one another, nor should they be.

Perspectives on the economic geography of the United States

Let us now use this matrix, and later a critique of its inadequacies, to see how well or how thoroughly we have studied the economic geography of the United States.[20] We should first define a submatrix in which the rows embrace those variables of interest to economic geography and the columns encompass all places in the U.S. By projecting the box backwards, we get historical depth.

Studies of this box *per se* have been done very well. The spatial distribution and associations of many variables have been mapped and analyzed. The character of the economic enterprises of most places is well known, as is the historical development of most of the major industries. Attempts of varying degrees of quality have been made to define the relatively homogeneous economic regions of the country both in the partitive sense of agricultural regions, manufacturing regions and the like, and in the holistic sense of real multivariate uniform economic regions. Spatial aspects of the economic growth of the country have been the subject of many investigations.

Yet serious limitations to a general understanding of the economic geography of the country should also be noted. We have already argued that an understanding of the spatial association of any single set of variables requires an evaluation of their actual covariance and theoretical relationship to many other sets of variables, since we are dealing with a system of which interdependence is the essence. Explicit and implicit hypotheses relating to such broader associations are restricted to something which varies between hard-nosed and soft-headed environmentalism. Similarly, more profound understanding of areal differentiation hinges upon comparative regional investigations. This literature is also limited. A third problem is that the model we have developed embraces most of the approaches conceived and undertaken by geographers, but not all; the model itself is limited. There are important geographic questions which the matrix we have developed does not show.

The discussion was initially phrased in terms of General Systems Theory. This theory tells us what some of these unanswered questions are. Any system, including the 'world-wide ecosystem of which man is the dominant part' can be viewed at a variety of levels, the first three of which are those of *static structure, connectivity of parts (functional organization)*, and *dynamic processes*. Figure 10 shows the ways in which the system of interest to geography may be viewed at the first of these levels, that of static structure – of frameworks and patterns in space and time. It says nothing at all about the second level, of interconnections across areas, connectivity of places, flows and interactions, let alone of the third, that of dynamic, interrelated processes. Studies of the economic geography of the United States at the second level are fewer in number and more limited in scope compared with those at the static level in spite of the early efforts of Platt and the later investigations of Harris and Ullman. The growing central place literature is undoubtedly the best example of the level at which the spatial organization of the U.S. economy is understood. This literature refers to a single sector, the distributive, and is generally confined to the local level of very small urban places. There is no understanding of the spatial organization of the U.S. economy that compares with our understanding of the static patterns, no functional regionalization to match the uniform.

There is no longer any real reason why this gap should exist, in spite of the complexity of the system which has to be clarified. What needs to be grasped is roughly as follows:

1   We live in a specialized society in which there is a progressively greater division of labor and scale of enterprise, accompanied by increasing degrees of regional specialization.

2   But in spite of the increasing diversity of people as producers, as consumers they are becoming more and more alike from one part of the country to another, consuming much the same 'basket of goods' wherever they may live as well as increasingly large baskets because of rising real incomes.

3   The physical problem in the economic system is therefore one of articulation – ensuring that the specialized products of each segment of the country are shipped to final consumers, seeing that consumers in every part of the country receive the basket of goods and services they demand and are able to purchase; bringing demands and supplies into equality over a period of time.

4   Articulation requires flows of messages, of goods and services, and of funds. The flows appear to be highly structured and channeled, with major metropolitan centers serving as critical articulation points, as follows: products move from their specialized production areas to trans-shipment or shipping points in the locally dominant metropolitan centres; a complete matrix of intermetropolitan product transfers takes place on a national basis, with each metropolitan center shipping out the specialized products of its hinterland, and collecting the entire range of specialized products from other metropolitan centers spread throughout the country to satisfy the demands of the consumers residing in the area it dominates; distribution then takes place from the metropolis to its hinterland through the medium of wholesale and retail contacts organized in the familiar central place hierarchy. In the reverse direction move both requests for goods and services, and funds to pay for goods and services received, so that the flows are not unidirectional.

The foregoing seems simple enough *but it is mostly unsupported by substantive studies of the spatial organization of the economy of the United States.* Here is a pressing need for careful analysis and synthesis. The amount we do not know at only the second level of viewing the system of interest to geographers is immense, without raising such third-level questions as the ways in which the complex spatial organization of the country is changing through time, and why. The challenge is great, and if these considerations constitute poorly developed or new approaches to the economic geography of the United States, it is towards their solution that we should be moving.

References

1   Reprinted from *Annals of the Association of American Geographers,* Vol. 54, No. 2 (March 1964), Washington, D.C.

2   *The Oxford Universal Dictionary on Historical Principles* (3rd ed., 1955).

3   As applied later in the paper, and as befits the purpose of the President's Program, the particular area of concern is the United States, but the remarks should apply generally.

4   The ideas are directly attributable to BERLINER, JOSEPH S., who developed them in his review of anthropology: 'The Feet of the Natives are Large: An Essay on Anthropology by an Economist,' *Current Anthropology,* Vol. 3 (1962), pp. 47–77.

5   The idea of General Systems Theory was originally formulated by VON BERTALANFFY, LUDWIG, 'General System Theory: A New Approach to Unity of Science,' *Human Biology,* Vol. 23 (1951), pp. 303–61. The journal *General Systems* is a basic source for all interested in General Systems Theory.

6   This contrasts with HARTSHORNE'S view that geography is a chorological science similar to the chronological sciences but contrasting with the sciences classified by categories of phenomena. See *The Nature of Geography,* Chapters 4, 5 and 9, and *Perspective on the Nature of Geography,* Chapters 2, 3, and 11. We are not alone in questioning Hartshorne's views, for a similar debate has been raging for some time in history. Anyone interested in this debate should refer to the journal *History and Theory.*

7   BLAUT, JAMES M. 'Object and Relationship,' *The Professional Geographer,* Vol. 14 (1962), pp. 1–7.

8   BOULDING, KENNETH E. 'General Systems Theory – The Skeleton of Science,' *Management Science,* Vol. 2 (1956), p. 197.

9   A caveat is appropriate at this point. EDWARD SHILS' remarks concerning sociology, which appear in the *Epilogue* to his monumental collection *Theories of Society* (with Parsons, Talcott, Naegele, Kasper D. and Pitts, Jesse R., and published by the Free Press in two volumes in 1961), might well have been written about the scientific status of contemporary geography:

> In so far as a science is a coherent body of empirically supported propositions which retain their stability within a particular theoretical framework, sociology is not a science today. The empirically verified propositions at a level of low particularity are many; as they rise toward generality they become fewer, not because the structure of any science requires it, but because of the deficient coherence of the analytical scheme that explicitly or implicitly guides these inquiries, and because the techniques of research have still not been sufficiently well-adapted to the observation of more abstractly-formulated variability. Nor, for that matter, has theory become sufficiently articulated and explicit. The gap between general theory and actual observation is still considerable.

This statement subsumes R. B. Braithwaite's views concerning the structure of a science, namely that a science properly includes several elements: (a) the facts observed and the simple inductive generalizations based upon these facts; (b) abstract logical constructs; and (c) scientific theories, which are initially stated as hypothesis, and only assume the status of valid and accepted theory when the simple inductive generalizations and the final deductions of the abstract logical constructs coincide. 'Coincidence' is achieved when a satisfactory level of explanation of the inductive generalizations from the deductive constructs is achieved. Nagel provides an excellent discussion of the four modes on scientific explanation, strictly logical, genetic, functional, and probabilistic. See Braithwaite, R. B., *Scientific Explanation,* Cambridge: Cambridge University Press (1953), and Nagel, Ernest, *The Structure of Science,* New York: Harcourt, Brace and World (1962). Adherence to these views we consider basic to this paper.

10   ACKERMAN, EDWARD A. 'Where is a Research Frontier?' Presidential Address, Denver, Colorado, September 4, 1963, *Annals of the Association of American Geographers,* Vol. 53 (1963), pp. 429–40.

11   MCINTOSH, ROBERT P. 'Ecosystems, Evolution and Relational Patterns of Living Organisms,' *American Scientist,* Vol. 51 (1963), pp. 246–67.

12   Ibid.

13   ULLMAN, EDWARD L. 'Geography as Spatial Interaction,' *Proceedings of the Western Committee on Regional Economic Analysis,* Revzan, David and Englebert, Ernest A. (Eds.), Berkeley (1954), pp. 1–13.

14   Op. cit., pp. 198–9.

15   I am indebted to discussions with Alex Orden for clarification of many concepts concerning systems and general systems theories.

16   This 'Geographic Matrix' differs from the matrix developed for anthropology by Berliner only in that the columns are *places* for geography and *cultures*

for anthropology. This difference perhaps indicates the kind of variability of major interest to the anthropologist vis-à-vis the geographer, and thus the different perspective in looking at the same systems taken by the two subjects.

17 HARTSHORNE, RICHARD 'On the Concept of Areal Differentiation,' *The Professional Geographer,* Vol. 14 (1962), pp. 10–12.

18 CLARK, ANDREW H. 'Praemia Geographiae: The Incidental Rewards of a Professional Career,' *Annals of the Association of American Geographers,* Vol. 52 (1962), p. 230, quoting Hartshorne.

19 In this grouping I relied upon JAMES, PRESTON E. and JONES, CLARENCE F., (Eds.), *American Geography: Inventory and Prospect,* by Syracuse University Press for the Association of American Geographers (1954).

20 The evidence supporting these remarks will be found in BERRY, BRIAN J. L. and HANKINS, THOMAS D., *A Bibliographic Guide to the Economic Regions of the United States,* a study prepared for the Commission on Methods of Economic Regionalization of the International Geographic Union, and published as Research Paper No. 87, Department of Geography Research Series, University of Chicago (1963).

III Geography and the Systems Approach

'System is *one* of the names of order,
the antonym of chaos.'

S. Beer,
Beyond the Twilight Arch

During the last century scientific knowledge has depended heavily upon the analytic method. Basically, this method consists of isolating all the variables that influence an object or event, determining the precise effect of each variable by means of controlled experiments and producing the general relationship between objects and variables. As verbal languages proved to be incapable of expressing the multitude of relationships involved in any situation, mathematics developed as a language capable not only of giving objectivity and precision to the various relationships, but also of being used as a deductive tool in its own right. Through the application of this analytic method mankind has witnessed an enormous expansion of technological ability and knowledge so that in the third part of this century he is capable not only of transplanting hearts, or travelling to the moon, but also of destroying all life on earth.

Geography, throughout most of the last century, has not been a 'leading' discipline in this expansion of knowledge: the tone of the essays in this book has indicated that it has tended to lag behind in scientific progress. However the essays have also indicated that strenuous efforts are being made to catch up with the main body of science by the wholesale adoption of analytical methodology (Section II). But one must be wary and take note of the history of previous methodologies. The analytical method represents only one approach to knowledge and, like all approaches, is not without its problems, particularly those stemming from the specialization of knowledge. The first problem concerns the narrowing of focus of any individual scientist; the second with the breakdown of scientific communication, not only between disciplines, but between exponents of the same discipline; the third with the compartmentalization of knowledge and the attendant difficulty of perceiving the relationships between the individual pieces of knowledge. Fortunately for scientific progress, time has proved to be a great healer in this respect, and communication has not broken down completely. Given the rapidity of technological change and the acceleration of the amount of knowledge, we cannot be so confident that the problems will continue to solve themselves.

255

The task of finding solutions is becoming even greater and more urgent. Moreover, it must be remembered that much of the work being published today may not be so productive as it could be. Stoddart (Chapter 3) and Chorley (Chapter 9), in particular, have shown how certain ideas which are currently important in the terms of modern scientific methodology made an early appearance in the literature, but lay dormant for years until a more modern revival occurred or, in more wasteful cases, a rediscovery or independent discovery took place. Indeed this has led Harvey to observe that 'one of our immediate aims should be to improve some semblance of order upon the chaos of our ideas about the spatial system before returning to the problem of establishing the nature of the order within the system itself.'[1]

Introduced by the use of a purely analytical methodology, these problems have led certain workers to eschew this approach. They have returned to those concepts of synthesis and systems that were so prevalent in the pre-analytical scientific era, focusing, in other words, on the whole rather than the individual parts (cf. the reversal of attitudes relating to space investigated by Blaut in Chapter 2). Now, the objective is to re-establish the holistic approach to knowledge, though it must be stressed that this is not a complete reversal. The pre-analytical reliance upon teleological explanations or directly perceived analogies is avoided, leaving in their place the rigour of modern analytic science.

The search for generalizations based on the whole rather than on individual parts is, therefore, a complementary method of modern science known as *systems analysis*. The broad framework of concepts lying behind it is called *General Systems Theory* and, as the quotation at the beginning of this section points out, it is one, though only one, of the methods employed by science. Since all systems, whether physical or human or combinations of both, consist of a set of objects and the relationships binding these objects together into some organization, it is not surprising that the approach is especially useful in dealing with functional aggregates. Indeed, now that the major focus of scientific enquiry has moved away from the study of objects or substances to the study of relationships and organizations, and all organizations are recognized as being particularly complex, systems analysis proves to be a particularly appropriate framework of study. By carefully defining the structure, functioning and general process change in any situation, these three major perspectives may be integrated into one system of study, as Berry has already revealed in Chapter 11. Moreover, the approach may also be said to be particularly relevant to those studies attempting to bridge the gap between the natural and social sciences (Stoddart, Chapter 14), the position long claimed, but rarely effectively occupied by geographers. (In view of these developments the concluding remarks of Matley in Chapter 5 may be noted.) Hence the objective of the systems approach is not to provide a unified substantive theory – rather it is to provide a synthesis of the functional and structural relationships, to emphasize the comparative study of systems and to point the way to future research into the existing gaps in our understanding. In this way the cross-fertilization of scientific disciplines will be preserved and the individual pieces of knowledge thrown up by purely analytical research will be inter-related, particularly those dealing with the functioning and organization of the whole system. In this section four essays represent examples of the development of this line of enquiry in geography, though at the very outset it must be repeated that the systems approach is not a replacement for the analytical method, but is an additional line of modern scientific enquiry designed to break down the barriers between interdisciplinary enquiries. It thus represents one of the major current research frontiers in geography.

The first paper of the section 'Where is a Research Frontier?' by Ackerman (Chapter 12) provides a general introduction to the need for geographers to develop systems analysis, and in doing so emphasizes many of the points raised by the authors of essays in Section II. Writing almost a personal confession of the errors of traditional geographical enquiry, Ackerman maintains that the core of the current methodological problem rests in the fact that geographers have tended to regard their discipline as an end in itself, forgetting one of the basic axioms of science: namely, that the course of science as a whole determines the progress of its parts. Ackerman is emphatic that if geography is ever going to shake itself out of its isolationist past it must rejoin the mainstream of scientific effort. If it fails to do this in the near future, other disciplines will cultivate the research frontiers appropriate to its perspectives.

Although the adaptation of modern scientific methodology represents a solution to the current problems of geographic enquiry, Ackerman is careful to observe that these ideas, techniques and concepts should not be used uncritically. Adoption of mathematical techniques and symbolic logic does not remove the discipline from the dangers of dead-end research or unproductive enquiry. Certain non-quantifiable ingredients must be added: namely, the creation of possibilities for inter-disciplinary communication; intuition based on original thought and experience; and a highly developed problem-solving ability. Only with these additions will the discipline be given life and direction.

In this attempt to illuminate the future direction of geographical enquiry, Ackerman dwells upon two fundamental problems of past enquiry – the failure to develop an adequate corpus of theory and the overzealous concentration upon areal differentiation. In company with most of the writers in this book, it is an article of faith to Ackerman that theory is the core of the discipline and represents the fundamental refuge of the specialist geographers, so he pays little attention to this issue. The emphasis upon areal differentiation is, however, a different matter. In many ways this approach is opposed to modern scientific method, for instead of searching for similarities it stresses the differences between areas, and leads to a static goal. As an alternative, Ackerman proposes that the empirical aspects of geography should be re-defined to deal with the spatial aspects of the inter-acting system of humanity and its natural environment. By focusing upon this system of relationships the inter-disciplinary communications of scientific findings will be preserved, and individual sub-systems will be placed in perspective. Moreover, the emphasis placed upon connectivity rather than the differentiation of any system means that particular attention will be paid to the pattern of flows within the system; this, to Ackerman, is the most important distinguishing feature of systems. Areal differentiation, he maintains, is only important in so far as it helps to define the connectivity of flows.

Ackerman's basic thesis is that once geographers adopt this redefinition of interest a more profitable dialogue with the students of the natural and behavioural environment would be maintained. Exploration of the common ground between these sciences would contribute to the growth of science as a whole and geography would not run the risk of the triviality of independent work. Examples of possible future research directions are provided, but it is particularly relevant to note one of Ackerman's major defences of the need for systems analysis. Already human society is being manipulated and radically altered by our knowledge of the system. If geographers are to contribute to the understanding of the changes that have occurred, expertise is needed in the field.

The second paper in this section, Chorley's 'Geomorphology and General Systems Theory' (Chapter 13), represents a great step forward in the development of modern scientific method in geography, for it was the first major paper devoted exclusively to a systems theme and to the framework of General Systems Theory. At this stage it is worth emphasizing again that this new integrative approach does not mean that the arguments in the first two sections of this collection of essays are now irrelevant. Systems analysis draws upon these ideas extensively and illuminates many of the fundamental issues. Thus, the points made in Section I about the difficulty of disentangling the factual interpretation of the world from the systematic framework of analysis employed by any worker are particularly relevant in this essay. By interpreting the Davisian cycle of erosion in systems terms, Chorley provides a valuable example of these points and incidentally demonstrates that knowledge of these frameworks of thought is as important to the interpretation of the theoretical-deductive system of enquiry as to any empirical-inductive one.

A major part of Chorley's paper is devoted to the description of the differences between the two major systematic frameworks – open and closed systems – and the application of these ideas to geomorphology. In a closed system definite and rigid boundaries occur, through which no material or energy can pass. The development of such systems leads to the eradication of the existing differences or gradients in the system, with the initial conditions determining the final equilibrium form. Open systems, on the other hand, are systems which need an energy supply for their maintenance and preservation and can maintain their organization and regularity of form by adjustments within the geometry of the system. Moreover, different initial conditions in such systems may lead to similar end results. By means of an exceptionally well balanced argument Chorley does not criticize the Davisian cycle as being totally irrelevant. He observes that the ability of the model to treat all landforms as an assemblage is particularly useful, but in the state of current knowledge the Davisian model is not a very productive line of investigation. Thus the Davisian system of thought has certain elements of a closed system framework which, given the early twentieth century bias towards the study of evolutionary history rather than the mechanism of change (cf. Stoddart, Chapter 3), led to a historically based, deterministic and almost irreversible sequential process of landform development. In this sort of system, time and initial conditions determine the ultimate equilibrium conditions or the position of the cycle at any point in time, though it may be noted that the concept of 'grade' was introduced to the cycle as an additional, though not particularly integrated, equilibrium concept. Instead, Chorley proposes that an open system framework is more applicable to geomorphology and would lead to several major advantages. The most important of these would be to focus attention on the multivariate nature of most problems, and on the relationships not only between form and stage, but between process and form (cf. comments of Blaut in Chapter 2).

In conclusion, therefore, it may be noted that Chorley's article demonstrates that it is not the Davisian model so much that needs to be criticized: rather it is the dominating position given to this model in geomorphology. By ignoring alternative lines of investigation, this consensus (cf. Lowenthal, Chapter 4) led to the isolation of geomorphology from the major lines of scientific enquiry, a position to which it is only gradually returning.

Unlike most of the papers in this book, the third article in this section, Stoddart's 'Geography and the Ecological Approach' (Chapter 14), was written as an additional contribution to a discussion on the ecological viewpoint initiated by Eyre. Unlike Eyre, Stoddart justifies the ecological approach not by empiricism (by the use of series of case studies to show the need for a synthesis between organisms and their habitat) but on methodological grounds, on the creation of a framework of concepts. He shows that the search for synthesis is not new, but geographers in particular have rarely achieved much success in this direction because results were either framed in a qualitative way or on a regional basis, while the true nature of the geographic relationships involved was obscured by naïve deterministic statements. Recognition of the ecosystem, Stoddart maintains, means that man, the environment and the flora and fauna are brought into one monistic framework, though it might be noted in the light of Matley's paper (Chapter 5) that Marxist geographers would probably disagree with this interpretation. This framework is structured in a particular way and it is the elucidation of this structure, together with the functional relationships of the whole system, that represents the focus of enquiry. It is admitted that a multitude of structures and relationships can be recognized in any ecosystem, but the potential value of the concept depends on the correct selection of the components to deal with the situation. Moreover, particular attention must be paid to the appropriate scale of investigation of the problem when these components are analysed.

Stoddart observes that the ecosystem is a form of general system. As such, its properties may be compared with the system derived by information and communication theorists, and from these comparisons fruitful lines of further investigation may be suggested. However, this does not lose sight of the fact that certain systems ideas were used profitably in the past. After all, analogous ideas lie behind the concept of 'climax' in vegetation studies and 'maturity' in soils. What is different is that the limitations of these concepts are recognized, for they represent 'closed system' ideas, whereas an ecosystem is essentially an 'open system'. Today a major research effort is devoted towards reformulating these concepts in open system terms.

By providing a general framework for any individual study the eco-system concept has considerable utility, but Stoddart stresses that it must not be forgotten that the research objective is to understand the functioning of the whole system. Gradually this knowledge is being attained, for the techniques of analytic scientific methods are being used to quantify the relationships between the components of an ecosystem. Hence, a precise and objective synthesis of the relationships is provided, with an emphasis upon the functional and dynamic aspects of the system rather than on the formal static relationships.

In the last paper of the section, 'Cities as Systems within Systems of Cities' (Chapter 15), Berry not only demonstrates the utility of the systems approach to the study of urban areas but also provides a vivid illustration of the direction in which modern scientific method, allied to the new computer technology, is moving. The basic contention of the paper is that studies dealing with the geography of urban areas have fallen into two main groups, one group providing inductive generalizations, the other providing logical constructs. Berry observes that if a sound urban theory is ever going to develop these two lines of enquiry must be integrated. In this process of integration inductive generalizations will be translated into theory and logical constructs will be quantified and tested in the real world. New empirical works and experiments are also likely to be generated as by-products of these developments.

Having established this broad frame of reference for his study, Berry devotes the rest of the paper to showing how this integration can be achieved in four major fields. One set deals with intra-urban characteristics, the other set deals with inter-urban characteristics. Both may be further divided into single and multi-feature distributions on the basis of the number of variables incorporated in the study, the former case being represented by the study of urban population densities and the population sizes of cities, the latter by the study of social areas and central place characteristics. In every case Berry shows the impact of the new quantitative methods by devising a set of structural equations for each of the four lines of enquiry. The single feature distributions are shown to possess very similar equations and are descriptively related to existing theoretical statements. By contrast, studies of multi-feature urban characteristics were rather different, for theories relating to the distribution of central places and social areas already existed. What was lacking was the rigorous testing of these theories. Berry uses the results of the latest research to demonstrate how this testing can be carried out. His own

work has provided a reasonable empirical summary of many of the aggregative patterns of central place systems, and it is worth noting that the hierarchy/
continuum debate has at last been resolved by consideration of scale factors.[2]
In social area studies Berry shows that the existing theories of intra-urban
social characteristics are not in as much opposed as they seem: the theories
are alternative and partial viewpoints of a complex reality. If the theories are
combined they provide an adequate description of real world patterns in
which local distortions can be placed in their true perspective.

The importance of this article does not only rest with these attempts at
integration. Berry concludes his article with a review of the relevance of
General Systems Theory concepts to the study of certain types of system, in
this case urban systems. Definitions of the concepts and their application to
the four major lines of investigation are given in the paper and need not be
repeated here. However, it does seem worth clearing up one point of possible
confusion. Although the concepts deal primarily with the organization of a
system and the processes that operate, this does *not* imply that any statement
is made about the particular nature of the causal mechanisms at work in any
system. They point out the general trends or generalizations that may be
derived from any particular system. Thus to observe that the stability of a
rank-size distribution of urban population sizes through time represents an
example of a morphostatic process does not make any statement about the
causes of this process – it merely describes the result and makes it easier to
compare the trend with those derived from other systems. It is, therefore, this
desire for comparability, this search for generalization, that represents the
objective of systems analysis. It is a methodology that focuses particularly
upon the organizational and process changes in the world, placing each
particular part in relation to all the others.

As a final point it must be admitted that systems ideas and General
Systems Theory have been subjected to a great deal of criticism and at least
one geographer has dismissed it as an 'irrelevant distraction.'[3] Such criticism
must be considered to be unfortunate because, with one or two exceptions,
the exponents of General Systems Theory do not consider their theory to be
a substantive theory in the way Chisholm describes. General Systems Theory
'is a point of view rather than a theory in the scientific sense, because there is
little in General Systems Theory that can be embodied into the sequence of
related propositions that one usually conceives a theory to be. More properly
... [it] ... is a certain approach to describing and investigating.'[4] In short,
General Systems Theory is not a substantive theory, but is a conceptual
theory, and its utility cannot be judged only by reference to the admittedly
limited homologies derived from empirical investigations in different sciences.
That certain critics find it an irrelevant distraction is, of course, their prero-
gative, but it is certainly not irrelevant. One should be wary of condemning
out of hand an approach aimed at synthesis, particularly in view of the consequences of the methodological isolation of geography during the past
thirty years. In addition we should remember that planners are using systems
analysis to alter and control our environment.[5] Unless we are aware of and

chronicle these methods, future generations of geographers may be condemned to naïve interpretations of the changes taking place in the world.

In conclusion it is worthwhile returning to the quotation from Wittgenstein used at the beginning of the Introduction, and stress that there is no single path to scientific understanding. All scientists search for order in the real world, whatever their disciplinary perspective, and the final test of any methodology is its productivity in organizing the search. Many workers have found the systems approach to be a fruitful source of ideas and concepts, and it is, moreover, a convenient aid to the clarification of thought. In such circumstances, the approach cannot be dispensed with lightly, unless a neo-empiricism is to emerge as the new commanding heights of the philosophy of geography.

References

1   HARVEY, D. 'The Development of Theory in Geography', *Journal of Regional Science,* Vol. 7 (December 1967), p. 215.
2   BERRY, B. J. L. *Market Centres and Retail Distribution,* Englewood Cliffs, New Jersey: Prentice Hall (1967).
3   CHISHOLM, M. 'General Systems Theory and Geography', *Transactions of the Institute of British Geographers,* No. 42 (December 1967), p. 51.
4   RAPPORT, A. 'Conceptualization of a System as a Mathematical Model' in Lawrence, J. R. (Ed.), *Operational Research and the Social Sciences,* London: Tavistock Publications (1966), p. 516.
5   It must be stressed that a scientific theory, a system of definitions and hypothesis, is capable of empirical verification; a conceptual model or framework is merely a tool for the organization of thought.
     ZETTERBERG, H. L. *On Theory and Verification in Sociology,* Totowa, New Jersey: Bedminster Press (3rd edition) (1965).

Selected bibliography for further reading

BOOKS

VON BERTALANFFY, L. *General System Theory: Foundations, Development, Applications,* New York: George Braziller (1968).
BLACK, G. *The Application of Systems Analysis to Government Operations,* New York: F. A. Praeger (1968).
BUCKLEY, W. *Modern Systems Research for the Behavioral Scientist,* Chicago: Aldine (1968).
BUCKLEY, W. *Sociology and Modern Systems Theory,* Englewood Cliffs, New Jersey: Prentice Hall (1967).
CHURCHMAN, W. *The Systems Approach,* New York: Delacorte Press (1968).
GLUCKMAN, M. (Ed.) *Closed Systems and Open Minds,* Edinburgh: Oliver and Boyd (1964).
KAPLAN, H. *Urban Political Systems,* New York: Columbia (1967).

KAPLAN, M. A. *System and Process in International Politics,* New York: J. Wiley (1964).

MCDANIEL, R. and HURST, M. E. ELIOT 'A Systems Analytic Approach to Economic Geography', *Association of American Geographers Commission on College Geography,* No. 8 (1968).

MCLOUGHLIN, J. B. *Urban and Regional Planning: A Systems Approach,* London: Faber (1969).

MESAROVIC, M. D. (Ed.) *Views on General Systems Theory,* New York: J. Wiley (1964).

QUADE, E. S. and BOUCHER, W. L. *Systems Analysis and Policy Planning,* New York: American Elsevier Publishing Co. (1968).

ZWICK, C. J. *Systems Analysis and Urban Planning,* Chicago: Rand Corporation (1963).

ARTICLES

AJO, R. 'An Approach to Demographical System Analysis', *Economic Geography,* Vol. 38 (1962), pp. 359–71.

BERRY, B. J. L. and WOLDENBERG, M. J. 'Rivers and Central Places: Analogous Systems', *Journal of Regional Science,* Vol. 7, No. 2, (1967), pp. 129–39.

BLALOCK, H. M. and BLALOCK, A. 'Toward a clarification of Systems Analysis in the Social Analysis,' *Journal of the Philosophy of Science,* Vol. 26 (1959), pp. 84–92.

BLAUT, J. M. 'Object and Relationship', *Professional Geographer,* Vol. 14 (1962), pp. 1–7.

FOOTE, C. D. and GREER-WOOLTEN, B. 'An Approach to Systems Analysis in Cultural Geography,' *Professional Geographer,* Vol. 20 (1968), pp. 86–92.

HUGHES, J. and MANN, L. 'Systems and Planning Theory,' *Journal of the American Institute of Planners,* Vol. (September 1969), pp. 330–3.

ISARD, W. *et al.,* 'On the Linkage of Socio-Economic and Ecologic Systems', *Papers—Regional Science Association,* Vol. 21 (1968), pp. 79–100.

KASPERSON, R. E. 'Environmental Stress and the Municipal Political System', in Kasperson, R. and Minghi, J., *The Structure of Political Geography,* Chicago Aldine Publishing Co. (1969), pp. 481–96.

MEDVEDKOV, Y. V. 'The Concept of Entropy in Settlement Pattern Analysis', *Papers—Regional Science Association,* Vol. 18 (1967), pp. 165–8.

QUASTLER, H. 'General Principles of Systems Analysis' in Waterman, T. H. and Morowitz, H. J. (Eds.), *Theoretical and Mathematical Biology,* New York: (1965).

12   Where is a research frontier?[1]

Edward Ackerman

Carnegie Institution of Washington, U.S.A.

Science in the last fifty years

I was born in the year 1911. The half century since that time has contained some of the most tremendous events in the history of the human race. Among them have been two world wars of unprecedented extent and violence, the near doubling of the total population of the world,[2] the rise of the great Communist states, and other events profoundly changing the course of human history. But pre-eminent among all are the growth of science and the growth of man's capacity to apply his mind to the problems of learning and discovery.

Many striking examples might be offered by the changes within science, and the changes wrought by science, in these 50 years. To me, a most striking illustration is a comparison of our knowledge of the universe in 1911 and now. In 1911 what men knew of space was confined to our own galaxy. Our solar system was thought to be near the center of that galaxy, whose shape was only dimly conjectured. Now we know that the sun and its secondary satellite, the earth, are far out on one arm of our vast, beautiful, spiral galaxy. We know also that there are at least a billion such galaxies within the space our telescopes have penetrated. Furthermore, we have seen the photographic record of objects five billion light-years away, moving away from us at half the speed of light. These 50 years have brought a more profound change in our knowledge of the cosmos than was achieved in all man's previous existence.

Although astronomy may stretch our minds most of all, there are other examples of advances in our learning and knowledge, of the deepest meaning and most comprehensive impact. Among them we might mention the general theory of relativity, the dismembering of the atom by nuclear physics, the discovery of the biochemical basis of heredity, the developments in engineering that made possible the Manhattan Project and the orbiting of men in space, and the chemical discoveries and developments in social organization that have promoted the world population explosion. It has been a truly epochal period, without any equal in history. Not least has been the final world acceptance of science as a tremendous social force.

As one views this panorama of glorious scientific achievement in the last 50 years, he cannot fail to be impressed by two things: the unity of scientific effort as it progresses; and great differences in the rates of progress among the subdivisions of science.

The first observation was skillfully described by the biologist Frank R. Lillie in 1915. 'Scientific discovery is a truly epigenetic process in which the germs of thought develop in the total environment of knowledge. Investigation of particular problems cannot be accelerated beyond well-defined limits; progress in each depends on the movement of the whole of science.'[3]

Lillie's observation must be considered in the light of the second point, the differentials of progress among separate subfields of the scientific community. The progress of 'science as a whole' at any given time in large measure is the progress of a relatively few subjects with growing points. As growing-point salients move, they furnish ground for practitioners in other disciplines to stand on and in turn push into new territory. This is what makes inter-communication among the sciences so important, and even more the proper choice of those with whom we communicate. To paraphrase an ancient observation, every scientist stands on the shoulders of giants. But one might add that it is important to stand on the shoulders of the right giant. The selection is as important as the standing. In the period between 1910 and the mid-1940's, physics and the mathematical disciplines stood out as examples of the giants.[4] Chemistry was of shorter stature in this comparison, biology considerably shorter, and geology less visible. Comparisons among the social sciences were more difficult, but perhaps anthropology, psychology, and economics deserve some distinction for their accomplishments in the pre-World War II period. However, the differences among subgroups within a field were in most cases as pronounced as differences between major fields.

The place of geography in the advancing front

We are, naturally, especially interested in the place that geography occupied in this advancing front of science. There is no reason to avoid frankness. I am sure that all but a few here would agree that our contributions have been modest thus far. We have not been on the forward salients in science, nor, until recently, have we been associated closely with those who have. The reasons are not difficult to find. During the early part of this 50-year period, in the 'teens and early twenties, our closest associations were with history and geology. Geological study of that period, and of the thirties, was not among the inspiring growing points in science. The history and the geology connections did not correct the predisposition of our scholars of the 'teens and early twenties to the deceptive simplicity of geographic determinism. This was perhaps one of the last appearances of the Newtonian view of the world.

As determinism began to fade and independent geography departments sporadically appeared in this country, geography turned to association with the social sciences of the period. 'Possibilism' in man's relation to the earth took the place of determinism. Because of the limitations of the social sciences and history at the time, these associations were only slightly more productive sources of inspiration than geology. It was only much later, indeed in the early fifties, when association with the social sciences bore its soundest fruits for geography. This was in the methods descended from mathematical statistics, first applied in biometrics, anthropometry, and econometrics. Their full application has not yet run its course.

Independence and separation of geography

I began my professional interest in geography at a time when the old moorings to geology were almost severed. The groping for solid footing among the social sciences was well under way.[5] The geographers who turned in the direction of the social sciences made a prescient choice of direction, but the difficulties confronting us were enormous, considering the methods then at our disposal. In the face of those difficulties, it was only natural that we became somewhat introspective. We tried to build a platform, as it were, from our own materials and to anchor it ourselves.[6] This search for a professional identity was, of course, found during other periods in the history of geography. It goes back at least to the nineteenth century German geographers. Alfred Hettner and others in Germany undertook influential studies from the early 1900's onward. But the succession of methodological appraisals in the United States that commenced with Harlan Barrows' 'Geography as Human Ecology'[7] in 1923 and continued for nearly forty years must certainly rank as one of the most intensive efforts toward this end.

Our search for a professional identity led to an intellectual independence and eventually to a degree of isolation against which a number of the rising younger generation of geographers have now reacted. In our search for a solid footing, a meaningful image of ourselves, many of us tended to separate ourselves from other sciences. Our principal interdisciplinary communications were with other sciences which also had problems of isolation, like cultural anthropology and geomorphology. In effect, some of us saw geography as an end in itself rather than in the broader context as a contributor to a larger scientific goal. Perhaps this is the fate of many specializations.

Insistence on the independence and separation in the 1930's and 1940's may seem a shockingly incorrect statement to some of you. Did not geography alone recognize its relations to both the physical sciences and the social sciences? Did not geography deal constantly with the data accumulated through the efforts of other disciplines? Indeed, was not geography even alert to analogous methods of inquiry from other disciplines? One can cite such major statements in the field as Barrows' 'Geography as Human Ecology,' and Sauer's 'The Morphology of Landscape'[8] as proof of this alertness. But I must note that both these statements came in the mid-twenties, and thereafter for at least 25 years an atmosphere of separatism and independence characterized the profession.[9] Furthermore, morphology was not a particularly happy choice as an analogue method, and the hint given by Barrows on ecology was never seriously followed up by his colleagues. For science at large, morphology already was becoming a somewhat sterile concept when we took to it, and the analytical methods of the twenties and thirties were not yet equal to the multivariate problems of ecology.[10] The concept that became dominant among us was that of 'areal differentiation,' derived from Hettner and introduced in the United States by Sauer.[11] This concept favored (although did not demand logically) a goal of investigation independent of the goals of other sciences. The same might be said of another important concept in the field, that of areal functional organization, introduced by Platt.[12] On the other hand, the work of Sauer and his disciples did find common ground with cultural anthropology, but it also was a somewhat isolated science until the 1940's.

In our desire to make our declaration of independence viable we neglected to maintain a view of the advancing front of science as a whole. We acted as though we did not believe in anything more than the broadest generalities about the universality of scientific method. In effect we neglected to appraise continuously the most profound current of change in our time. We neglected an axiom: The course of science as a whole determines the progress of its parts, in their greater or lesser degrees.

Influence of mathematical statistics

What did we miss in the course we took? For one thing, we missed early contact with developments in mathematical statistics, and early touch with the antecedents of systems analysis. The scholars whose thought influenced life (and social) systems concepts greatly, like R. A. Fisher and Karl Pearson in biology and anthropology, Alfred Lotka in biology, Sewall Wright in genetics, and L. L. Thurstone in psychology, were all active in the 1920's and 1930's. The flowering of the application of their techniques and concepts awaited the availability of electronic computers and mathematical progress in the late 1940's and early 1950's, but they provided forceful organizing advances in genetics and other biological fields, in physical anthropology, demography, psychology, and economics from 15 to 25 years earlier than in geography. We thus missed for a period the new thought their techniques generated, because the techniques were essential keys to communication of that thought.[13]

Within the last decade we have made good our initial failure to respond to these modern techniques. We have even felt the influence of physics, as a few have experimented with the application of physical analogues to the phenomena of distribution. Although not a few among us have been uneasy about their meaning, these techniques have already proved their power. Mathematical analysis is a recognized part of instruction in alert departments of geography. We can only welcome the growth of these methods, because they have been a notable and needed stimulus to the rigor of our thinking. Even more important, they increase our capacity to communicate precisely with workers in other fields of science.

Is the mathematicization of our discipline the way of our future? In a sense, yes. The year is not far off when a geographer will be unable to keep abreast of his field without training in mathematics. Furthermore, he will find it increasingly difficult to conduct meaningful research without such training. But here we must enter more than a word of caution. There is a great deal more to science than the application of mathematics, or of rigorous logic. We must take care to examine carefully the paths of research down which our computerized mathematical colleagues lead us, or perhaps push us. The danger of a dead end and nonsense is not removed by 'hardware' and symbolic logic. Before we go too far we should see what else there is about science at large that produces its 'growing points.' What determines how productive the use of statistics and hardware will be? In a few other fields scientists are facing problems of this kind that are somewhat out of control today. Recent attention to the scientific part of space exploration is an illustration.[14]

Nonquantifying attributes of science

Can we make any observation about the methods of science at large that will enable us to keep needed mathematicization under control? There are a great many definitions of science. I am sure that many of you are familiar with most of them. One definition I like is: 'Science is a quest for regularity underlying diverse events.' This quest proceeds through careful, verifiable observation and description; through the construction of hypotheses, to project reality into the unknown; testing of the hypotheses through the conduct of experiment or further observation;[15] replication of experiment and observation; and the building of a body of theory from verified hypotheses which in turn becomes the basis for new hypotheses, and new observations and experiments. Mathematical and statistical analysis have found their important place in this procedure because they aid in obtaining exact observation, and because they aid enormously in designing hypotheses that lead into the unknown.

I might stop here, and you would recognize this as a portrait of science. However, it is a portrait only of its skeleton. Three important additions provide the all-important life and direction that have figured wherever great strides in science have been made. I have already mentioned one: cross-disciplinary communication.

The second is what some men have described as the intuitive side of science. Warren Weaver has said, '. . . science is, at its core, a creative activity of the human mind which depends upon luck, hunch, insight, intuition, imagination, taste, and faith, just as do all the pursuits of the poet, musician, painter, essayist, or philosopher.'[16]

But there is more to it than this. The mind of the scientist, no less than that of the poet or musician, must be structured by thought and experience before it reaches the creative stage. Some persons are able so to structure their minds more easily than others. It has been said, for example, that

Irving Langmuir always saw matter, of whatever form, wherever he was, in terms of its molecular structure, thus opening the way automatically for his many remarkable insights. Every scientist does this in some degree. There is no doubt that there is such a thing as 'thinking geographically.' To structure his mind in terms of spatial distributions[17] and their correlations is a most important tool for anyone following our discipline. The more the better. If there is any really meaningful distinction among scientists, it is in this mental structuring. It is one reason why we should approach the imposition of analogues from other fields, as from physics, with the utmost care. The mental substrate for inspiration does differ from field to field.

A third important ingredient of science is a highly developed sense of problem. In my pleasant and valued association with Professor Charles Colby at the University of Chicago, I can remember his frequent reference to the cultivation of such a sense. I now realize how wise and perceptive his advice was. In my duties of the past five years at the Carnegie Institution of Washington, I have had to maintain current knowledge about research in several biological and physical sciences. In all of them this sense of problem is very keen where outstanding progress is being made. Herbert Simon has observed that science is essentially problem-solving.[18] This observation is so important that it deserves a few words of elaboration. A sense of problem, at its most meaningful, is really a sense of the hierarchy of problems in a broad field, and possibly in all science.

Every major field with which I am familiar has an easily recognized overriding problem. The overriding problems always lie behind the frontiers of investigation. They are remarkably few, and all fade into infinity in their ultimate forms. Indeed, the overriding problems of all science may be reduced to four: (1) the problem of the particulate structure of energy and matter, which physics treats; (2) the structure and content of the cosmos, which astronomy, astrophysics, and geophysics treat; (3) the problem of the origin and physiological unity of life forms; and (4) the functioning of systems that include multiple numbers of variables, especially life systems and social systems. Others might express these problems differently, but I believe that each of them is a beacon orienting research on the frontiers of the rapidly advancing fields.

Beneath each overriding problem are major second-level problems, and finally the problems translatable directly into experiment or observational investigation. For example, a major secondary problem related to the overriding one of the origin of life is the description of life in pre-Cambrian times. It is translated directly into a search in pre-Cambrian rocks for stable chemical compounds known to be indicators of life. In this way Philip Abelson and his collaborators at the Carnegie Institution have produced firm evidence of the existence of life at least 2.6 billion years ago.

The same relation among the hierarchy of problems can be seen in growing-point research in astronomy, geophysics, biology, and elsewhere. I do not mean that all research is so organized, or is distinguished by the sense of problem. Most commonly an appreciation of the hierarchy of problems is shared by relatively few in each field. It is indeed one of the most troublesome questions facing the administrator of public research funds in the nation at the present time.

Geography in the mirror of all sciences

By now my theme should be obvious: The geographer should seek his personal identity in the mirror provided by all sciences. How is this translated into future geographic progress? The development of a professional identity in geography has two aspects: the future development of the theoretical study of spatial distributions; and a reappraisal of the overriding problem recognized by our discipline.

The first, further development of the theoretical, is our true inner refuge as specialists. It is what helps to structure the mind 'geographically.' The more rigorously the structuring is done, the more likely the discipline will have a cutting edge that places it on a research frontier. However unrelated and esoteric it may seem, the cultivation of theoretical study of spatial distributions is basic.

If we have had any generally accepted overriding problem in the past, it is areal differentiation, a concept widely accepted and usefully employed, particularly by American and German geographers. Its rationale has been ably presented and skillfully defended by Richard Hartshorne. His most recent definition of areal differentiation as the 'accurate, orderly, and rational description and interpretation of the variable character of the earth surface'[19] still stands as a useful general guide to geographic method. A second preoccupation, but less widely held, was with the geographical expression of culture processes. I shall refer mainly to areal differentiation in the remarks of the next few paragraphs.

A new look at geography's overriding problem

At a time when the social sciences provided us with very little firm assistance, and we were stressing our independence, areal differentiation of the earth's surface did serve as an overriding problem. It is time that we recognize the limitations of this concept. Do we need something more for a purposeful selection of research problems leading us to significant research frontiers? If we look at the concept of areal differentiation carefully, we see that it did not often lead us to common ground with the other sciences. We see it also as ending in a somewhat static goal. In effect, it stressed a hierarchy of regions as our hierarchy of problems.

I suggest that we take a fresh look at the hierarchy of problems, ignoring for the moment some of our traditional points of view. I noted earlier that science is problem-solving. The problems that can be examined meaningfully depend on the methods which are available for their solution. As the centuries have gone on, men have steadily increased their capacity for problem-solving, but the truly important changes in methods of problem-solving have been remarkably few. They might read somewhat as follows: writing; Arabic numerals; analytical geometry and calculus; and the combination of techniques that comprise systems analysis. There was a time perhaps just after the Second World War, when the inclusion of systems analysis in such a list might have been considered controversial. That is no longer true. Systems, as you know, are among the most pervasive and characteristic phenomena in nature. Each human being, man or woman, is a system, that is, a dynamic structure of interacting, interdependent parts.[20] Perhaps that is less appealing than a poet's definition of a pretty girl, but it has meaning in that it relates the girl as a system to all other systems, such as a colony of ants, or a city, or a business corporation.

Systems analysis provides methods of problem-solving which might be said to have been created for geography, if there were not also many other uses for them. Geography is concerned with systems. Indeed, we may now state its overriding problem. It is nothing less than an understanding of the vast, interacting system comprising all humanity and its natural environment on the surface of the earth. This might be compared with Humboldt's statement of a century ago. 'Even though the complete goal is unobtainable, . . . the striving toward a comprehension of world phenomena remains the highest and eternal purpose of all research.'[21] It may also be compared with Hartshorne's definition of the purpose of geography as 'the study that seeks to provide scientific description of the earth as the world of man.'[22] Compare also Barrows' 'geographers . . . define their subject as dealing solely with the mutual relations between man and his natural environment. By "natural environment" they of course mean the combined physical and biological environments. . . . Thus defined, geography is the science of human ecology.'[23] All these statements have some similarity. However, the concept of the world of man as a vast interacting, interdependent entity permits us an effective orientation to a set of problems *at different levels* in a way that we have never had before.[24] Furthermore, it puts us in a context of sharp new problem-solving methods.[25] If we are willing, it also places us in association and in close communication with other sciences whose overriding problems are similar.

Viewed in this way one can see a host of beneficial results. We no longer are concerned about whether what we are doing is geography or not; we are concerned instead with what we contribute toward a larger goal, however infinite it may seem. As in other sciences, an overriding problem of infinite extent should be a challenge, not cause for resignation or despair. We no longer debate about whether geography can construct 'laws.' At the same time we do retain an identity by structuring our minds to handle spatial distribution patterns in all their complexity. But as we go about our task of analyzing spatial distributions and space relations on the earth we should keep in mind the question, 'What, if anything, do geographic observations and analyses tell us about systems generally, and the man – environment system particularly?'

Summary statement of concept and method

We might elaborate this position in summary manner: (1) The basic organizing concept of geography has three dimensions. They are: extent, density, and succession.[26] 'Spatial distribution and space relations' are a verbal shorthand for describing the dimensions of the concept. A theoretical framework for investigation may be developed from this basic concept, as observations confirm hypotheses.[27] (2) The universe treated by geographers is the worldwide man–natural environment system. Geographers share their overriding problem, an understanding of this system, with other sciences. (3) The worldwide system is composed of a number of sub-systems. The sub-systems assist in identifying a hierarchy of problems for research. (4) The techniques of systems analysis are of particular value to geographers in applying their organizing (space) concept to the analysis of sub-systems of the worldwide man–environment system. These techniques, because of their rigor, permit replications of analysis and comparability of results among different research investigations. They also state the results of geograhic research in terms comparable to those of other sciences using systems techniques, and therefore make such results of greater potential use in treating the overriding problem, or any sub-problem.

Systems methods are changing society

Events in the world of today make it absolutely essential that geographers adopt such a view if they have aspirations to the frontier of research. Not only do much sharper probes exist for examining man's activities, but society itself is responding to scientific change. It is being organized in ways that are more easily evaluated. The scientific revolution we have been going through is being accompanied by a revolution of rationalism in our economic structure. Indeed, it has been called a 'second Industrial Revolution,' with effects already very profound for all humankind. Industrial engineering years ago removed the individual decision making of the artisan. 'Cybernation,' or systems design and systems engineering,[28] are now rapidly moving individuality from 'middle management' decision. This development is part of the social problem of automation. Not least, systems design and engineering, through the nation's defense program, is having a dominant role in domestic political affairs and international relations. Research approaches have even been made toward understanding the process of human thought itself. Herbert Simon has said, 'We shall be able to specify exactly what it is that a man has to learn about a particular subject – ... how he has to proceed – in order to solve effectively problems that relate to that subject.'[29] And already a great deal is known about manipulating some aspects of society, like consumer demands, in a more or less controlled fashion. What we in the United States are experiencing is also going on in Europe and in Japan.

Quite a different form is found in the Soviet Union, but it is still certainly an aspect of rationalization. We may expect similar developments in other parts of the world. And we may expect systems engineering to play an increasingly large role in coping with the social and economic crises that technological change has brought.[30]

These events and trends have the profoundest significance for the future spatial distribution of human activities, and we could not hope to anticipate or understand that distribution without being fully abreast of what is taking place. On the other hand, there must be something that the study of spatial distributions can tell us about these phenomena. In brief, the methods that have created important salients on the frontier of the physical sciences are changing society itself, both directly and through their impact on the behavioral sciences.

Systems and geography's frontiers

We are, then, concerned not only with a vast interacting system, but with one that is being altered by knowledge of systems. We now come to the most difficult part of our determination. Recognition of the overriding problem is of little significance unless we relate it to the direction of everyday research, and, by extension, to the fields with which we seek common ground in the definition of problems. What does this tell us about our own frontiers?

The one thing that most distinguishes a system is the flow of information within it. 'Information' is not to be confused with the ordinary meaning of the word, for it refers here to any mechanism that holds together the interdependent, interacting parts of a system. This is an interesting and critical point as far as geography is concerned, because the *connectivity* within a system is its most important characteristic. Many geographers, on the other hand, have stressed *differences*, as exemplified in the term 'areal differentiation.' If you accept my proposal of the overriding problem for our science, it then follows that to choose a research problem without reference to the connectivity of the system is to risk triviality. What space relations tell us of connectivity in the system is significant to science as a whole. Areal differences are significant *only* insofar as they help to describe and define the connectivity or 'information' flow. We now see that the geographers who have been concerned with cultural and other processes have had an insight of significant direction in research. Eight such processes were suggested in the past – four physical and four cultural. Among the cultural you may remember demographic movement, organizational evolution, the resource-converting techniques, and the space-adjusting techniques. Among the physical, dynamics of the soil mantle, movement of water, climate, and biotic processes were suggested.[31]

A second important characteristic of a system is the existence of sub-systems within it. The pretty girl, if you like, can be broken down into an astonishing number of sub-systems, like any complex being. The same is true of other complex systems. This is another important and critical point, for we must make the proper selection of sub-systems for study if we are to maintain significance. We already have a clue in the past suggestions made about the importance of processes. It is the functional sub-systems that are generally the significant ones. Thus the systematic aspects of geography, insofar as they treat functions, are disposed to a high level of significance. Those geographers who have thought in terms of areal functional organization again have had a significant insight as to research direction.

However, not all types of region have equal significance for research. Political regions are territorial units with a high level of significance because they are functional. A watershed is an example of a physically determined region that is significant. On the other hand, the old concept of a 'geographic' region may have very little significance. We may need to review critically the significance of other types of regions within the context we are considering. The concept of a region is potentially valuable in systems study, but we should take care that the regional concepts we actually use are significant to the overriding system.

Selection of significant collaborating sciences

This brings us through the second level in the hierarchy of problems, down to a level where one must seek specific examples of significance. As geographers have long appreciated, the flow of 'information' within the man–natural environment system is indeed vast. Selection of a research problem at random again risks triviality, even though it may be entirely 'geographic' in conception. At this point one commences to be most actively concerned about clues from other sciences as to significant working problems.

Here we may go back to one of the first observations made in this discussion: The sciences differ enormously in their rates of progress. For example, not all divisions of the behavioral sciences or the earth sciences offer channels for productive communication. Without doubt we can benefit greatly from some collaborative definition of research problems with other sciences, but the co-operation must be selective. A good rule of thumb would be: Where systems analysis techniques are understood and incorporated at the working face of the discipline, a collaborative definition of problem may profitably be sought. In other words, co-operation is likely to be rewarding where methods made familiar in the physical sciences are now reaching into the neighbouring earth sciences and the behavioral sciences. Where the concepts and approaches using systems analysis methods are making inroads, a possible place of interest is suggested for geography. Relations with other sciences which at times have been loose, vague, and hard to define may thus become more meaningful.

The profession is becoming equipped gradually to take such a view in its fundamental research. The wind of change which we have felt for the last decade includes the application of some methods of systems analysis. Thus far they generally have been the application of more rigorous techniques to old geographic problems. Except for collaboration with economists and others of the 'regional science' group and the older collaboration between cultural geographers and cultural anthropologists, we thus far have done relatively little to explore common ground with other sciences on the definition of significant problems. In almost any direction we turn, interesting possibilities appear. Indeed, there are so many opportunities that the number of people undertaking geographic research seems remarkably few.

The relation of geography and the neighboring natural sciences is particularly interesting. By the neighboring natural sciences I mean studies that focus on the surface features of the earth, like soils, biotic features, and water movement. The logical point of contact of these sciences with the human part of the great man–land system is geography. In all of them there is increasing appreciation of the role of man. For example, it is realized that pollution has become a major feature in world hydrology; biological ecologists now admit that even the most 'inviolate' natural preserves will be affected by man, no matter what protection is given; and a few geomorphologists now recognize the significance of man as a part of geomorphic processes. We should be particularly alert to overtures from these neighboring sciences, like that of Geoffrey Robinson in geomorphology, who suggests that at least some geomorphologists are interested in a collaborative definition of problems.[32] We should continue to capitalize on a point of view that geography alone, until recently, has maintained among the sciences concerned with man: land is half of the man–land system.

There are signs that geography's position as a 'gateway' between the behavioral sciences and the earth sciences is being challenged somewhat by the behavioral sciences themselves. Economists, for example, in the last ten years have become increasingly concerned with natural resource development problems. To be sure, geographers helped to start them along these lines, but there is now a direct working relation between economics and hydrology. It is significant that the aspects of economics emphasizing a systems approach provided the important recent contributions to the study of resources.

Relations with the behavioral sciences

These events suggest that we need to maintain a comprehensive view of the frontiers in the behavioral sciences, and that we have a good clue to common interest in looking for those investigators who pursue a systems approach. It has been said, 'The behavioral sciences are diverse in subject matter and state of development, yet ideas and concepts circulate quite freely among them. . . .'[33] The quotation may be a slight overstatement, but it does represent an agreed-upon ideal in these sciences which we might well contemplate. How far do we join them in the shaping of goals and in the exchange of methods which we have commenced to use?

An illustration or two may direct our attention to possibilities. We have mentioned that a most important characteristic of a system is the flow of 'information,' broadly defined. 'Information' may be in the form of goods, people, messages containing data or ideas, or other dynamic phenomena. The geographer, by definition, looks at what spatial distributions tell concerning this information flow, or vice versa. Geographers have already attacked some of these problems of information flow successfully.[34] Probably the most important general question of this kind familiar to geographers is: 'What can we say about how people distribute themselves and their culture on the earth, given free choice?' Much of the work geographers have done thus far is within the context of economic constraints, but they also respond to their concepts of amenities, to neighborhood and other group attachments, to the diffusion of information, and perhaps to other factors. There is a wealth of significant problems here to examine. Attention to them can bring us into a common area with students of motivation in the behavioral sciences. This is a key area in behavioral science research. We may find eventually some interesting common ground with psychology, thus finally connecting with the inferences of M. G. Kendall twenty-five years ago.[35] Indeed, study of the brain is considered one of the most useful approaches to the study of systems generally.[36]

Geographers recently have been alert to non-economic 'information' flow studies. For example, a much respected pattern for geographical research with mathematical methods has been the diffusion of innovation studies by Torsten Hägerstrand in Sweden.[37] These studies, well known to American geographers, have stimulated diffusion research in this country. Such research is also a natural outgrowth of long-continued American interest in diffusion phenomena, followed particularly by cultural geographers.[38]

At the same time the interest in diffusion studies illustrates our past relations with other scientific subjects. American sociologists have been carrying on very similar work since the early 1940's, including some elaborately designed experiments.[39] As far as I can discover there was little cross-disciplinary communication on this remarkably similar path of research until about a year and a half ago. It is obvious that collaboration here between geography and sociology can be of value. This is of more than academic interest, for as Ullman has noted, 'the relative "stickiness" of society, the resistance of certain areas to spread of innovations and improvements,' has strong implications for public policy both nationally and internationally.[40]

Allow me another and more unusual example. An interesting offshoot in the behavioral sciences at the present time is the study of conflict theory, to which Kenneth Boulding of the University of Michigan and others have contributed. Looked at from the point of view of systems and their information flow, conflict theory is essentially a search for 'redundancy,' or the capacity to handle in channels multiple movements with the same destination. Boulding has suggested that the theory may be of interest in studying land use.[41] Here is an opportunity to help in exploring the overriding system through a fresh idea.

A common front with the behavioral sciences is important not only in framing significant research questions but also because of geography's long association with historical study. Increasingly, it looks as though history would acquire scientific meaning through the dimensions given it by behavioral science.

Conclusion

We emerge with four general points that could help to place our science on a research frontier. (1) Continue to strengthen quantifying methods. Attempt to add to them rigorous analytical approaches in our theory and habits of constructing hypotheses. (2) Recognize an earth-wide man–environment system as our overriding problem. We can seek significant research questions in the study of sub-systems at different levels, amenable to our spatial distribution analyses. (3) Choose our research problems in the light of the advancing frontier of the behavioral sciences, and with attention to systems-oriented study in the neighboring earth sciences. Finally, (4) supplement our present heavy commitments to studies within economic constraints and to morphology studies by other approaches. The rising interest in cultural geography is healthy, but we could diversify still more. I particularly commend to your attention political geography within the systems framework. It is concerned with regions that have true functional significance in the great man–land system.

Seeking and staying on a research frontier is a most exacting task. It is now very clear that, in this age of specialization, special knowledge and specialized concepts are not sufficient to hold a science on the frontier. The sense of overriding problem is essential, and so is a view of at least a part of the spectrum of all science. This does not mean that future accomplishment will be entirely by those who are mathematically sophisticated. For those of us not so endowed it is comforting to remember that A. A. Michelson, the first American to win the Nobel Prize, was, by his own admission, poorly prepared in mathematics. But he did have an extremely keen sense of the overriding problem in his field, a passion for exactness, and an alertness to the contributions of neighboring disciplines. There is an important place for a comprehensive view, but it must be a view based on something more than undergraduate and graduate courses. I believe the time is near when postgraduate training and a second doctoral degree may be the price for reaching a research frontier. In our plans for future professional action and in our advice to those in professional training, we must think about these matters before it is much later. If we do not, others will cultivate our frontier, for that is the way of science. If we do, perhaps we may come closer to justifying Charles Darwin's words, '. . . that grand subject, that almost keystone of the laws of creation, Geographical Distribution.'[42]

References

1   Address given by the Honorary President of the Association of American Geographers at its 59th Annual Meeting, Denver, Colorado, 4 September 1963.
      This paper makes no pretence to coverage of all the ways in which geography may be viewed, or practised. It discusses geography as a science. There is equal justification for placing some geographic scholarship among the humanities, as William L. Thomas has noted in a letter to me (24 June 1963). If such a distinction is needed, one might follow Howard Mumford Jones in his definition: 'The humanities are . . . a group of subjects devoted to the study of man as a being other than a biological product and different from a social or sociological entity' (Howard Mumford Jones, 'What Are the Humanities?,' *One Great Society* [1959], p. 17). Insofar as we encounter spatial distribution entities not amenable to the methods of science, and of interest to any serious scholar, our subject does have a humanistic content. But one may also question, as some scientists do, the appropriateness of these dividing lines between science and the humanities. As Marston Bates has provocatively said, '. . . science is only one of man's approaches to the understanding of the universe and of himself. By understanding, . . . I mean trying to make sense out of the apparent chaos of the outer world in terms of the symbol systems of the human mind. This might be considered the function of all art; and in that case I am led, half seriously, to call science the characteristic art form of Western civilization . . . the sciences and the humanities form a false dichotomy, because science is one of the humanities' (Marston Bates, 'Summary Remarks: Process,' *Man's Role in Changing the Face of the Earth* [1956], p. 1139). Compare also William Shockley, '. . . the practice of science is an art' (*Science,* Vol. 140 [1963], p. 384).
      Reprinted from *Annals of the Association of American Geographers,* Vol. 53, No. 4 (December 1963), Washington, D.C.

2   Estimated mid-1963 world populations: 3.25 billion; 1910 populations 1.7 billion (1963, extrapolated from United Nations data; 1910 extrapolations from estimates by W. F. Willcox and A. M. Carr-Saunders).

3   LITTLE, FRANK R. 'The History of the Fertilization Problem,' *Science,* Vol. 43 (1916), pp. 39–53.

4   The above comparison is not intended to reflect popular, or even professional, evaluations of the time. General appreciation of events in mathematics and physics during the late thirties, for example, did not come until the mid-forties. Yet they were sources of basic thought on methods that have affected all sciences.

5   Cf. BLAUT, J. M. 'Objective and Relationship,' *The Professional Geographer,* Vol. 14 (1962), pp. 1–7. 'In this respect we behaved like the social sciences: our philosophical weakness, like theirs, had its roots in chronically unsolved problems. Their problems concerned values, causes, and social wholes. Our problem, then as now, concerned the nature of our subject matter.'

6   The work of Carl Sauer and the 'California School' in collaboration with cultural anthropology was an exception.

7   BARROWS, HARLAN H. 'Geography as Human Ecology,' *Annals of the Association of American Geographers,* Vol. 13 (1923), pp. 1–14.

8   SAUER, CARL O. 'The Morphology of Landscape,' *University of California Publications in Geography,* Vol. 2 (1925), pp. 19–53.

9   The drive for the independent department typified this atmosphere at the time. Again the interest of the California group in cultural anthropology may be cited as an exception.

10 Barrows had true insight in stressing 'place relations,' but his concept of geography as human ecology set forth too ambitious a field. Neither qualitative nor quantitative methods of the time offered much solid ground for exploiting the ecological concept. At least in retrospect we can see the ecological concept of Barrows' time as incompletely formed (i.e., the adjustment of an organism to environment). It has now been replaced by the much more powerful monistic concept of an ecosystem, in which organism and environment are one interacting entity.

11 SAUER, op. cit., p. 20.

12 See PLATT, R. S. *Field Study in American Geography,* University of Chicago Department of Geography Research Paper 61, Chicago (1959), especially pp. 302–51.

13 An interesting demonstration of these techniques falling on sterile ground in geography occurred in 1938, when the mathematical statistician M. G. Kendall presented his paper 'The Geographical Distribution of Crop Productivity in England' before the Royal Statistical Society (*Journal of the Royal Statistical Society,* Vol. 102 [1939], pp. 21–62'). This study was an analysis of covariance among ten crops in the 48 English counties. Besides the interesting direct conclusions he drew, Kendall made some provocative observations about the similarity of statistical techniques for studying a psychological problem and for studying a geographical problem. However, the two geographers present, L. Dudley Stamp and E. C. Willats, devoted their comments on Kendall's paper mainly to its shortcomings in interpretating the observable landscape. So far as I know, there was no sequel in geographical study to Kendall's interesting exploration. I am indebted to Brian J. L. Berry for calling my attention to Kendall's paper.

14 Some scientists fear that space 'hardware' is causing an inefficient, even dangerous, misallocation of high-quality scientific talent in the United States. (See ABELSON, P. H., Testimony before the United States Senate Committee on Aeronautical and Space Sciences hearings on National Goals in Space, 10 June 1963).

15 The geographer may observe through field investigation; he may experiment with the use of statistical models (or idealized reality).

16 WEAVER, WARREN 'Science, Learning and the Whole of Life,' Address at 70th Anniversary Convocation, Drexel Institute of Technology (December 1961).

17 By 'spatial distributions,' 'earth-spatial distributions' is, of course, understood here. They are the parallel of distributional associations in other sciences.

18 SIMON, HERBERT *The New Science of Management Decision* (New York: 1960), p. 34. There are other similar statements, like that of Kuhn, T. S., who calls it 'puzzle-solving' (*The Structure of Scientific Revolutions,* Chicago (1962), pp. 35 ff).

19 HARTSHORNE, RICHARD *Perspective on the Nature of Geography,* Chicago (1959), p. 21. The concept of areal differentiation as Hartshorne explains, 'stems from Richtofen's synthesis of the views of Humboldt and Ritter, and has been most fully expounded in Hettner's writings' (ibid., p. 12).

20 A useful short categorization of systems is given by BOULDING, KENNETH E. in his 'General Systems Theory – The Skeleton of Science,' *Management Science,* Vol. 2 (1956), pp. 197–208. He distinguishes nine 'levels' of systems in increasing order of complexity. A social system is of the eighth order among his levels.

21 VON HUMBOLDT, ALEXANDER *Kosmos: Entwurf einer physichen Weltbeschreibung,* Vol. 1 (Stuttgart: 1845), p. 68. Quoted from Hartshorne, Richard *Perspective on the Nature of Geography,* p. 162.

22 HARTSHORNE *Perspective on the Nature of Geography,* p. 172.

23 BARROWS, op. cit., p. 3.

24   The closest approach to this in the geography of the '30's and '40's was in Robert S. Platt's view of geography 'as the science of regional process patterns of dynamic space relations,' PLATT, 'A Review of Regional Geography,' *Annals of the Association of American Geographers,* Vol. 47 (1957), p. 190. However, the appropriateness of formal systems concepts to geographic research is not mentioned by Platt.

25   A very gracefully stated description of the indivisible attribute (and others) of systems is given by BEER, SIR STAFFORD, in 'Below the Twilight Arch – A Mythology of Systems,' in *Systems: Research and Design* (Eckman, Donald P., Ed.), New York: Wiley (1961), pp. 1–25.

26   Extent is measurable as size, shape, and orientation. Density is shown by the amount of 'betweenness.' Simultaneity is a special case of succession.

27   Cf. BLAUT (op. cit., pp. 5–6), who interprets Hartshorne (op. cit., pp. 74–80, 133, 144–5) and states the organizing concept as 'areal integration.' I do not find Blaut's statement inconsistent with the statement given in this paragraph, but his does leave the epistemological problem of what space is. (Discussed by Blaut in his 'Space and Process,' *Professional Geographer,* Vol. XIII (July 1961), pp. 1–6.) In addition, the word 'integration' has a connotation of study technique that (to me) detracts from clarity.
       By extension I also do not find the statements of this paragraph inconsistent with Hartshorne's latest careful analysis of geographic concept and method (Hartshorne, op. cit.). It may be noted that Hartshorne, always precise in his definitions, has described the components of geographic study in a manner that allows them to fit the view of geography suggested here, and probably other views also.

28   See MICHAEL, DONALD N. *Cybernation: The Silent Conquest,* for a summary account of the social changes caused by systems engineering. Simon, op. cit., also describes them.

29   SIMON, op. cit., p. 34.

30   JOHNSON, E. A. has stated one aspect of this problem, from a national point of view: '. . . the increase in physical knowledge has made the future . . . uncertain, . . . we must plan much further ahead in a way that will provide much greater flexibility, whether this be in peaceful or military affairs, whether it be for the individual or for the country . . . our primary problem is to find a way to manage our very big systems affairs in this new situation. . . . We will have to examine our individual, group, and national values to see what it is we want to do in a rapidly changing world, and to see what we can do consciously to manipulate in our favor the real and perhaps hostile physical and world environment so that it will serve us better. This is a problem of big systems.' Johnson, E. A., 'The Use of Operations Research in the Study of Very Large Systems.' *Systems: Research and Design* (Eckman, Donald P., Ed.), pp. 52–93.

31   ACKERMAN *Geography as a Fundamental Research Discipline,* Chicago (1958), p. 28.

32   ROBINSON, GEOFFREY 'A Consideration of the Relations of Geomorphology and Geography,' *The Professional Geographer,* Vol. XV (1963), pp. 13–17.

33   Behavioral Sciences Subpanel, President's Science Advisory Committee, Strengthening the Behavioral Sciences (Washington: 20 April 1962), p. 13.

34   A number of reports on such research have appeared in the *Annals*; e.g., articles by Garrison, W. L. and others. Publications of the 'regional science' group also are illustrative.

35   See footnote 13 above.

36   BEER, op. cit., p. 19. 'The brain is itself the most resplendent system of them all. . . .' We may well reflect to what degree social reality reflects the structure of the brain.

37 HÄGERSTRAND, TORSTEN, 'The Propagation of Innovation Waves,' *Lund Studies in Geography*, Sweden: Lund (1952), and succeeding publications.

38 See, for example, KNIFFEN, FRED 'The American Covered Bridge,' *Geographic Review* (1951), p. 114.

39 See, for example, COLEMAN, JAMES S. 'The Diffusion of an Innovation among Physicians,' *Sociometry,* Vol. 20 (1957); DeFleur, Melvin and Larsen, Otto, *Flow of Information,* New York: Harper (1958); Rapoport, Anatol, 'Spread of Information through a Population with a Social Structure Bias,' *Bulletin of Mathematical Biophysics,* Vol. 15 (1953).

40 ULLMAN, EDWARD L. 'Geography Theory in Underdeveloped Areas,' *Essays on Geography and Economic Development* (Ginsburg, Norton S., Ed.), University of Chicago, Department of Geography Research Paper 62, Chicago (1960), pp. 26–32.

41 BOULDING, KENNETH *Conflict and Defense,* New York: Harper (1962), p. 1.

42 DARWIN, CHARLES letter to Hooker, Joseph Dalton (1845).

13 Geomorphology and general systems theory[1]

Richard J. Chorley
University of Cambridge, U.K.

ABSTRACT   An appreciation of the value of operating within an appropriate general systematic model has emerged from the recognition that the interpretation of a given body of information depends as much upon the character of the model adopted as upon any inherent quality of the data itself. Fluvial geomorphic phenomena are examined within the two systematic models which have been found especially useful in physics and biology – closed and open systems, for which simple analogies are given. Certain qualities of classic closed systems, namely the progressive increase in entropy, the irreversible character of operation, the importance of the initial system conditions, the absence of intermediate equilibrium states and the historical bias, permit comparisons to be made with the Davisian concept of cyclic erosion. The restrictions which were inherently imposed upon Davis's interpretation of landforms thus become more obvious. It is recognized, however, that no single theoretical model can adequately encompass the whole of a natural complex, and that the open system model is imperfect in that, while embracing the concept of grade, the progressive reduction of relief cannot be conveniently included within it. The open system characteristic of a tendency toward a steady state by self-regulation is equated with the geomorphic concepts of grade and dynamic equilibrium which were developed by Gilbert and later 'dynamic' workers and, despite continued relief reduction, it is suggested that certain features of landscape geometry, as well as certain phases of landscape development, can be viewed profitably as partially or completely time-independent adjustments. In this latter respect the ratios forming the bases of the laws of morphometry, the hypsometric integral, drainage density, and valley-side slopes can be so considered. The relative values of the closed and open systematic frameworks of reference are recognized to depend upon the rapidity with which landscape features can become

[1] Reprinted from *Geological Survey Professional Paper 500-B*. U.S. Government Printing Office, Washington D.C. (1962).
ACKNOWLEDGMENTS   The author would like to thank Professor Ludwig von Bertalnaffy of the University of Alberta and Dr. Luna B. Leopold of the United States Geological Survey for critically reading this manuscript and for making many valuable suggestions both regarding methodology and the application of general systems theory to geomorphology.

282

adjusted to changing energy flow, and a contrast is made between Schumm's (1956) essentially open system treatment of weak clay badlands and the historical approach which seems most profitable in treating the apparently ancient landscapes of the dry tropics.

Finally, seven advantages are suggested as accruing from attempts to treat landforms within an open system framework:

1   The focusing of attention on the possible relationships between form and process.

2   The recognition of the multivariate character of most geomorphic phenomena.

3   The acceptance of a more liberal view of changes of form through time than was fostered by Davisian thinking.

4   The liberalizing of attitudes toward the aims and methods of geomorphology.

5   The directing of attention to the whole landscape assemblage, rather than to the often minute elements having supposed historical significance.

6   The encouragement of geomorphic studies in those many areas where unambiguous evidence for a previous protracted erosional history is lacking.

7   The introduction into geography, via geomorphology, of the open systematic model which may prove of especial relevance to students of human geography.

Geomorphology and general systems theory

During the past decade several valuable attempts have been made, notably by Strahler (1950, 1952A, and 1952B), by Culling (1957, p. 259–61), and by Hack (1960, p. 81, 85–6; Hack and Goodlett, 1960), to apply general systems theory to the study of geomorphology, with a view to examining in detail the fundamental basis of the subject, its aims and its methods. They come at a time when the conventional approach is in danger of subsiding into an uncritical series of conditioned reflexes, and when the more imaginative modern work in geomorphology often seems to be sacrificing breadth of vision for focus on details. In both approaches it is a common trend for workers to be increasingly critical of operating within general frameworks of thought, particularly with the examples of the Davis and Penck geomorphic systems before them, and 'classical' geomorphologists have retreated into restricted historical studies of regional form elements, whereas, similarly, quantitative workers have often withdrawn into restricted empirical and theoretical studies based on process.

It is wrong, however, to confuse the restrictions which are rightly associated with preconceived notions in geomorphology with the advantages of operating within an appropriate general systematic framework. The first lead to the closing of vistas and the decrease of opportunity; the second, however, may increase the scope of the study, make possible correlations and associations which would otherwise be impossible, generally liberalize the whole approach to the subject and, in addition, allow an integration into a wider general conceptual framework. Essentially, it is not possible to enter into a study of the physical world without such a fundamental basis for the investigation, and even the most qualitative approaches to the subject show very strong evidence of operations of thought within a logical general framework, albeit a framework of thought which is in a sense unconscious. Hack (1960), for example, has pointed to the essential difference between the approaches to geomorphology of Gilbert and Davis, and in this respect the fundamental value of the adoption of a suitable general framework of investigation based on general systems theory becomes readily apparent.

Following the terminology used by Von Bertalanffy (1950 and 1960), it is possible to recognize in general two separate systematic frameworks wherein one may view the natural occurrence of physical phenomena; the closed system and the open system (Strahler, 1950, p. 675–6, and 1952A, p. 934–5). Hall and Fagan (1956, p. 18) have defined a system as '. . . a set of objects together with relationships between the objects and between their attributes.' In the light of this definition, it is very significant that one of the fundamental purposes of Davis's approach to landforms was to study them as an assemblage, in which the various parts might be related in an areal and a time sense, such that different systems might be compared, and the same system followed through its sequence of time changes. Closed systems are those which possess clearly defined closed boundaries, across which no import or export of materials or energy occurs (Von Bertalanffy, 1951). This view of systems immediately precludes a large number, perhaps all, of the systems with which natural scientists are concerned; and certainly most geographical systems are excluded on this basis, for boundary problems and the problems of the association between areal units and their interrelationships lie very close to the core of geographical investigations.

Another characteristic of closed systems is that, with a given amount of initial free, or potential, energy within the system, they develop toward states with maximum 'entropy' (Von Bertalanffy, 1951, p. 161–2). Entropy is an expression for the degree to which energy has become unable to perform work. The increase of entropy implies a trend toward minimum free energy (Von Bertalanffy, 1956, p. 3). Hence, in a closed system there is a tendency for levelling down of existing differentiation within the system; or, according to Lord Kelvin's expression, for progressive degradation of energy into its lowest form, i.e. heat as undirected molecular movement (Von Bertalanffy, 1956, p. 4). This is expressed by the second law of thermodynamics (Denbigh, 1955) which, in its classic form, is formulated for closed systems. In such systems, therefore, the change of entropy is always positive, associated with a decrease in the amount of free energy, or, to state this another way, with a tendency toward progressive destruction of existing order or differentiation.

Thus, one can see that Davis's view of landscape development contains certain elements of closed system thinking – including, for example, the idea that uplift provides initially a given amount of potential energy and that, as degradation proceeds, the energy of the system decreases until at the stage of peneplanation there is a minimum amount of free energy as a result of the levelling down of topographic differences. The Davisian peneplain, therefore, may be considered as logically homologous to the condition of maximum entropy, general energy properties being more or less uniformly distributed throughout the system and with a potential energy approaching zero. The positive change of entropy, and connected negative change of free energy, implies the irreversibility of events within closed systems. This again bears striking similarities to the general operation of the geomorphic cycle of Davis. The belief in the sequential development of landforms, involving the progressive and irreversible evolution of almost every facet of landscape geometry, in sympathy with the reduction of relief, including valley-side slopes and drainage systems, is in accord with closed system thinking. Although 'complications of the geographical cycle' can, in a sense, put the clock back, nothing was considered by Davis as capable of reversing the clock. The putting back of the clock by uplift, therefore, came to be associated with a release, or an absorption into the new closed system, of an increment of free energy, subsequently to be progressively dissipated through degradation.

Also, in closed systems there is the inherent characteristic that the initial system conditions, particularly the energy conditions, are sufficient to determine its ultimate equilibrium condition. This inevitability of closed-system thinking is very much associated with the view of geomorphic change held by Davis. Not only this, but the condition of a closed system at any particular time can be considered largely as a function of the initial system conditions and the amount of time which has subsequently elasped. Thus closed systems are eminently susceptible to study on a time, or historical, basis. This again enables one to draw striking analogies between closed-system thinking and the historical approach to landform study which was proposed by Davis.

Finally, it is recognized that closed systems can reach a state of equilibrium. Generally speaking, however, this equilibrium state is associated with the condition of maximum entropy which cannot occur until the system has run through its sequential development. In addition, it is impossible to introduce the concept of equilibrium into a closed-system framework of thought without the implication that it is associated with stationary conditions. The only feature of the cyclic system of Davis which employed the general concept of equilibrium was that of the 'graded' condition of stream channels and slopes which, significantly, Davis borrowed from the work of Gilbert, who had an entirely noncyclic view of landform development (Hack, 1960, p. 81). Characteristically, the concept of grade was the one feature of Davis's synthesis which seems least well at home in the cyclic framework, for it has always proved difficult to imagine how, within a closed system context, a graded or equilibrium state could exist and yet the associated forms be susceptible to continued change – namely, downcutting or reduction.

The foregoing is not meant to imply that it is unprofitable to consider any assemblage of phenomena within a closed system framework or, as Davis did, to overstress those aspects or phases which seem to achieve most significance with reference to the closed system model. It is important, however, to recognize the sources of partiality which result, not from any inherent quality of the data itself, but from the general systematic theory under which one is operating. In reality, no systematic model can encompass the whole of a natural complex without ceasing to be a model, and the phenomena of geomorphology present problems both when they are viewed within closed and open systematic frameworks. In the former, the useful concept of dynamic equilibrium or grade rests most uncomfortably; in the latter, as will be seen, the progressive loss of a component of potential energy due to relief reduction imposes an unwelcome historical parameter.

A simple, classic example of a closed system is represented by a mass of gas within a completely sealed and insulated container. If, initially, the gas at one end of the container is at a higher temperature than that at the other, this can be viewed as a condition of maximum segregation, maximum free energy, and, consequently, of maximum ability to perform work, should this thermal gradient be harnessed within a larger closed system. This is the state of minimum entropy. It is obvious, however, that this state of affairs is of a most transient character and that immediately an irreversible heat flow will begin toward the cooler end of the container. This will progressively decrease the segregation of mass and energy within the system, together with the available free energy and the ability of this energy to perform work, bringing about a similarly progressive increase of entropy. While the system remains closed nothing can check or hinder this inevitable leveling down of differences, which is so predictable that, knowing the initial energy conditions, the thermal conductivity of the gas and the lapse of time, one could accurately calculate the thermal state of the system at any required stage. Thus the distribution of heat energy and the heat flow within the system have a progressive and sequential history, the one becoming less segregated and the other ever-decreasing. Nor is it possible to imagine any form of equilibrium until all the gas has attained the same temperature, when the motion of the gas molecules is quite random and the static condition of maximum entropy obtains.

Open systems contrast quite strikingly with closed systems. An open system needs an energy supply for its maintenance and preservation (Reiner and Spiegelman, 1945), and is in effect maintained by a constant supply and removal of material and energy (Von Bertalanffy, 1952, p. 125). Thus, direct analogies exist between the classic open systems and drainage basins, slope elements, stream segments and all the other form-assemblages of a landscape. The concept of the open system includes closed systems, however, because the latter can be considered a special case of the former when transport of matter and energy into and from the system becomes zero (Von Bertalanffy, 1951, p. 156). An open system manifests one important property which is denied to the closed system. It may attain a 'steady state' (Von Bertalanffy, 1950; and 1951, p. 156–7), wherein the import and export of energy and material are equated by means of an adjustment of the form, or geometry, of the system itself. It is more difficult to present a simple mechanical analog to illustrate completely the character and operations of an open system but it may be helpful to visualize one such system as represented by the moving body of water contained in a bowl which is being constantly filled from an overhead inflow and drained by an outflow in the bottom. If the inflow is stopped, the bowl drains and the system ceases to exist; whereas, if the inflow is stopped and the outflow is blocked, the system partakes of many of the features of a closed system. In such an arrangement, changes in the supply of mass and energy from outside lead to a self-adjustment of the system to accommodate these changes. Thus, if the inflow is increased, the water level in the basin rises, the head of water above the outflow increases, and the outflow discharge will increase until it balances the increased inflow. At this time the level of water in the bowl will again become steady.

Long ago, Gilbert recognized the importance of the application of this principle of self-adjustment to landform development:

> The tendency to equilibrium of action, or to the establishment of a dynamic equilibrium, has already been pointed out in the discussion of the principles of erosion and of sculpture, but one of its most important results has not been noticed.
>
> Of the main conditions which determine the rate of erosion, namely, the quantity of running water, vegetation, texture of rock, and declivity, only the last is reciprocally determined by rate of erosion. Declivity originates in upheaval, or in the displacement of the earth's crust by which mountains and continents are formed: but it receives its distribution in detail in accordance with the laws of erosion. Wherever by reason of change in any of the conditions the erosive agents come to have locally exceptional power, that power is steadily diminished by the reaction of the rate of erosion upon declivity. Every slope is a member of a series, receiving the water and the waste of the slope above it, and discharging its own water and waste upon the slope below. If one member of the series is eroded with exceptional rapidity, two things immediately result: first, the member above has its own level of discharge lowered, and its rate of erosion is thereby increased; and second, the member below, being clogged by an exceptional load of detritus, has its rate of erosion diminished. The acceleration above and the retardation below diminish the declivity of the member in which the disturbance originated: and as the declivity is reduced, the rate of erosion is likewise reduced.
>
> But the effect does not stop here. The disturbance that has been transferred from one member of the series to the two which adjoin it, is by then transmitted to others, and does not cease until it has reached the confines of the drainage basin. For in each basin all lines of drainage unite in a main line, and a disturbance upon any line is communicated through it to the main line and thence to every tributary. And as a member of the system may influence all the others, so each member is influenced by every other. There is an interdependence throughout the system. (Gilbert, 1880, pp. 117–18).

This form-adjustment is brought about by the ability of an open system for self-regulation (Von Bertalanffy, 1952, pp. 132–3). Le Châtelier's Principle (originally stated for equilibrium in closed systems) can be expanded also to include the so-called 'Dynamic Equilibrium' or steady states in open systems:

> Any system in . . . equilibrium undergoes, as a result of a variation in one of the factors governing the equilibrium, a compensating change in a direction such that, had this change occurred alone it would have produced a variation of the factor considered in the opposite direction. (Prigogine and Defay, 1954, p. 262.)

A geomorphic statement of this principle has been given by Mackin (1948):

> A graded stream is one in which, over a period of years, slope is delicately adjusted to provide, with available discharge and with prevailing channel characteristics, just the velocity required for the transportation of the load supplied from the drainage basin. The graded stream is a system in equilibrium; its diagnostic characteristic is that any change in any of the controlling factors will cause a displacement of the equilibrium in a direction that will tend to absorb the effect of the change.

The cyclic adaptation of the concept of grade did not give sufficient importance to the factors, other than channel slope, which a stream system can control for itself, and in this respect Davis's ignorance of the significance of the practical experiments of Gilbert (1914) is most evident. A stream system cannot greatly control its discharge, which represents the energy and mass which is externally supplied into the open system. Neither can it completely control the amount and character of the debris supplied to it, except by its action of abrasion and sorting or as the result of the rapport which seems to exist regionally between stream-channel slope and valley-side slope (Strahler, 1950, p. 689). However, besides adjusting the general slope of its channel by erosion and deposition, a stream can very effectively and almost instantaneously control its transverse channel characteristics, together with its efficiency for the transport of water and load, by changes in depth and width of the channel. As Wolman (1955, p. 47) put it:

> The downstream curves on Brandywine Creek . . . suggest that the adjustment of channel shape may be as significant as the adjustment of the longitudinal profile. There is no way in which one could predict that the effect of a change in the independent controls would be better absorbed by a change in slope rather than by a change in the form of the cross section.

It may be, therefore, that a stream or reach may be virtually always adjusted (Hack, 1960, pp. 85–6), in the sense of being graded or in a steady state, without necessarily presenting the smooth longitudinal profile considered by the advocates of the geomorphic cycle as the hallmark of the 'mature graded condition.' The state of grade is thus analogous to the tendency for steady-state adjustment, it is perhaps always present and, therefore, this presence cannot be employed necessarily as an historical, or stage, characteristic. It is interesting that the concept of the vegetational 'climax,' which has often been compared to that of grade, has passed through a somewhat similar metamorphosis. The original idea of a progressive approach to a static equilibrium of the ecological assemblage (Clements, 1916, pp. 98–9) has been challenged by the open system interpretation of Whittaker (1955, p. 48), with an historical link being provided by the 'individualistic concept' of Gleason (1926–27; 1927), much in the same way as Mackin's concept of grade links those of Davis and Wolman.

The forms developed, together with the mutual adjustment of internal form elements and of related systems, are dependent on the flow of material and energy in the steady state. The laws of morphometry (Chorley, 1957) express one aspect of this relationship in geomorphology. In addition, adjustment of form elements implies a law of optimum size of a system and of elements within a system (Von Bertalanffy, 1956, p. 7). This is mirrored by Gilbert's (1880, pp. 134–5) symmetrical migration of divides and by Schumm's constant of channel maintenance (1956, p. 607), and is illustrated by Schumm's (1956, p. 609) contrast between basin areas of differing order.

Although a steady state is in many respects a time-independent condition, it differs from the equilibrium of closed systems. A steady state means that the aspects of form are not static and unchanging, but that they are maintained in the flow of matter and energy traversing the system. An open system will, certain conditions presupposed, develop toward a steady state and therefore undergo changes in this process. Such changes imply changes in energy conditions and, connected with these, changes in the structures during the process. The trend toward, and the development of, a steady state demands not an equation of force and resistance over the landscape, but that the forms within the landscape are so regulated that the resistance presented by the surface at any point is proportionate to the stress applied to it.

> Erosion on a slope of homogeneous material with uniform vegetative cover will be most rapid where the erosional power of the runoff is greatest. This non-uniform erosional process will in time result in a more stable slope profile which would offer a uniform resistance to erosion. (Little, 1940, p. 33.)

In this way the transport of mass and energy (i.e., water and debris) is carried on in the most economical manner. With time, landscape mass is therefore being removed and progressive changes in at least some of the absolute geometrical properties of landscape, particularly relief, are inevitable. It is wrong, however, to assume, as Davis did, that all these properties are involved necessarily in this progressive, sequential change. To return briefly to the analogy of the bowl. If the rush of water through the outflow is capable of progressively enlarging the orifice, the increasing discharge at the outflow, uncompensated at the inflow, will cause the head of water in the bowl to decrease. This loss of head will itself, however, constantly tend to compensate the inceasing outflow, but, if the enlargement of the outflow orifice proceeds,

C.R.G.—10

this is a losing battle and an important feature of the system will be the progressive and sequential loss of head. However, not all features of this system will reflect this progressive change of head, and, for example, the structure of the flow within the bowl will remain much the same while any head of water at all remains there. The dimensionless ratios between landscape forms, similarly, seem to express the steady state condition of adjusted forms from which mass is constantly being removed. The geometrical ratios which form the basis of the laws of morphometry, and the height-area ratios involved in the dimensionless, equilibrium hypsometric integral are examples of this adjustment:

> In late mature and old stages of topography, despite the attainment of low relief, the hypsometric curve shows no significant variations from the mature form, and a low integral results only where monadnocks remain ... After monadnock masses are removed, the hypsometric curve may be expected to revert to a middle position with integrals in the general range of 40 to 60 percent. (Strahler, 1952B, pp. 1129–30).

In a drainage basin composed of homogenous material, in which no monadnocks would tend to form, it seems possible, therefore, that the dimensionless percentage volume of unconsumed mass (represented by the hypsometric integral) may achieve a time-independent value. It has been suggested, however, that the construction of the hypsometric curve may be so inherently restricted as to make the hypsometric integral insensitive to variations of an order which would be necessary to recognize such an equilibrium state (Leopold, written communication, 1961). This steady state principle has been tentatively extended by Schumm (1956, pp. 616–7) to certain other aspects of drainage basin form:

> ... the form of the typical basin at Perth Amboy changes most rapidly in the earliest stage of development. Relief and stream gradient increase rapidly to a point at which about 25 percent of the mass of the basin has been removed, then remains essentially constant. Because relief ratio [the ratio between total relief of a basin and the longest dimension of the basin parallel to the principal drainage] elsewhere has shown a close positive correlation with stream gradient, drainage density, and ground-slope angles, stage of development might be expected to have little effect on any of these values once the relief ratio has become constant.

In the steady state of landscape development, therefore, force and resistance are not equated (which would imply no absolute form change), but balanced in an areal sense, such that force may still exceed resistance and cause mass to be removed. Now, as has been pointed out, removal of mass under steady-state conditions must imply some progressive changes in certain absolute geometrical properties of a landscape, notably a decrease in average relief, but by no means all such properties need respond in this simple manner to the progressive removal of mass. The existence, for example, of the optimum magnitude principle for individual systems, or subsystems, implies that if the available energy within the system is sufficient to impose the optimum magnitude on that system, this magnitude will be maintained throughout a period of time and will not always be susceptible to a progressive, sequential change. Thus, Strahler (1950) has indicated that erosional slopes which are being forced to their maximum angle of repose by aggressive basal stream action will, of necessity, retain this maximum angle despite the progressive removal of mass with time.

Total energy is made up of interchangeable potential energy and flux, or kinetic, energy (Burton, 1939, p. 328) and even if the potential energy component decreases within an open system due to its general reduction, in other words along with a continual change in one aspect of form (i.e., relief), the residual flux energy may be of such overriding importance as to effectively maintain a steady state of operation. In practice the steady state is seldom, if ever, characterized by exact equilibrium, but simply by a tendency to attain it. This is partly due to the constant energy changes which are themselves characteristic of many open system operations, but the steady state condition of tendency toward attainment of equilibrium is a necessary pre-requisite, according to Von Bertalanffy (1950, p. 23; and 1952, pp. 132–3), for the system to perform work at all. Now, once a steady state has been established, the influence of the initial system conditions vanishes and, with it, the evidence for a previous history of the system (Culling, 1957, p. 261) (i.e., was our bowl full or empty at the start?) Indeed, in terms of analyzing the causes of phenomena which exhibit a marked steady-state tendency, considerations regarding previous history become not only hypothetical, but largely irrelevant. This concept contrasts strikingly with the historical view of development which is fostered by closed-system thinking. Wooldridge and Linton (1955, p. 3) have gone so far as to say that:

> Any such close comprehension of the terrain can be obtained in one way only, by tracing its evolution.

An even more extreme statement of the same philosophy has been made by Wooldridge and Goldring (1953, p. 165):

> The physical landscape, including the vegetation cover, is the record of *processes* and the whole of the evidence for its evolution is contained in the landscape itself.

The whole matter hinges on the rapidity with which landscape features become adjusted to energy flow, which may itself be susceptible to rapid changes, particularly during the rather abnormal latest geologic period of earth history. Obviously, most existing features are the product of both past and reasonably contemporary energy conditions, and the degree to which these latter conditions have gained ascendancy over the former is largely a function of the ratio between the amount of present energy application and the strength (whatever this may mean) of the landscape materials. Thus, the geometry of stream channels (Leopold and Maddock, 1953) and the morphometry of weak clay badlands (Schumm, 1956) show remarkable adjustments to contemporary processes – on whatever time level the action of these processes may be defined (Wolman and Miller, 1960) – whereas, at the other end of the energy/resistance scale, erosion surfaces cut in resistant rock and exposed to the low present energy levels associated with the erosional processes of certain areas of tropical Africa can only be understood on the basis of past conditions. Between these two extremes lies the major part of the subject matter of geomorphology including considerations of slope development, and it is here where the apparent dichotomy between the two systematic approaches to the same phenomena, termed by Bucher (1941; see also Strahler, 1952A, pp. 924–5) 'timebound' and 'timeless,' is most acute. In a related context, the problem of timebound-versus-timeless phenomena becomes especially obvious when rates of change and the ability to adjust are underestimated, as when vegetational assemblages have been correlated with the assumed stages of geomorphic history in the folded Appalachians by Braun (1950, pp. 241–2) and in Brazil by Cole (1960, pp. 174–7).

One can appreciate that in areas where good evidence for a previous landscape history still remains, the historical approach may be extremely productive, as exemplified by the work of Wooldridge and Linton on southeastern England. However, in many (if not most) areas the condition is one of massive removal of past evidence and of tendency toward adjustment with progressively contemporaneous conditions. It is an impossibly restricted view, therefore, to imagine a universal approach to landform study being based only upon considerations of historical development.

Another characteristic of the open system is that negative entropy, or free energy, can be imported into it – because of its very nature. Therefore, the open system is not defined by the trend toward maximum entropy. Open systems thus may maintain their organization and regularity of form, in a continual exchange of their component materials. They may even develop toward higher order, heterogeneity, hierarchical differentiation and organization (Von Bertalanffy, 1952, pp. 127–9). This is mirrored in geomorphology by the characteristic development of interrelated drainage forms, and goes along with a concept of progressive segregation (Von Bertalanffy, 1951, pp. 148–9). This, to a minor extent, militates against the general view of adjustment previously discussed, insofar as, with time, rates of interactions between form elements in an open system may tend to decrease. Therefore, it is quite reasonable to assume that mutual adjustments of form within geomorphic systems might be more difficult of accomplishment and delayed where the relief, through its influence over the potential energy of the system, is low rather than where there is a higher potential energy in the system.

Steady-state conditions can be interrupted by a disturbance in the energy flow or in the resistance, leading to form adjustments allowing a new steady state to be approached. These adjustments, however, do imply a consumption of energy and there is a 'cost of transition' from one steady state to another (Burton, 1939, pp. 334, 348). A particular geomorphic instance of this dissipation might be presented by the phenomenon of 'over-shooting' where active, but sporadic, processes are operating on weak materials, as instanced when the failure of steep slopes reduces them to inclinations very much below their repose angles, and by the excessive cutting and subsequent filling of alluvial channels associated with flash floods.

The dynamic equilibrium of the steady state manifests itself in a tendency toward a mean condition of unit forms, recognizable statistically, about which variations may take place over periods of time with fluctuations in the energy flow. These periods of time may in some instances be of very short duration, and the fluctuations of transverse stream profiles are measurable in the days, or even minutes, during which changes of discharge occur. These constant adjustments to new steady-state conditions may be superimposed on a general tendency for change possibly associated with the reduction of average relief through time. This general relief change, however, does not imply a sympathetic change of all the other features of landscape geometry. As has been demonstrated by Strahler (1958) and Melton (1957), for example, drainage density is controlled by a number of factors of which relief is only one. Recent work seems to be indicating that relief (naturally including considerations of average land slope) probably has only a relatively small influence over drainage density, which may be masked or neglected altogether by the other more important factors (for example, rainfall intensity and surface resistance) which are not so obviously susceptible to changes with time. Denbigh, Hicks and Page (1948, p. 491) have pointed out that:

> Quite large changes of environment may take place, without the need for more than a small internal readjustment.

Horton (1945) did not believe, as did Glock (1931), that drainage density could be employed as a measure of landscape 'age,' and, indeed, it is not difficult to entertain the possibility that certain features of landscape geometry may be relatively unchanging, in actual dimensional magnitude as well as in dimensionless ratio, throughout long periods of erosional history.

For many landscape units, changes on either level are slow, or in some instances nonexistent. Under steady state conditions, therefore, corresponding local morphometric units will, as regards their form and magnitude, tend to crowd around a very significant mean value, imparting to a geomorphic region its aspects of uniformity. Strahler's (1950, p. 685) 'law of constancy of slopes' is an expression of one phase of this adjustment. It is interesting that the general principle of the operation of a steady state condition was intuitively recognized long ago by Playfair (1802, p. 440):

> The geological system of Dr. Hutton resembles, in many respects, that which appears to preside over the heavenly motions. In both we perceive continual vicissitude and change, but confined within certain limits, and never departing far from a certain mean condition, which is such, that in the lapse of time, the deviations from it on one side, must become just equal to the deviations from it on the other.

Often the achievement of exact equilibrium in nature occurs only momentarily as variations about the mean take place (Mackin, 1948), and in these instances the existence of the steady state can only be recognized statistically (Strahler, 1954). In the study of landscape, the steady state condition indicated by discrete, close and recognizable statistical groupings of similar units, is characteristic of regions of uniform ratios between process and surface resistance.

Davis's view of landscape evolution was that the passage of time, of necessity, imprinted recognizable, significant and progressive changes on every facet of landscape geometry. The recognition, however, that landscape forms represent a steady-state adjustment with respect to a multiplicity of controlling factors obliges one to take a less rigid view of the evolutionary aspects of geomorphology. When a geometrical form is controlled by a number of factors, any change of form with the passage of time is entirely dependent upon the net result of the effect of time upon those factors. Some factors are profoundly affected by the passage of time, others are not; some factors act directly (using the term in the mathematical sense) upon the form, others inversely; some factors exercise an important control over form aspects, others a less important one. Thus, if a particular geometrical feature of landscape is primarily controlled by a factor the action of which does not change greatly with time, or if the changes of factors having direct and inverse controls tend to cancel out the net effect of the changes, then the resulting variation in geometry may itself be small – perhaps insignificant.

A last important characteristic of open systems is that they are capable of behaving 'equifinally'—in other words, different initial conditions can lead to similar end results (Von Bertalanffy, 1950, p. 25; and 1952, p. 143). Davisian (closed system) thinking is instinctively opposed to this view, and the immediate and facile assumption, for example, that most breaks of stream slope are only referable to a polycyclic mechanism is an illustration of the one cause – one effect mentality. The concept of equifinality accentuates the multivariate nature of most geomorphic processes and militates against the unidirectional inevitability of the closed system cyclic approach of Davis. The approach contrasts strikingly with that of Gilbert:

> Phenomena are arranged in chains of necessary sequence. In such a chain each link is the necessary consequent of that which precedes, and the necessary antecedent of that which follows ... If we examine any link of the chain, we find it has more than one antecedent and more than one consequent ... Antecedent and consequent relations are therefore not merely linear, but constitute a plexus; and this plexus pervades nature. (Gilbert, 1886, pp. 286–7.)

To sum up, the real value of the open system approach to geomorphology is:

Firstly, that it throws the emphasis on the recognition of the adjustment, or the universal tendency toward adjustment, between form and process. Both form and process are studied, therefore, in equal measure, so avoiding the pitfall of Davis and his more recent associates of the complete ignoring of process in geomorphology:

> In a graded drainage system the steady state manifests itself in the development of certain topographic form characteristics which achieve a time-independent condition ... Erosional and transportational processes meanwhile produce a steady flow (averaged over a period of years or tens of years) of water and waste from and through the land-form system ... Over the long span of the erosional cycle continual adjustment of the components in the steady state is required as relief lowers and available energy diminishes. The forms will likewise show a slow evolution.
>
> Applied to erosion processes and forms, the concept of the steady state in an open system focuses attention upon the relationship between dynamics and morphology. (Strahler, 1950, p. 676.)

The relation between process and form lies close to the heart of geomorphology and, in practice, the two are often so intimately linked that the problem of cause and effect may present the features of the 'hen and the egg.' Approach from either direction is valuable, however, for knowledge of form aids in the understanding of process, and studies of process help in the clearer perception of the significant aspects of form.

> The study of form may be descriptive merely, or it may become analytical. We begin by describing the shape of an object in the simple words of common speech: we end by defining it in the precise language of mathematics; and the one method tends to follow the other in strict scientific order and historical continuity ... The mathematical definition of a 'form' has a quality of precision which was quite lacking in our earlier stage of mere description ... [employing means which] are so pregnant with meaning that thought itself is economized; ...
>
> We are apt to think of mathematical definitions as too strict and rigid for common use, but their rigour is combined with all but endless freedom ... we reach through mathematical analysis to mathematical synthesis. We discover homologies or identities which were not obvious before, and which our description obscured rather than revealed: ...
>
> Once more, and this is the greatest gain of all, we pass quickly and easily from the mathematical concept of form in its statical aspect to form in its dynamical relations: we rise from the conception of form to an understanding of the forces which gave rise to it; and in the representation of form and in the comparison of kindred forms, we see in the one case a diagram of forces in equilibrium, and in the other case we discern the magnitude and the direction of the forces which have sufficed to convert the one form into the other ...
>
> ... Every natural phenomenon, however simple, is really composite, and every visible action and effect is a summation of countless subordinate actions. Here mathematics shows her peculiar power, to combine and generalize ...

A large part of the neglect and suspicion of mathematical methods in ... morphology is due ... to an ingrained and deep-seated belief that even when we seem to discern a regular mathematical figure in an organism ... [the form] which we so recognize merely resembles, but is never entirely explained by, its mathematical analogue; in short, that the details in which the figure differs from its mathematical prototype are more important and more interesting than the features in which it agrees; and even that the peculiar aesthetic pleasure with which we regard a living thing is somehow bound up with the departure from mathematical regularity which it manifests as a peculiar attribute of life ... We may be dismayed too easily by contingencies which are nothing short of irrelevant compared to the main issue; there is a *principle of negligibility* ...

If no chain hangs in a perfect catenary and no raindrop is a perfect sphere, this is for the reason that forces and resistances other than the main one are inevitably at work ..., but it is for the mathematician to unravel the conflicting forces which are at work together. And this process of investigation may lead us on step by step to new phenomena, as it has done in physics, where sometimes a knowledge of form leads us to the interpretation of forces, and at other times a knowledge of the forces at work guides us towards a better insight into form. (Thompson, 1942, pp. 1026–9.)

Secondly, open-system thinking directs the investigation toward the essentially multivariate character of geomorphic phenomena (Melton, 1957; Krumbein, 1959). It is of interest to note that the physical, and the resulting psychological, inability of geographers to handle successfully the simultaneous operation of a number of causes contributing to a given effect has been one of the greatest impediments to the advancement of their discipline. This inability has prompted, at worst, a unicausal determinism and, at best, an unrealistic concentration upon one or two contributing factors at the expense of others. Davis' preoccupation with 'stage' in geomorphology has been paralleled, for example, by an undue emphasis on the part of some economic geographers upon the factor of 'distance' in many analyses of economic location.

Thirdly, it allows a more liberal view of changes of form with time, so as to include the possibility of non-significant or nonprogressive changes of certain aspects of landscape form through time.

Fourthly, while not denying the value of the historical approach to landform development in those areas to which the application of this framework of study is appropriate, open-system thinking fosters a less rigid view regarding the aims and methods of geomorphology than that which appears to be held by proponents of the historical approach. It embraces naturally within its general framework the forms possessing relict facets, those indeed which form the basis for the present studies of denudation chronology, under the general category of the 'inequilibrium' forms of Strahler (1952B). There is no uniquely correct method of treatment for a given body of information, and Postan (1948, p. 406) has been at pains to demonstrate the purely subjective distinction which exists between alternative explanations of phenomena on an immediately causal or generic basis, as against an historical or biographical one:

> For the frontier they draw separates not the different compartments of the universe but merely the different mental attitudes to the universe as a whole. What makes the material fact a fit object for scientific study is that men are prepared to treat it as an instance of a generic series. What makes a social phenomenon an historical event is that men ask about it individual or, so to speak, biographical questions. But there is no reason why the process should not be reversed; why we should not ask generic questions about historical events or should not write individual biographies of physical objects. Here Spinoza's argument still holds. The fall of a brick can be treated as a mere instance of the general study of falling bricks, in which case it is a material fact, and part and parcel of a scientific enquiry. But it is equally possible to conceive a special interest in a particular brick and ask why that individual brick behaved as it did at the unique moment of its fall. And the brick will then become an historical event. Newton must have been confronted with something of the same choice on the famous day when he sat under the fabulous apple tree. Had he asked himself the obvious question, why did that particular apple choose that unrepeatable instant to fall on that unique head, he might have written the history of an apple. Instead of which he asked himself why apples fell and produced the theory of gravitation. The decision was not the apple's but Newton's.

Davis was metaphorically struck by landscape and chose to write a history of it.

Fifthly, the open-system mentality directs the study of geomorphology to the whole landscape assemblage, rather than simply to the often minute elements of landscape having supposed evolutionary significance.

Sixthly, the open-system approach encourages rigorous geomorphic studies to be carried out in those regions – and perhaps these are in the majority – where the evidence for a previous protracted erosional history is blurred, or has been removed altogether.

Lastly, open system thinking when applied to geomorphology, has application within the general framework of geography; for geomorphology has always influenced geographical thinking to a great, and possibly excessive, degree (as, for example, that of Whittlesey, 1929; Darby, 1953; Beaver, 1961). Open-system thinking is characteristically less rigidly deterministic in a causative and time sense than the closed-system approach. The application of this closed-system approach to problems of human geography is extremely dangerous because, of its nature, it directs the emphasis toward a narrow determinism, and encourages a concentration upon closed boundary conditions, upon the tendency toward homogeneity and upon the levelling down of differences. Open-system thinking, however, directs attention to the heterogeneity of spatial organization, to the creation of segregation, and to the increasingly hierarchical differentiation which often takes place with time. These latter features, are, after all, hallmarks of social, as well as biological evolution.

Selected bibliography for further reading

BEAVER, S. H. 'Technology and geography,' *The Advancement of Science* (1961), Vol. 18, pp. 315–27.

BRAUN, E. L. *Deciduous forests of eastern North America,* Philadelphia, Pa.: Blakiston Co. (1950).

BUCHER, W. H. 'The nature of geological inquiry and the training required for it,' *American Institute of Mining Metallurgy Engineers Technical Publication 1377* (1941), 6 pp.

BURTON, A. C. 'The properties of the steady state compared tot hose of equilibrium as shown in characteristic biological behavior,' *Journal Cellular Comparative Physiology*, Vol. 14 (1939), pp. 327–49.

CHORLEY, R. J. 'Illustrating the laws of morphometry,' *Geological Magazine,* Vol. 94 (1957), pp. 140–9.

CLEMENTS, F. E. 'Plant succession: an analysis of the development of vegetation,' *Carnegie Institution, Washington, Pub. 242 (1916).*

COLE, M. M. 'Cerrado, Caatinga, and Pantanal: distribution and origin of the savanna vegetation of Brazil', *Geographical Journal,* Vol. 126 (1960), pp. 168–79.

CULLING, W. E. H. 'Multicyclic streams and the equilibrium theory of grade,' *Journal of Geology,* Vol. 65 (1957), pp. 259–74.

DARBY, H. C. 'On the relations of geography and history,' *Transactions of the Institute of British Geographers,* No. 19 (1953), pp. 1–11.

DENBIGH, K. G. 'The principles of chemical equilibrium; with applications in chemistry and chemical engineering,' Cambridge, England: Cambridge University Press (1955).

DENBIGH, K. G., HICKS, M., and PAGE, F. M. 'The kinetics of open reaction systems': *Faraday Society Transcripts,* Vol. 44 (1948), pp. 479–91.

GILBERT, G. K. *Report on the geology of the Henry Mountains,* 2nd ed. (1880), Washington, D C.: Government Printing Office, 170 pp.

GILBERT, G. K. 'The inculcation of the scientific method by example,' *American Journal of Science,* 3rd series, Vol. 31 (1886), pp. 284–99.

GILBERT, G. K. 'The transportation of debris by running water,' *U.S. Geological Survey Professional Paper 86* (1914).

GLEASON, H. A. 'The individualistic concept of the plant association': *Bulletin Torrey Botany Club,* Vol. 53 (1926–7), pp. 7–26.

GLEASON, H. A. 'Further views on the succession-concept', *Ecology,* Vol. 8, (1927), pp. 299–326.

GLOCK, W. S. 'The development of drainage systems,' *Geographical Review,* Vol. 21 (1931), pp. 475–82.

HACK, J. T. 'Interpretation of erosional topography in humid temperate regions,' *American Journal of Science,* Vol. 258-A (1960), pp. 80–97.

HACK, J. T. and GOODLETT, J. C. 'Geomorphology and forest ecology of a mountain region in the central Appalachians,' *U.S. Geological Survey Professional Paper 347* (1960).

HALL, A. D. and FAGEN, R. E. 'Definition of system,' *General Systems Yearbook,* Vol. 1 (1956), Ann Arbor, Mich., pp. 18–28 (mimeographed).

HORTON, R. E. 'Erosional development of streams and their drainage basins: hydrophysical approach to quantitative morphology,' *Geological Society of America Bulletin,* Vol. 56 (1945), pp. 275–370.

KRUMBEIN, W. C. 'The "sorting out" of geological variables, illustrated by regression analysis of factors controlling beach firmness,' *Journal of Sedimentary Petrology,* Vol. 29 (1959), pp. 575–87.

LEOPOLD, L. B. and MADDOCK, T. (JR.) 'The hydraulic geometry of stream channels and some physiographic implications,' *U.S. Geological Survey Professional Paper 252* (1953).

LITTLE, J. M. *Erosional topography and erosion,* San Francisco, Calif.: A. Carlisle and Co. (1940).

MACKIN, J. H. 'Concept of the graded river,' *Geological Society of America Bulletin,* Vol. 59 (1948), pp. 463–512.

MELTON, M. A. 'An analysis of the relation among elements of climate, surface properties, and geomorphology,' *Office of Naval Research Project NR 389–042, Technical Report* 11, Dept. Geol., Columbia University (1957).

PLAYFAIR, J. *Illustrations of the Huttonian theory of the earth,* 1802, Facsimile reprint, Champagne, Ill.: University of Illinois Press (1956).

POSTAN, M. 'The revulsion from thought,' *The Cambridge Journal,* Vol. 1 (1948), pp. 395–408.

PRIGOGINE, I. and DEFAY, R. *Chemical thermodynamics,* London: Longmans, Green and Co. (1954).

REINER, J. M. and SPIEGELMAN, S. 'The energetics of transient and steady states, with special reference to biological systems,' *Physical Chemistry Journal,* Vol. 49 (1945), pp. 81–92.

SCHUMM, S. A. 'Evolution of drainage systems and slopes in badlands at Perth Amboy, New Jersey,' *Geological Society of America Bulletin,* Vol. 67, pp. 597–646.

STRAHLER, A. N. 'Equilibrium theory of erosional slopes, approached by frequency distribution analysis,' *American Journal of Science,* Vol. 248 (1950), pp. 673–96, 800–14.

STRAHLER, A. N. 'Dynamic basis of geomorphology,' *Geological Society of America Bulletin,* Vol. 63 (1952), pp. 923–38.

STRAHLER, A. N. 'Hypsometric (area-altitude) analysis of erosional topography', *Geological Society of America Bulletin,* Vol. 63 (1952), pp. 1117–42.

STRAHLER, A. N. 'Statistical analysis in geomorphic research,' *Journal of Geology,* Vol. 62 (1954), pp. 1–25.

STRAHLER, A. N. 'Dimensional analysis applied to fluvially dissected landforms: *Geological Society of America Bulletin,* Vol. 69 (1958), pp. 279–300.

THOMPSON, D'ARCY W. *On growth and form,* Cambridge, England: (1942).

VON BERTALANFFY, L. 'The theory of open systems in physics and biology,' *Science,* Vol. 111 (1950), p. 23–9.

VON BERTALANFFY, L. 'An outline of general system theory,' *Journal of British Philosophy of Science,* Vol. 1 (1951), pp. 134–65.

VON BERTALANFFY, L. *Problems of Life,* London: Watts and Co. (1952).

VON BERTALANFFY, L. 'General system theory', *General Systems Yearbook,* Vol. 1, Ann Arbor, Michigan, pp. 1–10 (mimeographed).

VON BERTALANFFY, L. 'Principles and theory of growth,' Chapter 2 in *Fundamental aspects of normal and malignant growth,* edited by Nowinski, W. W., Amsterdam: Elsevier Pub. Co. (1960), pp. 143–56.

WHITTAKER, R. H. 'A consideration of the climax theory: the climax as a population and pattern,' *Ecological Monographs,* Vol. 23 (1955), pp. 41–78.

WHITTLESEY, D. 'Sequent occupance,' *Annals of the Association of American Geographers,* Vol. 19 (1929), pp. 162–5.

WOLMAN, M. G. 'The natural channel of Brandywine Creek,' Pennsylvania: *U.S. Geological Survey Professional Paper 271* (1955).

WOLMAN, M. G. and MILLER, J. P. 'Magnitude and frequency of forces in geomorphic processes,' *Journal of Geology,* Vol. 68 (1960), pp. 54–74.

WOOLDRIDGE, S. W. and GOLDRING, F. *The Weald,* London: Collins (1953).

WOOLDRIDGE, S. W. and LINTON, D. L. *Structure, surface and drainage in south-east England,* London: G. Philip and Son Ltd. (1955).

14 Geography and the ecological approach: the ecosystem as a geographic principle and method[1]

D. R. Stoddart[2]

University of Cambridge, U.K.

Dr Eyre's brief but interesting paper on 'Determinism and the ecological approach in geography'[3] seeks to demonstrate that by adopting an ecological viewpoint, geographers will rid themselves of naïve determinism and misinterpretation in both physical and human geography. This theme is, however, of wider importance than Eyre would indicate, for the significance of the ecological approach in geography is not merely that studies of vegetation and soils add 'cohesion and distinctiveness' to geographical work, nor that 'a more ecological approach enhances the prestige of geographers within the academic world'[4], but that ecological concepts provide a research method which geography so sadly lacks. Geography, as derived by Hettner and Hartshorne from the writings of Kant, occupies what Schaefer[5] calls an 'exceptionalist' position, that of a branch of knowledge with a unique integrating function, synthesizing more specialized fields in space as history does in time. With this idea Eyre apparently agrees; for him, 'geography either stands or falls as an integrating discipline'.[6] Schaefer's paper a decade ago marked the beginning of the retreat from this aloof and rather superior position; recent workers, from Ackerman to Bunge, take the position that geography in its aims and methods is essentially no different from other branches of science, and that it is precisely because of its Kantian heritage that the subject has managed to isolate itself from virtually every major development in the field of scientific thought since 1859.

 The fundamental contribution of ecology to the geographer, therefore, is in providing a methodology. British geographers, in their quest for the region and its spirit, exemplified in the work of Herbertson,[7] held up as an ideal a concept which defies rational analysis and which should have died with the biological vitalism which inspired it. American geographers, following Hartshorne in his quest for 'areal differentiation', committed themselves to what was, at worst, an exercise in the classification of areas, involving as an afterthought problems of organization and function. Neither view of geography provided an analytical tool of sufficient power to lead to new insights

301

and new approaches. In recent years, geographers have become aware of the potentialities of the *ecosystem concept* in geographical work. The idea of the ecosystem is implicit in much of Eyre's paper; this note briefly outlines the properties and applications of the ecosystem idea in geography, and indicates some of the great potentialities which the concept possesses as a research tool in geography.

Properties

The term *ecosystem* was formally proposed by the plant ecologist Tansley in 1935,[8] as a general term for both the biome ('the whole complex of organisms – both animals and plants – naturally living together as a socio-logical unit'[9]) and its habitat. 'All the parts of such an ecosystem – organic and inorganic, biome and habitat – may be regarded as interacting factors which, in a mature ecosystem, are in approximate equilibrium: it is through their interactions that the whole system is maintained.'[10] Fosberg[11] has developed the definition as follows:

> 'An ecosystem is a functioning interacting system composed of one or more living organisms and their effective environment, both physical and biological. .... The description of an ecosystem may include its spatial relations; inventories of its physical features, its habitats and ecological niches, its organisms, and its basic reserves of matter and energy; its patterns of circulation of matter and energy; the nature of its income (or input) of matter and energy; and the behaviour or trend of its entropy level.'

Properties of biological ecosystems have been recently outlined by Evans,[12] Whittaker,[13] and Odum[14] while the whole terrestrial ecosystem has been termed the *ecosphere*, derived from ecosystem and biosphere, by Cole.[15]

The ecosystem concept has four main properties which recommend it in geographical investigation. First, it is *monistic*: it brings together environment, man, and the plant and animal worlds within a single framework, within which the interaction between the components can be analysed. Hettner's methodology, of course, emphasizes this ideal of unity, and some synthesis was achieved in the regional monographs of the French school, but the unity here was aesthetic rather than functional, and correspondingly difficult to define. Ecosystem analysis disposes of geographic dualism, and with it the problem of determinism which Eyre discusses; the emphasis is not on any particular relationship, but on the functioning and nature of the system as a whole. Thanks very largely to the work of Hartshorne, the monism–dualism controversy is no longer a live issue in the west, but in the U.S.S.R., where Anuchin[16] has attempted to put forward monistic ideas on the unity of geography, he has been violently attacked.[17]

Secondly, ecosystems are *structured* in a more or less orderly, rational and comprehensible way. The essential fact here, for geography, is that once structures are recognized they may be investigated and studied, in sharp contrast to the transcendental properties of the earth and its regions as organisms or organic wholes.[18] Much geographical work in the past has been concerned with the framework of systems, and the current concern with geometry of landforms, settlement patterns and communication networks may be interpreted on this level. As an example of a structural investigation in biology, reference may be made to the work of Hiatt and Strasburg[19] on the food web and feeding habits of over 200 species of fish in coral reefs of the Marshall Islands in the Pacific. Observation showed that the fish could be classified into five trophic groups, which were related in a rather complex manner. These relationships formed the structure shown in Fig. 11, which includes all levels from plankton and algae to sharks and other carnivores.

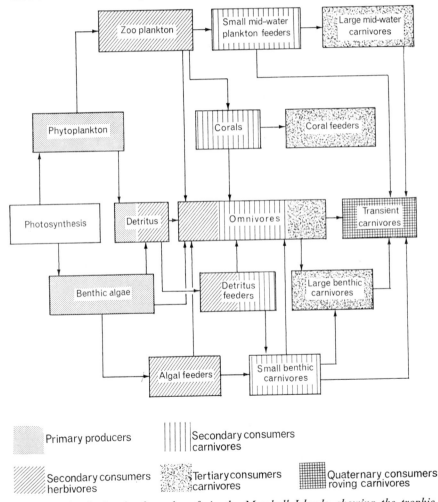

Fig. 11. The food web of coral reefs in the Marshall Islands, showing the trophic structure in a qualitative manner. After Hiatt and Strasburg.

Thirdly, ecosystems *function*; they involve continuous through-put of matter and energy. In geographic terms, the system involves not only the framework of the communication net, but also the goods and people flowing through it. Once the framework has been defined, it may be possible to quantify the interactions and interchanges between component parts, and at least in simple ecosystems the whole complex may be quantitatively defined. Odum and Odum[20] in a pioneering study, again on a Marshall Island coral reef, attempted to quantify the major trophic stages in the coral reef community – the primary producers, the herbivores and the carnivores. Figure 12A shows a biomass pyramid for a measured quadrat near the seaward edge of a reef; Figure 12B is a mean biomass pyramid generalized from quadrats

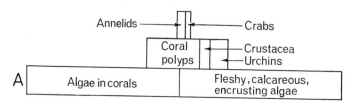

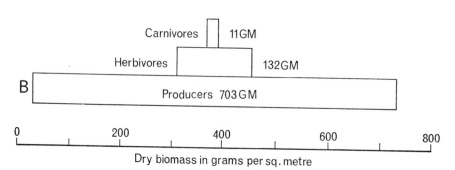

Dry biomass in grams per sq. metre

Fig. 12. Biomass pyramids, showing the dry weight of living materials in quadrats on the reef of Eniwetok Atoll in the Marshall Islands. A–a quadrat on the reef edge; B–average biomass for the reef. Gross trophic structure is here shown in a quantitative way. After Odum and Odum.

across a whole reef flat. While the details of the interpretation, particularly the trophic status of the corals, is open to question, the Odums have certainly demonstrated the possibility of quantifying the gross structural characteristics of small ecosystems. Equally remarkable is Teal's study[21] of a salt marsh ecosystem in Georgia. Teal constructed a food web for the salt marsh, and then measured standing crop, production and respiration for each of its components. Figure 13 shows in diagrammatic form the energy flow through this ecosystem, and the part played by each component of the web, with an energy input (light) of 600,000 kcal/sq.m./yr.

Fourthly, ecosystems are a type of general system, and the ecosystem

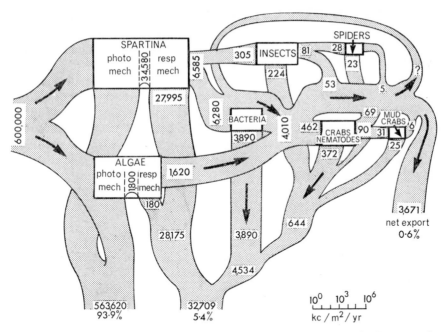

Fig. 13. Energy-flow diagram for a salt marsh in Georgia. After Teal. The numerals refer to kcal/sq. m./yr.

possesses the attributes of the general system. In general system terms, the ecosystem is an open system tending towards a steady state and obeying the laws of open-system thermodynamics. Many of the properties of such systems have been implicitly recognized in the past – for example, the idea of climax in vegetation, of maturity in soils, and of grade in geomorphology – but most of these conceptions have been, in effect, the application of classic thermodynamic ideas to closed system situations. With the development of open system thermodynamics,[22] many of these older ideas are being reinterpreted in a dynamic rather than a static manner. Whittaker[23] has thus revised Clements' views on succession and climax; Jenny, and more recently Nikiforoff,[24] have done the same for soils; and Chorley[25] and others have reinterpreted landforms in open system terms. Ecosystems in a steady state possess the property of self-regulation (action and reaction), and this is similar in principle to a wide range of mechanisms such as homeostasis in living organisms, feedback principles in cybernetics, and servomechanisms in systems engineering.[26] Systems such as ecosystems, moreover, can be conceived on different levels of complexity, and it is the task of the geographer to search out aspects of reality which are significant at the level at which the system is conceived. Systems, in fact, possess many of the structural properties of theoretical models, and a first approximation to system structure may be reached in a model-building manner, by selection, simplification, and ordering of data at a series of levels.[27] Thus systems may be built at the framework level (e.g., settlement hierarchies or transport nets) or as simple cybernetic systems (e.g., the mechanism of supply and demand, and of Malthus' doctrine), or at the more complex level of social systems and living organisms. Often in the case of more complex systems, a system has been conceived at a very much lower level of complexity, in the hope of gaining insight into problems where the data are too involved or the techniques inadequate. In geography, for example, the study of human groups, highly complex systems, has often been carried out on the level of 'clockwork' systems, such as simple deterministic, cause-effect relationships. The potential value of a system clearly depends on the correct selection of components when the system is structured, and this normally presupposes considerable experience with the problems or data involved.[28]

Applicability

The ecosystem concept is in origin a biological idea, and most of its applications, including those already quoted, have been from the non-human world. Some attempts have been made, however, to describe fairly complex ecosystems in which man may play some part. Fosberg,[29] for example, after many years' work on coral atolls, attempted a general qualitative description of the coral atoll ecosystem, in terms of the media involved, the inflow of energy, primary productivity and successive elaboration, transformation and decomposition of its animal and plant community, excretion and accumulation of matter, and total turnover of matter and energy. Fosberg subsequently convened a symposium[30] to discuss the role of man in the isolated ecosystems of islands, in which the discussion ranged from man's own appraisal of his ecological status to more detailed discussion of the effects of overpopulation on island life. Islands in fact provide small laboratories for the testing and analysis of relatively simple and well-defined ecosystem structures. Thus Sachet[31] has described the effects of the introduction of pigs on the ecology of Clipperton Island; vegetation was severely checked by crabs until men introduced pigs, which ate crabs and allowed the vegetation to grow again. The pigs have recently been killed, and the ecological readjustments are awaited. In a similar situation, Stoddart[32] has shown how coral islands in the Caribbean are in equilibrium with major storms, and are even built up during hurricanes, but when man replaces the natural vegetation by coconut plantations, the storms begin to cause catastrophic erosion. A classic study of an island ecosystem involving man is that of Thompson[33] on the interaction of man, plants and animals in Fiji.

Most ecosystems involving man are more complex than the salt marsh and coral reef systems already descibed (Figures 11–13), and attempts to describe ecosystems at such complex levels are likely to be difficult until experience is gained with relatively simple or restricted systems. Fosberg's focus on islands is one way out of this problem; another, which has received considerable attention recently, is to concentrate on primitive human and sub-human groups, in the hope of obtaining insight into the structure and function of more complex organizations. For example, Schaller's extraordinary study[34] of the mountain gorilla, *Gorilla gorilla berengei*, its territoriality, population structure, ecology and behaviour, and DeVore's[35] of the baboon, demonstrate the intriguing possibilities of primate geography. Among geographers, Sauer has been pre-eminent in the study of the ecology of man in the Pleistocene; and Daryll Forde, in a classic volume, studied the ecology of some two dozen primitive peoples.[36] Most of these studies, however, were conducted on traditional lines, and not within an explicit system framework; with some of the simpler groups it should be possible to delineate ecosystems with as much precision as in the non-human world.

The power of ecosystem analysis to pose new problems in geography, and hence to seek new answers, is demonstrated by Clifford Geertz's discussion of shifting cultivation and wet rice cultivation in Indonesia.[37] Geertz points out that most discussions of shifting cultivation emphasize its negative characteristics,[38] but that it is more profitably viewed in its system characteristics in relation to the tropical forest it replaces. Both are highly diverse systems, in which matter and energy circulate rapidly among the vegetation components and the topmost soil layer; the soil itself plays little part in this energy flow, and may often be impoverished. Burning is seen as a means of channelling the nutrients locked up in the vegetation into certain selected crop plants; the general ecological efficiency is lowered, but the yield to man increased. In a well-developed shifting cultivation system both structure and functions are comparable to those in the tropical forest, but the equilibrium is more delicately poised. By contrast, in wet rice cultivation, the ecosystem structure is quite different, the productivity is high, and the system equilibrium is more stable. The analysis is given in qualitative terms, but points the way to several lines of quantitative investigation, with clear import for land-use planning and rural reform programmes.

Apart from Geertz's work, there are few specific system-building studies in geography. The ecologist Dice, after working on natural communities, has produced a survey of ecosystem properties which may serve as a programme for human ecosystem research;[39] Chorley has carried systems analysis into geomorphology;[40] systems theory is being used in many branches of land-use planning, for example in the study of water resources;[41] Brookfield has briefly noted the potential of ecosystem studies;[42] but only Ackerman, in a major paper,[43] has pointed to systems analysis as geography's great research frontier.

Problems

It may be objected that the study of ecosystems in geography is either (*a*) not new, or (*b*) 'not geography'. In a sense, it is true, the study of systems is implicit in most geographic work; in economics system-building goes back to Smith and Ricardo, and in human and physical geography elements of systems are even older. The study of the ecosystem, however, requires the explicit elucidation of the structure and functions of a community and its environment, with the ultimate aim of the quantification of the links between the components. The ecosystem is a type of general system, defined as a 'set of objects together with relationships between the objects and between their attributes.'[44] Partaking in general system theory, the ecosystem is potentially capable of precise mathematical structuring within a theoretical framework, a very different matter from the tentative and incomplete descriptions of highly complex relationships which too often pass for geographical 'synthesis'. The charge that ecosystems study is 'not geography' lies in the fact, presumably, that the ecosystem definition does not explicitly define the earth's surface as a field of operation. 'Ecology is the study of environmental relationships; geography is the study of space relationships,' states Davies,[45] but he goes on to add that 'what is not clear is where the one stops and the other starts.' The study of space relationships, if it is to be more than mere nominal-scale classification of areas, must involve system-building, while the limits of the ecosystem may be set at any desirable areal extent. So flexible is the ecosystem concept, moreover, that it may be employed at any level from the drop of pond water to the universe, and is currently being employed in the study of the artificial ecosystem within space capsules and interplanetary rockets.[46] Within any areal framework the ecosystem concept will give point to enquiry, and thus highlight both form and function within a spatial setting. Simplistic ideas of causation and development, or of geographic dualism, are in this context clearly irrelevant; ecosystem analysis gives the geographer a tool with which to work.

Potentialities

The value of systems analysis lies not only in its emphasis on organization, structure and functional dynamics. By its general system properties, it brings geography back into the realm of the natural sciences, and allows us to participate in the scientific revolutions of this century from which the Kantian exceptionalist position excluded us. Perhaps the most significant implications of the ecosystem approach in geography is that systems may be linked with information theory, and thus with the whole new world of cybernetics, communication, and related mathematical techniques. Ecosystems are ordered arrangements of matter, in which energy inputs carry out work. Remove the energy input and the structure will break down until the components are randomly arranged (maximum entropy), which is the most probable state. Brillouin[47] has shown that order, or negative entropy, in systems, corresponds to information. First attempts have been made, as a result, to apply information theory to ecosystem analysis[48] and to interpret ecosystems in terms of cybernetics,[49] but the geographical implications of this have yet to be assessed. New mathematical techniques are being applied to geographical problems – for example, Gould's use of game theory in economic development studies in Ghana[50] and Kansky's application of network analysis to transportation patterns.[51] These are, however, essentially tools; they do not provide a methodology. This systems analysis does, and using it, geography can no longer stand apart in its isolated 'integrating' position.

Conclusion

Geography therefore stands today in much the same position as in 1859, following the publication of *On the origin of species*. In the century which has elapsed geography has borrowed many diverse ideas from biology, and attempts have been made to restate the nature of geography in ecological terms, most notably by Barrows in 1923.[52] Barrows, however, stated his position in deterministic terms, and succeeded in frightening off both geographers and sociologists, leaving the field which he delineated in the possession of neither.[53] Many of the insights gained from biology were applied in geography in an over-simplified and incautious way, and soon lost their power to stimulate fresh insight. The emergence of the ecosystem idea as a tightly-knit interacting complex of man and nature – clearly enough stated in the third chapter of Darwin's *Origin* – awaited the development of the growing body of systems theory. In the last few years it has begun to be applied by geographers both as a research tool and as a methodological instrument offering an alternative to that of Kant and Hettner. It links geography with the mainstream of modern scientific thought, in systems analysis and related disciplines, and opens up as yet unexplored possibilities in the application to geography of the whole field of information theory and communication techniques. In the ecosystem concept ecology makes its most profound and powerful contribution to geography.

References

1   Reprinted from *Geography*, Vol. 50 (1965), The Geographical Association Sheffield, U.K.
2   The author records his thanks to R. J. Chorley for his comments on and discussion of this paper.
3   EYRE, S. R. 'Determinism and the ecological approach to geography', *Geography*, Vol. 49 (1964), pp. 369–76.
4   Ibid., p. 374.
5   SCHAEFER, F. K. 'Exceptionalism in geography: a methodological examination', *Annals of the Association of American Geographers*, Vol. 43 (1953), pp. 226–49.
6   EYRE, S. R. op. cit., p. 376.
7   HERBERTSON, A. J. 'The higher units: a geographical essay', *Scientia*, Vol. 14 (1913), pp. 203–12.
8   TANSLEY, A. G. 'The use and abuse of vegetational concepts and terms', *Ecology*, Vol. 16 (1935), pp. 284–307.
9   TANSLEY, A. G. *Introduction to Plant Ecology*, London (1946), p. 206.
10 Ibid., p. 207.

11   FOSBERG, F. R. 'The island ecosystem' in Fosberg, F. R. (Editor), *Man's Place in the Island Ecosystem, a Symposium,* Honolulu (1963), pp. 1–6, reference on p. 2.

12   EVANS, F. C. 'Ecosystem as the basic unit in ecology', *Science,* Vol. 123 (1956), pp. 1127–8.

13   WHITTAKER, R. H. 'Ecosystem', *McGraw-Hill Encyclopaedia of Science and Technology,* Vol. 4, New York (1960), pp. 404–8.

14   ODUM, E. P. *Ecology,* New York (1963).

15   COLE, L. 'The ecosphere', *Scientific American,* Vol. 198, No. 4 (April 1958), pp. 83–92.

16   ANUCHIN, V. A. *Teoreticheskiye Problemy Geografii,* Moscow (1961); reviewed by Baranskiy, N. N., *Soviet Geography,* Vol. 2, No. 8 (1961), pp. 81–4.

17   KALESNIK, S. V. 'About "monism" and "dualism" in Soviet geography', *Soviet Geography.* Vol. 3, No. 7 (1962), pp. 3–16.

18   VIDAL DE LA BLACHE, P. 'Le principe de la géographie générale', *Annales de Géographie,* Vol. 5 (1895–6), pp. 129–42; Herbertson, A. J., op. cit.; Stevens, A. 'The natural geographical region', *Scottish Geographical Magazine,* Vol. 55 (1939), pp. 305–17.

19   HIATT, R. W. and STRASBURG, D. W. 'Ecological relationships of the fish fauna on coral reefs of the Marshall Islands', *Ecological Monographs,* Vol. 30 (1960), pp. 65–127.

20   ODUM, H. T. and ODUM, E. P. 'Trophic structure and productivity of a windward coral reef community on Eniwetok Atoll', *Ecological Monographs,* Vol. 25 (1955), pp. 291–320.

21   TEAL, J. M. 'Energy flow in the salt marsh ecosystem of Georgia', *Ecology,* Vol. 43 (1962), pp. 614–24.

22   DENBIGH, K. G. *The Thermodynamics of the Steady State,* London (1951).

23   WHITTAKER, R. H. 'A consideration of climax theory: the climax as a population and pattern', *Ecological Monographs,* Vol. 23 (1953), pp. 41–78.

24   JENNY, H. *Factors of Soil Formation,* New York (1941); Nikiforoff, C. C., 'Reappraisal of the soil', *Science,* Vol. 129 (1959), pp. 186–96.

25   CHORLEY, R. J. 'Geomorphology and general systems theory', *Professional Paper,* United States Geological Survey, No. 500-B (1962).

26   WIENER, N. 'Cybernetics or control and communication in the animal and the machine', *Actualités scientifiques et industrielles,* Vol. 1053 (1948), pp. 1–194.

27   CHORLEY, R. J. 'Geography and analogue theory', *Annals of the Association of American Geographers,* Vol. 54 (1964), pp. 127–37.

28   On the levels of systems, BOULDING, K. 'General systems theory – the skeleton of science', *General Systems,* Vol. 1 (1956), pp. 11–17.

29   FOSBERG, F. R. 'Qualitative description of the coral atoll ecosystem', *Atoll Research Bulletin,* No. 81 (1961), pp. 1–11; also in *Proceedings,* Ninth Pacific Science Congress, Vol. 4 (1962), pp. 161–7.

30   FOSBERG, F. R. (Editor), *Man's Place in the Island Ecosystem, a Symposium,* Honolulu (1963).

31   SACHET, M.-H. 'History of change in the biota of Clipperton Island', in Gressitt, J. L. (Editor), *Pacific Basin Biogeography, a Symposium,* Honolulu (1963), pp. 525–34.

32   STODDART, D. R. 'Storm conditions and vegetation in equilibrium of reef islands', *Proceedings,* Ninth Conference on Coastal Engineering (Lisbon), New York (1964), pp. 893–906.

33   THOMPSON, L. 'The relations of man, animals and plants in an island community (Fiji)', *American Anthropologist,* Vol. 51 (1949), pp. 253–67.

34   SCHALLER, G. B. *The Mountain Gorilla: Ecology and Behavior,* Chicago (1963).

35   DE VORE, I. and WASHBURN, S. L. 'Baboon ecology and human evolution', in

Howell, F. C. and Bourlière, F. (Editors), *African Ecology and Human Evolution,* London (1964), pp. 335–67.

36 DARYLL FORDE, C. *Habitat, Economy and Society,* London (1934).

37 GEERTZ, C. *Agricultural Involution; the Process of Ecological Change in Indonesia,* Berkeley (1963).

38 GOUROU, P. *The Tropical World,* London (1953).

39 DICE, L. R. *Natural Communities,* Ann Arbor (1952); Dice, L. R., *Man's Nature and Nature's Man: the Ecology of Human Communities,* Ann Arbor (1955).

40 CHORLEY, R. J. op. cit. (1962).

41 MCKEAN, R. N. *Efficiency in Government through Systems Analysis, with emphasis on Water Resources Development,* New York (1958).

42 BROOKFIELD, H. C. 'Questions on the human frontiers of geography', *Economic Geography,* Vol. 40 (1964), pp. 283–303.

43 ACKERMAN, E. A. 'Where is a research frontier?', *Annals of the Association of American Geographers,* Vol. 53 (1963), pp. 429–40.

44 HALL, A. D. and FAGEN, R. E. 'Definition of system', *General Systems,* Vol. 1 (1956), pp. 18–28.

45 DAVIES, J. L. 'Aim and method in zoogeography', *Geographical Review,* Vol. 51 (1961), pp. 412–17, reference on p. 415.

46 KONECCI, E. B. 'Space ecological systems', in Schaefer, K. E., (Editor), *Bioastronautics,* New York (1964), pp. 274–304.

47 BRILLOUIN, L. *Science and Information Theory,* New York (1962); Brillouin, L., *Scientific Uncertainty and Information,* New York (1964).

48 MARGALEF, D. R. 'Information theory in ecology', *General Systems,* Vol. 3 (1958), pp. 36–71.

49 PATTEN, B. C. 'An introduction to the cybernetics of the ecosystem: the trophic-dynamic aspect', *Ecology,* Vol. 40 (1959), pp. 221–31.

50 GOULD, P. R. 'Man against his environment: a game theoretic framework', *Annals of the Association of American Geographers,* Vol. 53 (1963), pp. 290–7.

51 KANSKY, K. J. 'Structure of transportation networks: relationships between network geometry and regional characteristics', *Research Papers,* Department of Geography, University of Chicago, Vol. 84 (1963).

52 BARROWS, H. H. 'Geography as human ecology', *Annals of the Association of American Geographers,* Vol. 13 (1923), pp. 1–14.

53 SCHNORE, L. F. 'Geography and human ecology', *Economic Geography,* Vol. 37 (1961), pp. 207–17.

15   Cities as systems within systems of cities[1]

Brian J. L. Berry[2]

Center for Urban Studies, the University of Chicago, U.S.A.

Introduction

This paper[3] examines some of the ways in which understanding of cities and sets of cities has been advanced during the first decade of Regional Science. Three channels that lead toward development of sound urban models are explored and relevant implications drawn. By models we mean *symbolic* models, not those of the *iconic* or *analogue* kinds.[4] Further, the symbolic models of interest are those that provide idealized representations of properly formulated and verified scientific theories relating to cities and sets of cities perceived as spatial systems. Any scientific theory logically comprises two parts:

a.   Simple inductive generalizations drawn from observable facts about the world.

b.   Abstract logical constructs.

It is the coincidence of deductions drawn from the logical constructs and inductive generalizations drawn from fact that makes for a valid scientific theory. Ten years ago urban studies were in an either/or situation; either inductive generalizations or logical constructs existed, the former as likely as not produced by urban geographers and the latter by urban economists. As the word *model* became fashionable, both called their products models, but neither had models of theories in the strict sense.

The importance of the last decade has been that the two *have* met through the medium of Regional Science. Moreover, the meeting came just when quantitative methods of analysis, facilitated by rapid developments in computer technology, began a technological revolution that has wrought havoc throughout the sciences. What more shattering change could there be than one which facilitates the large-scale studies that lead to specification of strength of belief in inductive generalizations, allow objective testing of the degree of coincidence between inductive generalizations and deductions from logical constructs and ease replication? The technological advance has meant more, however: virtual elimination of the once lengthy gap between problem formulation and evaluation of results; sharpening of the questions asked;

initiation and completion of experiments of a size unthinkable under earlier technical conditions; and many more.

The meeting, then, was timely. Inductive generalizations could be eased toward theory; logical constructs could be faced with the ultimate test of reality and new kinds of empiricism and experimentation could be developed. These are the three channels discussed in this paper. Examples are presented in an expository rather than a rigorous manner, since each has been elaborated elsewhere. The conclusions of the paper are that urban models are the same kinds of models as appear in other kinds of systems inquiry. Urban theory therefore may be viewed as one aspect of General Systems Theory. Viable avenues for future urban research might therefore be identified by looking at those other aspects of General Systems Theory that are relatively well advanced, to see how they reached this more developed position.

Inductive generalizations in search of a theory

Two of the better-known generalizations concerning cities are the rank-size relationship for sets of cities and the inverse-distance relationship for population densities within cities. Both had been observed many times when they were formalized as empirical 'rules' a decade or so ago, the former as the *rank-size rule* by G. K. Zipf and the latter as the *negative exponential density distance relationship* by Colin Clark. Yet as Isard noted in 1956, 'How much validity and universality should be attributed to the rank-size rule is, at this stage, a matter of individual opinion and judgment.'[5] Further, although Clark argued that the negative exponential 'appears to be true for all times and places studied,' he provided no theoretical rationale for his observations, only specified that they might have something to do with transport costs.[6] During the past decade both inductive generalizations have been brought closer to the status of scientific models, with the range of their validity carefully specified.

DISTRIBUTION OF CITY SIZES[7]

The rank-size rule says that for a group of cities, usually the cities exceeding some size in a particular country

(1) $$P_r^q. = P_1/r$$

where

P_1 is the population of the largest or first-ranking city
P_r is the population of the city of rank r

and

q is a constant.[8]

Whence it follows that

(2) $$\text{Log } r = \log P_1 - q.\log P_r$$

so that a plot of rank against size on double logarithmic paper should give a straight line with a slope of $-q$.

Another way of expressing the foregoing is that the frequency distribution of cities by size seems to be highly skewed in the shape of a reversed-*J*. A whole series of probability distributions, among them the lognormal and the Yule, have a similar reversed-*J* shape, each bearing a general family resemblance through their skewness. Each is, in fact, the steady-state distribution of the same simple stochastic process. Could it be that rank-size regularities of city sizes also result from such a stochastic process? The tenor of arguments provided in the past decade is that stochastic processes do indeed provide such a framework, and both the Yule distribution and the lognormal have been proposed as the basis of rank-size regularities.[9] The two are in fact so similar that each could obtain when the cumulative distribution of cities by size forms a straight line on lognormal probability paper, and which is applicable to the particular case depends upon whether a closed or an expanding system of cities is being considered.

Consider the transition matrix of a stochastic process in which the rows and columns are specified by city-size groups. If the probability density function of each size-class of cities is approximately the same,[10] then the steady state of the stochastic process will be lognormal if the set of cities existing at the beginning of the process is the same as the set achieving the steady state at the end. If, however, the smallest size class is augmented by new cities at a fairly steady rate throughout the process, the steady state is that of the Yule distribution. If growth of cities within the set can be said to occur in small independent increments, with probabilities of growth the same for each size class (growth is the result of 'many factors operating in many ways' and occurs such that if city sizes for time period 1 are plotted against sizes for time period n the resulting scatter of points is homoscedastic with a slope of -1), then the basic conditions of such a stochastic process can be said to have been met. One or other constraint leads to the lognormal or the Yule; in the former case a closed system of cities must exist, whereas in the latter the system must go on growing at a steady rate by addition of cities at the lowest level.

A recent study shows that the rank-size regularity applies throughout the world for countries which are highly developed with high degrees of urbanization, for large countries, and for countries such as India and China which, in addition to being large, also have long urban traditions; conversely, 'primate cities' or some stated degree of primacy obtains if a country is very small, or has a 'dual economy.'[11] Moreover, additional studies have recently shown that many distributions with some degree of primacy take on more of a rank-size form as level of development and degree of urbanization increase.[12] By virtue of size and complexity, then, countries with rank-size distributions appear to satisfy the condition of 'many factors operating in many ways' and increasing complexity of a space economy certainly brings the city-size distribution closer to rank-size. A rank-size regularity is not found when few factors mold the urban system in a few simple ways: in small countries, where economies of scale accrue in a single 'primate city'; or in 'dual economies,' where one or a few exogenous colonial cities of great size are superimposed upon an indigenous urban system of smaller places. In such cases, growth patterns cannot be summarized in the form of a stochastic process of the simple kind just outlined.[13] For all large, complex systems of cities which exist in the world, however, aggregate growth patterns do conform to such a stochastic process, so that one macroscopic feature of these systems is a rank-size regularity of city sizes. The regularity may, in turn, be 'explained'[14] by the stochastic process.

URBAN POPULATION DENSITIES[15]

No city has yet been studied for which a statistically significant fit of the expression

$$(3) \qquad d_x = d_0 e^{-bx}$$

does not obtain. In this equation, derived empirically by Colin Clark,

> d_x is population density d at distance x from the city center
> d_0 is central density, as extrapolated into the city's central business district

and

> b is the density gradient, so that, of course

$$(4) \qquad \operatorname{Ln} d_x = \ln d_0 - bx$$

Muth and Alonso have provided a satisfactory 'explanation'[16] of the observed regularity recently in terms of the rent-transport cost trade-off of individuals in different stages of the family cycle at different income levels and at different distances from the city center.[17] Apparently, the bid-rent function is steeper for the poorer of any pair of households with identical tastes in the American city, so the poor live toward the city center on expensive land consuming little of it; and the rich at the periphery consuming much.[18] The negative exponential shape of the decline stems from the nature of the production function for housing and the shape of the price-distance function.[19] Expression 3 is thus an equation of some generality that can be derived as a logical implication of the theory of the urban land market.

This being so, a variety of conclusions may be drawn. For example, the population residing at distance m from the city center is

$$(5) \qquad P_m = \int_0^m d_0 e^{-bx} (\pi 2x) \, dx$$

which becomes

$$(6) \qquad P_m = 2d_0 \pi b^{-2} [1 - e^{-bm}(1 + bm)]$$

This implies that the population pattern of an urban area can be described by two parameters alone, b and d_0. Winsborough has called the former a measure of the *concentration* of the city's population and the latter an index of its *congestion*.[20]

Now for any set of cities and for any particular city through time, another empirical expression holds:[21]

$$(7) \qquad\qquad b = aP^{-c}$$

Thus, b is in turn a function of city size, and a is the intercept. Central density d_0 is, on the other hand, apparently a function of the form of the city as established at the particular stage at which it grew, and is thus directly related to the city's age.[22] Knowing the population of a city and its age, it is possible to predict fairly closely the pattern of population densities within it.

In any system of cities for which the rank-size regularity obtains, the population P of a city of rank r, P_r, is a function of only P_1 and q (Equation 1). Hence, b must likewise be a function of P_1 and q (Equations 1 and 7). The distribution of population within cities is a function of the position of these cities within the entire system of cities, and age. If the larger system is Yule in form, age is simply the generation of the underlying stochastic process at which the city entered the system, so that congestion d_0 as well as concentration b is given within the framework of the larger system. The preceding statement can thus be modified to read: the distribution of population within cities is a function of the position of these cities within the entire system of cities at some point in time, and of the period of time for which they have been within the system.

Logical constructs in search of a test: Central Place Theory[23]

The preceding two models account for the size and the distributional characteristics of urban populations, but they say nothing of the locations of the cities concerned. Three sets of reasons for cities have been advanced, each with locational parameters more or less explicit: cities as strategic locations on transport routes; cities as the outcome of local concentrations of specialized economic activities; and cities as 'central places' performing retail and service functions for surrounding areas. Only the latter is of interest here.

Central Place Theory was formulated by Walter Christaller as a 'general purely deductive theory' designed 'to explain the size, number and distribution of towns' for reasons that also made it 'the theory of urban trades and institutions.'[24] A decade ago this theory was perhaps the only one concerning systems of cities that was at all well developed.[25] At that time, although many empirical studies of central places had been completed, the fact that no satisfactory test of the theory had been made largely reflects the fact that investigators looked for examples of theoretical implications drawn simply for exemplification by Christaller under the assumption of an isotropic plain. There was a lively debate as to whether certain of the most fundamental theoretical implications – for example, that of a hierarchy of central places – had any empirical validity. It has only been during the last decade that such questions have been settled. A thorough review of most aspects of the topic is to be found in *Central Place Studies: A Bibliography of Theory and Applications*, the first of the Regional Science Research Institute's Bibliography Series, and so will not be repeated here.[26] Subsequently to the Bibliography, the various postulates of the theory were drawn together in a model, however, and since the model appears to have some generality (implications drawn from the model have been verified independently, for example) it will be presented here.[27]

The model applies to systems of central places in which the elements are viewed aggregatively. A set of inequalities supplements the model, however, and these empirically derived expressions link aggregate patterns to local arrangements of central places under specified conditions of population density by specifying expectations as to the steps of the central place hierarchy. Random variations from ideal steplike patterns of central places in a series of local areas, combined with logical changes in location of the steps according to population density, interact to produce the regularities which may be observed in the aggregate. The definitions, identities, structural equations, and implications of the model follow without lengthy comment.

Definitions:

P_t the total population served by a central place

P_c population of the central place

P_r rural population and population of lower level centers served by the central place

A area of the trade area

Q_t population density of the area served

Q_r population density of those parts of the area served lying outside the central place

T number of central functions performed by the center, and since central functions enter in a regular progression and can be ranked from $1 \ldots T$ in decreasing order of ubiquity, also the highest level central function performed by the center.

E number of establishments providing the T types of business

D_m maximum distance consumers will travel to a central place of size T, *or* the range of good T

Identities:

$$(8) \qquad\qquad P_t = P_c + P_r$$

$$(9) \qquad\qquad P_t = AQ_t$$

$$(10) \qquad\qquad P_r = AQ_r$$

$$(11) \qquad\qquad A = kD_m^q$$

Figure 14 shows identity (9) in five distinct study areas in the United States.[28] In each case total population and total area served slope upwards to the right on double logarithmic paper with a slope of $+1$. Differences between study areas are simply a function of population densities.

Structural Equations:[29]

(12) $$\text{Log } P_c = a_1 + b_1 T$$

(13) $$\text{Log } D_m = a_2 + b_2 T$$

(14) $$\text{Log } E = a_3 + b_3 \log P_t$$

These structural equations hold in any study area (that is, at any level of density), and relate, by means of the intercept a and regression coefficient b, the population of a market center to the variety of central functions performed for surrounding areas, the drawing power of the center to its offerings, and the number of separate establishments performing the T functions (E exceeds T for all except the smallest villages and hamlets) to the total population served to account for non-basic demands for goods and services from the population P_c as well as basic demands generated by the population of the area served P_r.

Implications:

(15) $$P_c = P_t^s w^{-s} Q_t^{-s}$$

where

$$w = k\{\log^{-1} [qb_1^{-1}(a_2 - a_1 b_2)]\}$$

and

$$s = (b_1)/(qb_2)$$

As total population served increases, the central population comes to assume an increasing proportion of the total, but this tendency varies inversely with population densities.

(16) $$A = wP_c^x$$

where

$$x = s^{-1}$$

Area served increases exponentially with size of center.

(17) $$E = mQ_t^{b_3} P_c^{b_3}$$

where

$$\text{Log } m = a_3 + b_3 \log w$$

Total number of establishments varies exponentially with both size of center and total population densities.

These and similar structural equations and implications have now been verified in several studies[30] and appear to be a reasonable summary of many of the aggregate features of central place systems. Each has particular implications within the framework of Central Place Theory as well, particularly as it has been generalized. However, a set of inequalities is needed in combination with Figure 14 to lay out the steps of the central place hierarchy as it is found in local areas at different levels of density. These inequalities were established empirically by factor analyses of the functional structure of central places in each of several study areas, to determine the hierarchy

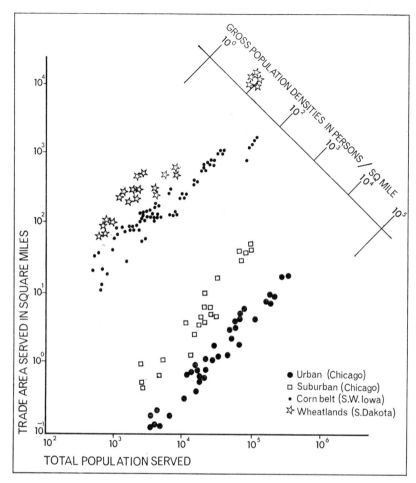

Fig. 14. Relationships between population and area served at different levels of population density.

individually within each of those areas, and then by discovering, unexpectedly, that limits to each of the levels varied consistently across the set of study areas as population density varied. With the second subscripts v referring to villages, t to towns, and c to cities, these inequalities are

(18) $$\text{Log } A_{tv} < 10.4 - 2.67 \log P_t$$

(19) $$\text{Log } A_{tt} < 9.3 - 2.067 \log P_t$$

(20) $$\text{Log } A_{tc} < 22.25 - 4.75 \log P_t$$

They are inserted into Figure 15, and in the case of the corn belt study area the individual observations are identified as they were classified in the factor analysis.[31]

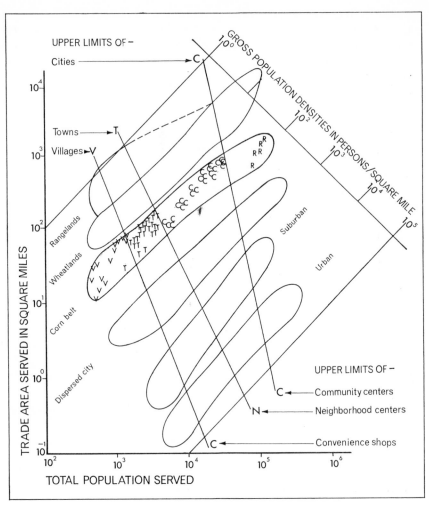

Fig. 15. Consistent upper limits have been identified for successive levels of central place at different levels of population density.

Innovation under technical impetus: social area analysis

The decade has seen a variety of innovations, most of them facilitated by rapid developments in computer technology, making possible kinds of research that could never have been contemplated prior to these developments. Beginnings are to be seen in the construction of urban simulators that will facilitate study of cities and sets of cities in laboratory-type experimental situations.[32] The most successful attempts so far have been those of Chapin in studies of land development[33] and Morrill in analyses of changing central place patterns,[34] although this statement is not intended to denigrate attempts along these lines in current urban transportation and economic studies. Out of these studies, particularly those undertaken in Chicago, Pittsburgh, and the Penn-Jersey region, and also those of the RAND Corporation and Resources for the Future Inc., will surely emerge models of some predictive power and experimental capability. Another paper at these meetings considers this topic, however, recognizing what may be the most significant new dimension to urban research added during the past decade. We will concentrate here on another topic, the new empiricism of the decade, stimulated by advancing computer technology and consequent diffusion of multivariate analysis throughout the social sciences. We focus here on one form of multivariate analysis, *factor analysis*, and briefly review how, in the form of *social area analysis*, it has facilitated studies of the internal structure of cities.

Social area analysis is one approach to the classic problem of urban ecology,[35] the succinct description of the location of residental areas by type within cities in terms meaningful to persons interested in social differentiation and stratification. Over the years several constructs have been developed in this context:[36] Hurd's concept of urban growth proceeding according to two patterns, central growth and axial growth; Burgess' concentric zone hypothesis of the location of residential areas by type, stemming from the nature of a growth process that proceeds outwards from the city center, accompanied by waves of residential invasion and succession; Hoyt's emphasis upon the axial growth of higher income neighborhoods outward from the city center along some sector; and Harris and Ullmans' notions of the multiple nucleation of the city. Both the social area analysts and their critics[37] have emphasized the difficulty of testing these hypotheses with the wide variety of socioeconomic data available, for example, from censuses. Which variables should be used in the test? Will the story told by different but presumedly related variables be the same? What in fact are the stories told about the structure and differentiation of urban neighborhoods by the wide range of census data available?

Factor analysis can provide answers to questions of the latter kinds. Let us review the basic features of the method. Consider a data matrix $_nX_m$ in which are recorded the data for n observations (say, census tracts) over m variables (census variables). If the column vectors of X are normalized and standardized to yield $_nZ_m$, then $n^{-1}Z^TZ = {}_mR_m$, which is, of course, the correlation matrix of the m variables. Since the column vectors of X were standardized, R is the variance-covariance matrix of Z, and the trace of R, equaling m, is the total variance of the m variables.

Now assume that each of the m variables is regressed in turn upon the $m - 1$ remaining. For each is then available a coefficient of determination expressing how much of its variance is held in common with the $m - 1$ other variables; in factor analysis these coefficients of determination are called *communalities*, and denoted h^2. For each variable, then, in its standardized form $1.0 - h^2 = u^2$ is the proportion of variance unique to the variable. A diagonal matrix U^2 can thus be formed with individual u^2's along the diagonal. It follows that $[R - U^2]$ has communalities on its diagonal, and the trace of $[R - U^2]$ is the total common variance of the m

variables. This total common variance plus the trace of U^2 equals m, the total variance.

Principal axes factor analysis provides a procedure whereby a matrix $_mA_r$ may be found such that

(21) $$[R - U^2] = AA^T$$

(22) $$A^TA = A$$

The dot product of each row vector of A yields one of the communalities, and the inner product of any pair of row vectors reproduces a correlation. The array Λ is a diagonal matrix, which implies that inner products of pairs of column vectors of A are zero. Such vectors are thus orthogonal (uncorrelated). The dot product of each column vector yields an eigenvalue λ. Since the sum of the eigenvalues must equal the sum of the communalities, these eigenvalues represent another way of parceling up the total common variance, the one (communalities) relating to the amount of the total common variance contributed by the association of any one variable with all other variables, the other (eigenvalues) to that part of the total attributable to one of the column vectors of A. These independent column vectors are the factors of factor analysis; the principal dimensions of variation underlying the original body of variables m.

Individual elements of A are factor loadings, the correlation coefficients between the original variables and each of the underlying common dimensions. The property of orthogonality of the dimensions is useful, because it means that each of the dimensions accounts for a different slice of the common variance, which slices are additive in any reconstitution of the whole; such additivity was not a property of the original intercorrelated m variables. Each dimension summarizes, then, one pattern of variation – one story told by the original m variables. A further step which is useful is to form

(23) $$_nS_r = [(A^TA)^{-1}A^TZ^T]^T$$

In the matrix S, the individual s_{ij} are factor scores of the original observations on each of the new dimensions formed by the analysis. The matrix S expresses all associations and common patterns found in X, but in a simpler form.

Factor analysis of census data for a whole series of U.S. cities by social area analysts has led to the conclusion that three dimensions are all that are required to summarize the stories told by the characteristics recorded for census tracts by the census. Study of the correlations between the original variables and the three dimensions has also revealed remarkably stable patterns from one city to another. One factor was consistently highly correlated with income, education, occupation, and wealth. A second was related to family structure, fertility, type of household, and position of women in the labor force. Finally, a third was associated with ethnic and racial structure of the population, age and sex composition, and measures of deterioration and blight. Speculation about the meaning of these regularities led social area analysts to identify the first as depicting variations in the *social rank* of individuals and families, the second as representing variations in the *urbanization* or *family status* of neighborhoods, and the third as resulting from *segregation*. The factor scores of tracts on these three dimensions could be used to characterize neighborhoods, since the three dimensions appear to be those responsible for the basic features of urban differentiation and stratification.

If the latter statement is true, then the three dimensions should enable research workers to test some of the classical constructs concerning such urban differentiation and stratification. A first study along these lines has revealed that factor scores of tracts with respect to social rank are differentiated in a sectoral fashion, as they should be if Hoyt's concepts apply, and that factor scores on urbanization and family status are differentiated in a concentric fashion, as they should be if Burgess's ideas are valid.[38] However, spatial variations in segregation show no regularity, but are specific to each case. As Hurd had speculated much earlier, concentric and axial patterns are therefore additive and independent sources of urban differentiation from city to city, with spatial variations specific to each city added by the third dimension of segregation.

It is clear that although social area analysts began simply in a 'look-see' manner, with later work facilitated by advancing computer technology, their work has now laid the bases for a spatial model of the internal socio-economic pattern of cities in which the relevance and role of the traditional concepts is clear.[39] [40]

A systems framework

The previous findings point in one direction: that cities and sets of cities are *systems* susceptible of the same kinds of analysis as other systems and characterized by the same generalizations, constructs, and models. *General Systems Theory* provides a framework for such inquiry into the nature of systems; indeed Boulding calls it the skeleton of science. Further, *Information Theory* has come to the fore as one of the foundations of general systems theory, contributing the two complementary ideas of *entropy* and *information* to the vocabulary of general systems research.[41] Entropy is achieved in the steady state of a stochastic process and is at its maximum if this process is unconstrained. Information is a measure of the order present if some systematic pressures for organization constrain the operation of the stochastic process.

Curry[42] has shown that, given Z settlements, with Z_i having a population i, the numbers of ways people can be distributed among settlements is

$$(24) \qquad P = Z! / \prod_{i=0}^{n} Z! \qquad (0 \leqslant i \leqslant n)$$

and in a large system the entropy E is given by

(25)
$$E_{def} = \log P = Z \log Z - \Sigma Z_i \log Z_i$$

E is maximized when

(26)
$$Z_i = (Z/N)e^{-(i/N)}$$

in which equation N is the mean population per settlement, or $N = n/Z$. Now if S is the size of the largest city,

(27)
$$Z_{i \leq s} = S(1 - e^{-(i/N)})$$

in which case

(28)
$$E_{max} = Z \log (eN)$$

and the most probable state of the system (that is, the state in which maximum entropy is found) is one in which, given the size of the largest city, the probability of the $(q + 1)$ st city having a population that is a given ratio of the qth city is a constant. Under these conditions the sum of the logarithms is a maximum, and of course it is the conditions satisfied when the rank-size rule for cities obtains. If a system of cities assumes rank-size, then, entropy has been maximized and it has assumed its most probable steady state.[43]

On the other hand, organization exists due to pressures for order in central place systems. If the per cent change in establishments in central places is constant with each addition of new business types, then[44]

(29)
$$dE/E \, dT = k$$

Integrating yields

(30)
$$\text{Log } E = k_1 T = c_1$$

If similar percentage ratios exist for the sizes of central places P_c then

(31)
$$\text{Log } P_c = k_2 T + c_2$$

From equations 30 and 31

(32)
$$T = K_1 \log E - C_1$$

(33)
$$T = K_2 \log P_c - C_2$$

Now the equation

(34)
$$I = K \log (\text{number of states})$$

has been identified as one measure of macroscopic negentropy, the inverse of entropy. It follows that the number of business types T is an index of the amount of information present in a set of establishments located in central places, or of the population of those places. This is consistent with the use of types of functions to identify and classify the central place hierarchy. Many attempts have been made to assess the 'centrality' of central places. It would seem that number of types of business – information content –

provides such an index. In southwestern Iowa, very strong fits to Equations 32 and 33 are found:

(35) $$T = 55.46 \log E - 58 \qquad (r^2 = 0.96)$$

(36) $$T = 50.00 \log P_c - 105 \qquad (r^2 = 0.91)$$

indicating that where urban centers are almost exclusively central places, necessary empirical bases for these arguments are to be found. It will be apparent that the above equations are compatible with those presented earlier for central place systems in the third section. Lösch and Christaller postulate such constant percentage relationships also with the addition of *levels* to the regular hierarchy ($k = 3$, $k = 4$, $k = 7$ networks and their implications); related measures of information should therefore exist for the order maintained by the steplike nature of the hierarchy.

It is not difficult to extend similar arguments to the situation within cities.[45] For example, urban population densities settle down to a most probable state in which densities are ranked with distance from the city center. Conversely, the model of central place systems also applies, indicating that certain aspects of urban life are constrained from reaching their most probable state.

Maruyama[46] has speculated about an apparent contradiction of the second law of thermodynamics in social phenomena, including those of cities. According to the second law, an isolated system will most probably trend to its most probable state, even if it begins in an inhomogeneous state. He points out that cybernetics, the study of equilibrating systems, considers many cases of self-regulation such that deviations are counteracted and the system is brought back towards its equilibrium, usually a most-probable state under constraint. But many instances can be cited in which feedback does not lead to self-correction toward some preset equilibrium (morphostasis). Rather, progressively greater contrasts appear, as between Myrdal's 'rich lands and poor' or with progressively greater centralization of urban functions in fewer larger cities, or when the 'growth of a city increases the internal structuredness of the city itself' (Maruyama's words). These are all examples of deviation *amplifying* processes (morphogenesis), which run counter to the second law.

Whether or not a system trends toward maximum entropy because processes working are deviation-correcting, or toward maximum information because the processes are deviation- and therefore structure-amplifying, apparently depends upon the nature of the causal relationships at work and of their feedback characteristics. Maruyama concludes that any system, together with the subsystems into which it may be partitioned, contains many examples of both deviation-correcting and deviation-amplifying processes. One subsystem may be becoming more highly organized; another may be approximating its most probable state. To understand the system as a whole demands that each of the subsystems be understood, as well as the relationships between them.[47]

So be it in the urban field. It is clear that cities may be considered as systems – entities comprising interacting, interdependent parts. They may be studied at a variety of levels – structural, functional, and dynamic – and they may be partitioned into a variety of subsystems. The most immediate part of the environment of any city is other cities, and sets of cities also constitute systems to which all the preceding statements apply. For systems of cities the most immediate environment is the socio-economy of which they are a part, and so forth.

Conclusions

Whereas progress has been made in understanding various facets of these systems and subsystems, we stand much as we did a decade ago for other facets. In a systems framework we should no longer worry about apparent contradictions between the kinds of conclusions reached for different subsystems, that is, between the distribution of city sizes and the functional arrangement of market centers in a hierarchy, however, for the difference is understood to be one of the relative balance of entropy-approximating or order-generating processes in various parts of the system. In contradistinction, however, we have very little understanding of how to put these different patterns together in more general models that are broad in scope. Sound models of partial kinds are providing the building blocks, but maximum progress during the next decade awaits the architectural systematizer.

References

1   Paper presented at the Annual Meeting of the Regional Science Association, December 1963.
        Reprinted from *Papers and Proceedings of the Regional Science Association,* Vol. 13 (1964), University of Pennsylvania, U.S.A.
2   *Editor's note:* A paragraph relating to the scope of the meeting at which this paper was presented has been omitted from this reprint. Berry was at pains to point out that this paper should not be regarded as a concise summary of the whole field of urbanization as was the following work by HAUSER, P. and SCHNORE, L. *The Study of Urbanization,* New York: John Wiley (1965). This volume (33) includes review papers by historians, geographers, political scientists, sociologists, economists, and the like.
3   Throughout this list of references, bracketed numbers refer to the Selected Bibliography at the end of the article on pp. 328–30.
4   ACKOFF (1) elaborates these terms.
5   See ISARD (34) in connection with a discussion of empirical regularities.
6   See (12) for review comments.
7   BERRY (8) lists the relevant literature in some detail. Subsequent contributions include those of Bell (5), Friedmann (29) and Ward (51).
8   If the entire population were urban, then $P_t = P_i \Sigma r^{-q}$. See WEISS (52).
9   SIMON (48), BERRY and GARRISON (7), THOMAS (49), DACEY (25) and WARD (51).
10   That is, so that the 'law of proportionate effect' holds.

11   BERRY (8).

12   BELL (5), FRIEDMANN (29).

13   Unless the process works, for example, to a random power of size, as with the log-lognormal, see THOMAS (49).

14   NAGEL (42) discusses the various modes of scientific explanation, and the role of explanation in science.

15   BERRY (12) lists the relevant literature. Also see WINSBOROUGH (54).

16   See (2) and (41), also footnote 12.

17   Ibid.

18   ALONSO (2).

19   MUTH (41).

20   WINSBOROUGH (54).

21   BERRY (12), WEISS (52), NEWLING (43).

22   WINSBOROUGH.

23   BERRY and PRED (9). Later studies include (10), (11), (13), (22). See also the parallel speculations of Rashevsky (46), (47).

24   CHRISTALLER (21).

25   BERRY and PRED (9).

26   Ibid.

27   See (10) and (14).

28   See (10) or (11) for details.

29   Only a sample of the structural equations necessary to facilitate the present discussion is given here.

30   See (10), also subsequent studies as yet unpublished by Karaska, Pitts, Murdie, and others.

31   The factor analytic results are presented in (14) and (10).

32   GARRISON (30) has one of the first presentations.

33   See (20).

34   See (38) and (39).

35   BELL (6) has an excellent review.

36   See the review by ANDERSON (3).

37   The Duncans write 'students of urban structure have lived for some time with the uncomfortable realization that their theories, or rather their abstract, schematic descriptions of urban growth and form are not very susceptible to empirical testing' (26).

38   ANDERSON (3).

39   This, in spite of criticism (26), has been the *accumulative* result.

40   It is worthwhile to note some of the other contributions made possible by factor analysis: (a) more general urban typologies (40); clear-cut evidence of the hierarchy of central places as an additive class system (10), (14); (c) multivariate regionalization (31); (d) metropolitan structure (32).

41   BERTALANFFY (16), (17), BOULDING (19), BEER (4).

42   CURRY (23). Other cases he examines are the spacing of nearest neighbors (see also Dacey [24]), the spacing of nearest neighbors of the same size, and the percentage manufacturing in an urban labor force.

43   CURRY points out that entropy in the same system, constrained such that persons had to be allocated in threes, as families, would be $H^1 = Z \log (eN/3)$. Hence, a measure of order is $R = 1 - H^1/E_{max}$.

44   ODUM (44), BERRY (11).

45   MEIER (37).

46   Reference 35 is a review statement of Maruyama's ideas, and contains other references of interest.

47   MARUYAMA provides an example of the operation of deviation amplifying mutual causal processes in a two-dimensional spatial distribution, and his discussion of systems, subsystems, and feedback is phrased in terms of cities.

Selected bibliography for further reading

1   ACKOFF, R. L. *Scientific Method. Optimizing Applied Research Decisions,* New York: John Wiley & Sons, Inc. (1961).

2   ALONSO, W. 'A Theory of the Urban Land Market,' *Papers and Proceedings of the Regional Science Association,* Vol. 6 (1960), pp. 149–58.

3   ANDERSON, T. R. and EGELAND, J. E. 'Spatial aspects of social area analysis,' *American Sociological Review,* Vol. 26 (1961), pp. 392–8.

4   BEER, S. 'Below the twilight arch – a mythology of systems,' in Eckman, D. F. (Ed.), *Systems: Research and Design,* New York: John Wiley & Sons, Inc. (1961).

5   BELL, G. 'Change in city size distribution in Israel,' *Ekistics,* Vol. 13 (1962), p. 98.

6   BELL, W. 'Social areas: typology of urban neighborhoods,' in Sussman, M. B. (Ed.), *Community Structure and Analysis,* Riverside, N.J., Crowell-Collier Press (1959).

7   BERRY, B. J. L. and GARRISON, W. L. 'Alternate explanations of urban rank-size relationships,' *Annals of the Association of American Geographers,* Vol. 48 (1958), pp. 83–91.

8   BERRY, B. J. L. 'City size distributions and economic development,' *Economic Development and Cultural Change,* Vol. 9 (1961), pp. 573–88.

9   BERRY, B. J. L. and PRED, A. *Central Place Studies: A Bibliography of Theory and Applications,* Philadelphia: Regional Science Research Institute (1961).

10   BERRY, B. J. L. *Comparative studies of central place systems,* final report of project NONR 2121–18, Office of Naval Research, Geography Branch, U.S. Department of the Navy (1961).

11   BERRY, B. J. L., BARNUM, H. G. and TENNANT, R. J. 'Retail location and consumer behavior,' *Papers and Proceedings of the Regional Science Association,* Vol. 9 (1962), pp. 65–106.

12   BERRY, B. J. L., SIMMONS, J. W. and TENNANT, R. J. 'Urban Population Densities, Structure and Change,' *The Geographical Review,* Vol. 53 (1963), pp. 389–405.

13   BERRY, B. J. L. *Commercial Structure and Commercial Blight,* Department of Geography Research Paper No. 85, University of Chicago (1963).

14   BERRY, B. J. L. and BARNUM, H. G. 'Aggregate Patterns and Elemental Components of Central Place Systems,' *Journal of Regional Science,* Vol. 4 (1964), pp. 35–68.

15   BERRY, B. J. L. 'Research Frontiers in Urban Geography,' in Hauser, P. and Schnore, L. (Eds.), *The Study of Urbanization,* New York: J. Wiley (1965).

16   VON BERTALANFFY, L. 'General System Theory: A New Approach to the Unity of Science,' *Human Biology,* Vol. 23 (1951), pp. 303–61.

17   VON BERTALANFFY, L. 'General System Theory,' *General Systems,* Vol. 1 (1956).

18   VON BERTALANFFY, L. 'General System Theory: A Critical Review,' *General Systems,* Vol. 7 (1962).

19   BOULDING, K. 'General Systems Theory – The Skeleton of Science,' *Management Science,* Vol. 2 (1956), pp. 197–208.

20   CHAPIN, F. S. and WEISS, S. F. *Factors Influencing Land Development,* Chapel Hill: University of North Carolina Press (1962).

21   CHRISTALLER, W. *Die zentralen Orte in Süddeutschland,* Jena: Gustav Fischer (1933).

22   CLAVAL, P. *Géographie Générale des Marchés,* Besançon (1962).

23   CURRY, L. 'Explorations in Settlement Theory: The Random Spatial Economy, Part I,' *Annals of the Association of American Geographers,* Vol. 54 (1964), pp. 138–46.

24 DACEY, M. F. and TUNG, T. H. 'The Identification of Point Patterns, I,' *Journal of Regional Science,* Vol. 4 (1963).

25 DACEY, M. F. 'Another Explanation for Rank-Size Regularity,' Philadelphia (1962).

26 DUNCAN, B. and DUNCAN, O. D. 'The Measurement of Intra-city Locational and Residential Patterns,' *Journal of Regional Science,* Vol. 2 (1960), pp. 37–54.

27 DUNCAN, B. 'Variables in Urban Morphology,' in Burgess, E. W. and Bogue, D. J. (Eds.), *Contributions to Urban Sociology,* Chicago: University of Chicago Press (1963).

28 FRIEDMANN, J. R. P. 'The Spatial Structure of Economic Development in the Tennessee Valley,' *Department of Geography Research Paper* No. 39, University of Chicago (1955).

29 FRIEDMANN, J. R. P. 'Economic Growth and Urban Structure in Venezuela,' *Cuadernos de la Sociedad Venezolana de Planificacion,* special issue (1963).

30 GARRISON, W. L. 'Toward a Simulation Model of Urban Growth and Development,' *Proceedings of the IGU Symposium in Urban Geography, Lund, 1960,* Lund: Gleerup (1962).

31 GINSBURG, N. *An atlas of economic development,* Chicago: University of Chicago Press (1961).

32 HATTORI, K., KAGAYA, K. and INANAGA, S. 'The Regional Structure of Surrounding Areas of Tokyo,' *Chirigaku Hyoron* (1960).

33 HAUSER, P. and SCHNORE, L. F. (Eds.), *The Study of Urbanization,* New York: J. Wiley (1965), in press.

34 ISARD, W. *Location and Space Economy,* New York, The Technology Press of the Massachusetts Institute of Technology and John Wiley & Sons, Inc. (1956).

35 MARUYAMA, M. 'The Second Cybernetics: Deviation Amplifying Mutual Causal Processes,' *American Scientist,* Vol. 51 (1963), pp. 164–79.

36 MCINTOSH, R. 'Ecosystems, Evolution and Relational Patterns of Living Organisms,' *American Scientist,* Vol. 51 (1963), pp. 246–67.

37 MEIER, R. L. *A Communications Theory of Urban Growth,* Cambridge, Mass., The M.I.T. Press (1962).

38 MORRILL, R. L. 'Simulation of Central Place Patterns over Time,' *Proceedings of the IGU Symposium in Urban Geography, Lund,* 1960, Lund: Gleerup (1962).

39 MORRILL, R. L. 'The Development of Spatial Distributions of Towns in Sweden: An Historical-Predictive Approach,' *Annals of the Association of American Geographers,* Vol. 53 (1963), pp. 1–14.

40 MOSER, C. A. and SCOTT, W. *British Towns: A Statistical Study of their Social and Economic Differences,* Edinburgh, Oliver and Boyd (1961).

41 MUTH, R. F. 'The Spatial Structure of the Housing Market,' *Papers and Proceedings of the Regional Science Association,* Vol. 7 (1961), pp. 207–20.

42 NAGEL, E. *The Structure of Science,* New York: Harcourt, Brace and World, Inc. (1961).

43 NEWLING, B. *The Growth and Spatial Structure of Kingston, Jamaica,* Ph.D. dissertation, Northwestern University (1962).

44 ODUM, H. T., CANTLON, J. E. and KORNICKER, L. S. 'An Organizational Hierarchy Postulate for the Interpretation of Species – Individual Distributions, Species Entropy, Ecosystem Evolution, and the Meaning of the Species – Variety Index,' *Ecology,* Vol. 41 (1960), pp. 395–9.

45 PIERCE, J. R. *Symbols, Signals and Noise,* New York: Harper and Row, Publishers (1961).

46 RASHEVSKY, N. 'Outline of a Mathematical Approach to History,' *Bulletin of Mathematical Biophysics* (1953).

47   RASHEVSKY, N. 'Some Quantitative Aspects of History,' *Bulletin of Mathematical Biophysics* (1953).

48   SIMON, H. A. 'On a Class of Skew Distribution Functions,' *Biometrika*, Vol. 42 (1955), pp. 425–40.

49   THOMAS, E. N. 'Additional Comments on Population-Size Relationships for Sets of Cities,' in Garrison, W. L. (Ed.), *Quantitative Geography*, New York, Atherton Press (1967).

50   VAN ARDSOL, M. D., CAMILLERI, S. F. and SCHMID, C. F. 'The Generality of Urban Social Area Indexes,' *American Sociological Review*, Vol. 23 (1958), pp. 277–84.

51   WARD, B. *Greek regional development*, Athens: Center for Economic Research (1962).

52   WEISS, H. K. 'The Distribution of Urban Population and an Application to a Servicing Problem,' *Operations Research*, Vol. 9 (1961), pp. 860–74.

53   WINGO, L. *Transportation and Urban Land*, Resources for the Future, Inc. (1961).

54   WINSBOROUGH, H. H. 'City Growth and City Structure,' *Journal of Regional Science*, Vol. 4 (1963).

IV Geography and Behaviour

'Our calculations have outrun conception;
We have eaten more than we can digest.'

P. B. Shelley, *In Defence of Poetry*

Geographers have not been particularly successful in dealing with behaviour. For much of the early part of this century interest in behaviour revolved around the extent to which man's actions were conditioned by his environment. Description was often by analogy, and problems were rarely formulated in any testable form. The determinists – advocates of a rigidly dependent relationship – may be seen as the lineal descendents of those individuals interpreting Darwin's work as a simplified cause-effect hypothesis (Stoddart, Chapter 3); their intellectual opponents, the possibilists, were heavily imbued with doctrines relating to man's free will. Fortunately the detailed discussion of these differences in interpretation, together with the half-way position adopted by many other geographers, need not concern us now. Within the last decade the debate has fizzled out; it is now a matter for the history of geography. Increasing awareness of the multitude of links between man and his environment as mediated through his perceptive process,[1] and recognition of the futility of any single feature explanations, means that any determinist or possibilist philosophy represents a dangerously naïve view of the world. Given our knowledge of the disrepute that the stimulus-response school brought to the field of animal behaviourism,[2] it is time we took comfort in the loss of relevance of the debate, though one important methodological point must be emphasized, an issue relating to the way in which evidence was presented. The determinist point of view explicitly excluded consideration of alternative events or courses of action, but even the possibilists failed to deal with the problem. Although aware of the importance of probability concepts, as Lukermann has recently stressed,[3] such issues were not explicitly utilized in any analytical framework. Evidence was still presented in a deterministic manner, not in the sense of a physical determinism, but in the sense of deriving one specific answer from any set of antecedents. Alternatives were a subjective possibility, not a specific and quantifiable probability. The consequence was that it was the course of evolution that was followed in any study, not the mechanism of change. Only within the last decade have geographers started to re-formulate their evidence on a probabilistic basis.

Perhaps of greater contemporary relevance to the geographical study of behaviour has been the emphasis placed on behavioural products rather than behaviour per se. This emphasis upon behavioural products, for instance upon completed migrations, completed journeys-to-shop, etc., even by such eminent workers as Hägerstrand[4], means that geographers have been one step behind in the reality of behaviour. By concentrating upon the spatial pattern of behaviour, rather than upon the process, the decision rules governing the behavioural process have rarely been commented on, let alone subjected to precise analysis. A consequence has been the tendency to apply hindsight to situations. There is a danger in emphasizing variables that might have been irrelevant to any contemporary situation, to attribute motives or perceptions that did not exist, and to forget about the multitude of equally probable events co-existing in any situation. In addition one must be constantly aware of imputing causal associations to distributions co-existing in area (casual associations), something in extreme form that links back to the rigid determinist stance of the Semple-Huntingdon era. Perhaps a more persistent problem has been the tendency to conceptualize issues in terms of optimizing behaviour (via a rational economic man) when 'satisfizing' principles operate in reality.[5] (One must be careful here and note that a so-called 'satisfizing' behavioural pattern – one in which an individual substitutes leisure time for the time taken to find the cheapest goods – may be considered optional to the individual. The 'optimizer-satisfizer' distinction is probably only useful if kept on the operational level of economic return.) Implicit in many empirical studies, optimality has been overstressed, particularly in theoretical and predictive studies. Practically all the major location theorists – Christaller, Lösch, Weber, Isard, etc. – made explicit use of optimal and rational assumptions, as have the first rash of transportation models. As economics is the most advanced theoretical discipline in the social sciences the emphasis is not surprising. Just as the biological analogy influenced late nineteenth century views of the world (Stoddart, Chapter 3) so economic formulations have left their mark upon the human geography of the mid-twentieth century. Again the importance of Section I must be stressed. We need explicit studies of the intellectual and ideological setting against which theories were formulated.

A third but related problem is the lack of correspondence between many of the empirical studies dealing with behaviour, whether behaviour per se or behavioural products. In most cases behaviour is used as another explanatory variable in the interpretation of the area under investigation; it is the description of the area that is important, for behaviour is just one aspect of the area. Given this objective and the inevitable limitations of the data (only recently have census authorities provided information on spatial movements apart from migration and even now the spatial basis is hardly above criticism), it is not surprising that the results derived from various study areas are not comparable. Unlike the scientist working in a laboratory, the temporal and spatial variations in which the geographer deals have led to different operational interpretations. The lack of correspondence rules means that precise generalizations have been few and far between. Not only does this mean that few behavioural laws have been derived, but even the attempts to effectively integrate different aspects of behaviour have not been particularly successful. Part of this problem has been dealt with by standardizing the sources of variation, though by far too few people,[6] while the sophisticated techniques displayed by Berry (Chapter 20) offer another possible avenue of research.

Attempts to derive behavioural laws do not seem, however, to have much possibility of success, unless a distinction is drawn between spatial behaviour and behaviour in space.[7] It is this that provides the fourth problem that must be considered. Today many geographers are not scrutinizing the specific behavioural patterns of any area because they recognize that these patterns depend upon the structure of the area (behaviour in space). Instead, they are searching for the rules governing spatial behaviour; they are experimentally searching for the postulates that describe the behavioural process. So far there has been a heavy dependence upon the work of other behavioural scientists, and operational models are still in their infancy. The search has mainly been directed at the elucidation of the relevant concepts and principles. Two examples, Wolpert (Chapter 18) and Golledge and Brown (Chapter 19), provide an indication of the directions taken by this line of research.

Always aware of the disintegrative tendencies of the discipline of geography, it seems important to conclude these preliminary observations on an optimistic note before turning to the individual studies. The behavioural approach seems to be encouraging the integration of studies of individual and aggregative behaviour. For much of this century aggregative research – especially that expressed in an ecological framework – has become more and more technically sophisticated, without being able to raise the level of its explanation. The dangers of inference have been constantly expressed.[8] By contrast, a subjective appreciation of the role of the individual has often led to an unreasoned denial of the possibility of generalization in such matters. Indeed, this fascination with the role of the particular (the ideographic approach) has frustrated the generalizing effort of a generation of geographers. Now a fusion is possible via probability analysis. The role of the unique factor, the unexpected action, the latent event made manifest, can be accommodated via the frank admission of our ignorance of all the details of any particular process. There is no need to deal with change in time or variation in space as if there is only *one* single result out of a multiplicity of causes or possibilities that is worthy of academic study. If the discipline is to progress, alternatives must be specifically built into any study. This is particularly relevant in view of the technological advances of our society. One of the consequences has been the creation of more and more choices, and geography may have an important role in evaluating the spatial implications of these choices. However, it is not only in the practical field that this development is important. The specific introduction of choice, of alternatives into any scheme, also means that the methodological dualism between individual and aggregative studies has at last been resolved. Both types of study can be incorporated into a single framework of analysis.

The first essay in this section, 'Chance and Landscape', by L. Curry

(Chapter 16), provides a concise and eloquent advocacy of the principle of indeterminacy. Although ranging widely across the whole spectrum of physical and human geography and possessing relevance to all the papers in this book, it provides a necessary backcloth to the analytical study of behaviour. In addition its effectiveness in certain circles will be increased by its concentration upon the philosophical rather than the technical implications of probability analysis.

Wary of indiscriminate criticism and possessing an admirable open-minded attitude, Curry is careful to admit that probability analysis is not the only possible methodological approach. Alternative forms can be recognized and in certain circumstances they might be more useful. Indeed, he even goes so far as to suggest that the probability approach may be a fad, though this seems a rather 'tongue in cheek' observation in the light of his evidence. What is important here is the observation that it is upon the utility of the results achieved in geography that the case for this methodological interpretation rests.

Five major issues are dealt with in the paper. The first is that ignorance is explicitly admitted by the investigator, perhaps particularly important in geography, a discipline dealing with complex processes at a macro scale. Secondly, the role of chance is not ignored but is incorporated into the analysis. Thirdly, by looking at the development process as a series of contingent events rather than as a simple temporal sequence, alternative events, or lines of action, are explicitly incorporated (cf. Chapter 4). Linked with this is the fourth point, namely that a probability formulation not only makes it possible to deal with time-independent processes (the steady state concept links back to system analysis) but also makes possible a reasoned appreciation of the periodicity of change. The potential fluctuations in any system are not seen as cataclysmic events (perhaps the nineteenth century debate on the origin of river valleys is worth noting here[9]), but as alternatives within the system; alternatives that may be abnormal or aberrant to the normal state but possibilities none the less within physical and human distributions. Finally, the description of small-scale processes within the same framework as large-scale processes may now be possible, as for instance via 'representative' or 'summation' man, although, as Curry briefly observes, there are problems in both approaches. Given the interlinking of these ideas within one methodological framework it is not surprising that Curry ends his article with a plea for geographers to place 'uncertainty' in the centre of their analysis.

Olsson and Gale's succinct study 'Spatial Theory and Human Behavior' (Chapter 17) reviews many of the problems of existing spatial theories. They show that although such theories have modified traditional economic views by explicitly incorporating spatial variables, one unfortunate consequence of this intellectual borrowing has been the implicit acceptance of the behavioural assumptions of economic theory. Despite the constant preoccupation by geographers with the initial assumption of spatial homogeneity, this is probably the least important problem. The 'profit maximization' and 'perfect knowledge' assumptions represent issues of more fundamental concern. In addition the deterministic frame of reference of most models imposes serious limitations upon extant theories.

Although serious criticisms can be levelled at existing theories, Olsson and Gale do not imply that these ideas should be thrown aside, particularly as little analytical work has been carried out on a new set of behaviouristic premises. Moreover, they stress that as the results from deterministic models may be the same as those derived from stochastic models, valuable insights into the creation of new models may be provided. For instance, probability inputs into the distance-decay functions of gravity models ensure that multiple rather than single answers will be forthcoming, while the norms provided by existing theories provide a useful first stage framework upon which to investigate chance factors. However, despite these mitigating points, the fact remains that stochastic frameworks provide more useful and comprehensive approaches, involving less rigidity in concept and technique. Probably of greatest importance is the recognition that patterns in the real world are not unique cases, but are the tangible realization of one event out of whole series of alternatives. Three other relevant criteria are also seen to lie behind the ideal stochastic model. The first is that any investigator should accept the limitations of his knowledge about behaviour and should not act as if he knew all the relevant facts. If he did, perhaps a deterministic model would be a viable alternative. Secondly, it is also accepted that a multitude of variables affects any situation so that a multivariate framework is necessary to reduce the dimensionality of the data. Thirdly, space and time need to be integrated into any study if a process study is going to be successfully formulated.

The establishment of these criteria lead Olsson and Gale into a study of the utility of Markov frameworks, though they pay particular attention to the problems of these methods. (It must be noted that this criticism is designed to be productive not carping). By drawing attention to the problems involved the authors hope that it will speed up the construction of an operational model of spatial behaviour.

The third paper of the section, Wolpert's examination of the 'Behavioral Aspects of the Decision to Migrate' (Chapter 18), stems from a dissatisfaction with the explanatory power of existing migration models. Analogous to the theories reviewed by Olsson and Gale, he shows that migration models are mainly deterministic in approach and have been shown to give increasingly inadequate interpretations of migration patterns in the United States. Wolpert's proposed solution is not the fitting of more powerful mathematical curves, but rather the adoption of a behavioural approach. This preliminary, but stimulating study mainly consists of a conceptual analysis of the factors lying behind any migration decision and is concluded by a short discussion of the translation of these ideas into an operational framework.

In his paper Wolpert is at pains to distinguish migration flows from other spatial flows such as mail or goods; the distinction lies in the ability of the migrant to act as an agent generating his own flow. Three concepts, 'place utility', 'action space' and 'life cycle', are considered to be relevant to migration generation and the effect of these concepts upon migration patterns are discussed. Apart from the lucid elaboration of these concepts and their implications, it is notable that Wolpert explicitly deals with the problem of non-movers. For a long time this has been a stumbling block in migration studies. Wolpert, however, is able to accommodate them into his system of study by treating them as lagged movers, as individuals postponing the decision to migrate – perhaps for all time. This ensures that any investigator deals with a single dimension of movement. A brief, but none the less valuable, comment is also made about the prospect of looking at migration in informative theory terms. By dealing with migration as a process involving the reduction of uncertainty and stressing the goals and search behaviour involved, important extensions of the theory may be derived.

An integral part of Wolpert's discussion deals with the distinction noted above between spatial behaviour and behaviour in space. Although not expressed in such a succinct way this does represent one of the first references to the distinction in geographical literature. Hence the study has a general methodological importance transcending specific migration problems and provides an overview of future research directions.

The fourth paper in this section, 'Search, Learning and the Market Decision Process' by Golledge and Brown (Chapter 19), concentrates upon the market decision process in studies of spatial behaviour. The author maintains that previous work by geographers has conceptualized behaviour as a stereotyped spatial pattern, not as the end product of an on-going process. Hence only equilibrium patterns were dealt with spatially. Golledge and Brown, drawing heavily upon the learning theories of educationalists and psychologists, view spatial behaviour as a process in which two critical phases occur. The first is a search phase in which individuals search among a set of alternative markets for the most satisfactory reward – random as well as motivated patterns can be accommodated here. The second is the so-called 'habitual response' phase, in which the most favourable response derived from these alternatives replaces the search activity and behaviour settles down into its habitual pattern. Although conceptualized as a two-phase activity, the requirements of a process formulation necessitates constant revaluation of the alternatives available to any individual. A final static response pattern applicable to all situations is never derived in this type of study. Even in the habitual response phase the probability exists of a change from one market to another.

Instead of leaving the discussion of the process on a verbal level, Golledge and Brown describe it in terms of a simple Markov chain model. The self-confessed limitations of the simple Markov model, particularly the need to vary the probability of choosing any market through time, leads to a consideration of three types of learning models, though one of the authors has recently extended this analysis.[10] Although an important contribution to the study of spatial behaviour, it must be noted that the authors conclude by admitting that the factors underlying the elements of their model are not known and await further work.

The final paper in this section by Berry is the 'Interdependency of Spatial Structure and Spatial Behavior' (Chapter 20). Unlike the previous two papers it is an example of the alternative approach to the spatial study of behaviour, namely an equilibrium study of behaviour in space. The article provides a valuable technical refinement of many of the concepts developed in Chapter 12, but it also makes an important statement of the way in which geographers are able to replace the subjectivity of past regional generalizations by an objective integration of the structural and behavioural approaches to the study of area. Also of note is the fact that, like the other papers in this section, Berry's work is close to the work of other social scientists, in this case to Lewin and Rumnel.[11]

In substantive terms the paper shows that behaviour, in the sense of the actual movement of goods and people, is usefully regarded as being related to the potentials for interaction. These depend in turn upon the spatial structure of any area. The first stage of analysis consists of reducing the dimensionality of the structural variables, via factor analysis, to a set of underlying bases. Each observation or area has a set of scores on these bases. The distance between any 'dyad' (or pairs of observations) as measured by these scores, provides an index of interaction potential. The second stage follows the same procedure for the behavioural variables. Once these preliminary stages are completed the behavioural and structural bases are related, via canonical analysis, to provide the basic patterns of independency for each observation. This may be interpreted as a regionalization procedure incorporating structural and behavioural elements. In order to clarify the issues involved in the combination of two previously disparate lines of study, Berry provides a case study of the Indian sub-continent in which three basic patterns of interdependency are revealed. Having reached such a level of technical sophistication, it is worth noting that in the conclusion Berry mentions that the next stage of analysis consists of a search for the generating mechanisms involved. Given the pre-occupation of the other essays in this section with the rules of spatial behaviour, their candid admission of their problems and their mapping out of the directions of future research, it may not be as long as we think before geographical studies of structure, behaviour in space, and spatial behaviour may be combined into a general process formulation. (cf. Chapter 9). This is the research task of tomorrow; these five studies represent signposts on the way towards that goal.

References

1*a*.   KATES, R. W. and WOHWILL, J. (Eds.), 'Man's response to the physical environment,' *Journal of Social Issues,* Vol. 22 (October 1966).

  b.   LOWENTHAL, D. (Ed.), 'Environmental perception and behavior', *University of Chicago Research Paper* No. 109 (1967).

2   KOESTLER, A. *The Ghost in the Machine,* London: Hutchinson (1967).

3   LUKERMANN, F. 'The "calcul des probabilités" and the école française de géographie', *Canadian Geographer,* Vol. 9, (1965), pp. 128–38.
4   HÄGERSTRAND, T. (translated by Pred, A.) *The Diffusion of Innovations,* Chicago University Press (1967).
5   For a discussion of this point see SIMON, H. A. *Models of Man,* New York: J. Wiley (1957).
6   BERRY, B. J. L. *Market centres and retail distribution,* Englewood Cliffs, New Jersey: Prentice Hall (1966).
7   RUSHTON, G. 'Analyses of spatial behavior by revealed space preference', *Annals of the Association of American Geographers,* Vol. 59 (1968), pp. 391–400.
8   ROBINSON, W. S. 'Ecological Correlations and the Behavior of Individuals', *American Sociological Review,* Vol. 15 (1950), pp. 351–7.
See also:
a.   GOODMAN, L. A. 'Some alternatives to ecological correlation', *American Journal of Sociology,* Vol. 64 (1959), pp. 610–25.
b.   BLALOCK, H. M. *Causal inferences in non-experimental research,* Chapel Hill: University of Caroline Press (1964).
9   BECKINSALE, R. P., CHORLEY, R. J. and DUNN, A. J. *The history of the study of landforms,* Vol. 1, London: Methuen (1964).
10   GOLLEDGE, R. G. 'The geographical relevance of some learning theories' in Cox, K. R. and Golledge, R. G. (Eds.), 'Behavioral Problems in Geography: A Symposium', *Northwestern University Studies in Geography* No. 17 (1969).
11*a.*   LEWIN, K. *Field Theory in Social Science,* London: Tavistock (1952).
    b.   RUMMEL, R. J. 'Dimensions of Conflict Behaviour Within and Between Nations,' *General Systems Yearbook,* No. 8, (1963), pp. 1–50.
    'A Social Field Theory of Foreign Conflict', *Peace Research Society Papers,* Vol. 9 (1966), pp. 131–50.

Selected bibliography for further reading

BOOKS

BLUMER, H. 'Collective Behaviour,' in Gittler, J. B. (Ed.), *Review of Sociology,* New York: J. Wiley (1957).
BOULDING, K. E. *The Image,* Ann Arbor: University of Michigan (1956).
BOWDEN, L. W. 'Diffusion of the Decision to Irrigate', *University of Chicago Research Paper,* No. 97 (1965).
BROWN, L. A. 'Diffusion Processes and Location: A Conceptual Framework and Bibliography', *Regional Science Bibliography Series,* No. 4 (1968), University of Pennsylvania Regional Science Institute.
CLARK, L. H. *The Life Cycle and Consumer Behaviour,* New York (1955).
COHEN, J. *Behaviour in Uncertainty,* London (1964).
COMBS, A. W. and SNYGG, D. *Individual Behaviour: A Perceptual Approach to Behaviour,* New York: Harper and Row, (1959).
COX, K. and GOLLEDGE, R. 'Behavioral Problems in Geography', *Northwestern University Studies in Geography,* No. 17, (1969).
CYERT, R. M. and MARCH, J. G. *A Behavioural Theory of the Firm,* Englewood Cliffs, New Jersey: Prentice Hall (1963).
DUIJKE, H. C. J. and FRIJDA, N. H. *National Character and National Stereotypes,* Amsterdam: New Holland Publishing Co. (1960).
EMERY, F. E. and OESER, O. A. *Information, Decision and Action: A Study of the Psychological Determinants in Changes in Farming Techniques,* Melbourne (1958).

FESTINGER, L. *et al., Conflict, Decision and Dissonance,* California: Stanford (1964).

HÄGERSTRAND, T. (translated by Pred, A.), *The Diffusion of Innovations,* Chicago University Press (1967).

HULL, C. L. *A Behaviour System,* New Haven: Yale University Press (1952).

NICOSIA, F. M. *Consumer Decision Processes,* Englewood Cliffs, New Jersey: (1966).

PARSONS, T. and SHILS, E. A. (Eds.), *Toward a General Theory of Action,* New York: Harper and Row (1951).

PRED, A. 'Behaviour and Location', *Lund Studies in Geography,* Series B, No. 27 (1967).

ROGERS, E. M. *The Diffusion of Innovations,* New York: The Free Press (1962).

ROSEN, R. *Optimality Principles in Biology,* London (1967).

SHELLY, M. W. and BRYAN, G. L. *Human Judgments and Optimality,* New York (1964).

SIMON, H. A. *Administrative Behaviour,* New York: Free Press (1957).

SIMON, H. A. *Models of Man: Social and Rational,* New York: J. Wiley (1957).

SPROUT, H. and SPROUT, M. *The Ecological Perspective on Human Affairs,* New Jersey: Princeton University Press (1965).

THEIL, H. *Economics and Information Theory,* Amsterdam: North Holland Publishing Company (1957).

WAGNER, R. L. and MIKESELL, M. W. *Readings in Cultural Geography,* Chicago: University of Chicago Press (1962).

ARTICLES

ALCHIAN, A. A. 'Uncertainty, evolution and economic theory', *Journal of Political Economy,* Vol. 58 (1960), pp. 211–21.

BACHRACH, P. and BARATZ, M. S. 'Decisions and Non Decisions: An Analytic Framework', *American Political Science Review,* Vol. 57 (September 1963), pp. 632–42.

BLAUT, J. M. 'Microgeographic Sampling', *Economic Geography,* Vol. 35 (1959), pp. 79–88.

BROEK, J. O. M. 'National Character in the Perspective of Cultural Geography', *Annals of the American Academy of Political and Social Science,* Vol. 370 (1967), pp. 8–15.

BUTTIMER, A. 'Social Space an Interdisciplinary Perspective', *Geographical Review,* Vol. 59 (1969), pp. 417–26.

COX, K. (Ed.), 'Studies in Political Interaction', *East Lakes Geographer,* Vol. 4 (December 1968).

CURRY, L. 'Central Places in the Random Spatial Economy', *Journal of Regional Science,* Vol. 7, No. 2 (1967), pp. 217–38.

CURRY, L. 'The Random Spatial Economy', *Annals of the Association of American Geographers,* Vol. 54 (1964), pp. 138–46.

CURRY, L. 'The Climatic Resources of Intensive Grassland Farming', *Geographical Review,* Vol. 52 (1962), pp. 138–46.

DACEY, M. F. and ISARD, W. 'On the projection of individual behaviour in regional analysis', *Journal of Regional Science,* Vol. 4, No. 1 (1962), pp. 1–34.

DACEY, M. F. 'A County Seat Model for the Areal Pattern of an Urban System', *Geographical Review,* Vol. 56, No. 4 (1966), pp. 527–42.

DANIELSSON, A. 'The Locational Decision from the point of view of the Individual Company', *Ekonomisk Tidskrift,* Vol. 66 (1964), pp. 47–87.

GOODMAN, L. A. 'Some Alternatives to Ecological Correlation', *American Journal of Sociology,* Vol. 64 (1959), pp. 610–25.

GOODMAN, L. A. 'Ecological Regression and the Behaviour of Individuals', *American Sociological Review,* Vol. 18 (1953), pp. 663–4.

GOULD, P. R. 'Man Against the Environment: A Game Theoretic Framework', *Annals of the Association of American Geographers,* Vol. 53 (1963), pp. 290–7.

GOULD, P. R. 'Joshua's Trumpet: The Crumbling Walls of the Social and Behavioural Sciences', *Geographical Review*, Vol. 55 (1965), pp. 599–602.

GREENHUT, M. L. 'The Decision Process and Entrepreneurial Returns', *Manchester School of Economic and Social Studies*, Vol. 34 (1966), pp. 247–67.

HARVEY, D. W. 'Behavioural Postulates and the Construction of Theory in Human Geography', *Seminar Papers, Series A, No. 6, Department of Geography, University of Bristol.*

HUFF, D. L. 'A Topographical Model of Consumer Space Preferences', *Papers – Regional Science Association*, Vol. 6 (1960), pp. 159–73.

KIM, K. W. 'Limits of Behavioural Explanation in Politics', *Canadian Journal of Economics and Political Science*, No. 31 (August 1965), pp. 322–4.

LANSING, J. B. and MARANS, R. W. 'Evaluation of Neighbourhood Quality', *Journal of the American Institute of Planners*, Vol. 35, No. 3 (May 1969), pp. 195–9.

LUCE, R. D. 'Psychological Studies of Risky Decision Making', in Strother, C. B. (Ed.), *Social Science Approaches to Business Behaviour*, London: Irwin Inc., Tavistock Publications (1962).

MARBLE, D. F. *Three Papers on Individual Travel Behaviour in the City'*, Illinois: Northwestern University Press (1967).

OSGOOD, C. E. 'Behaviour Theory and the Social Sciences', *Behavioural Science*, Vol. 1 (1956), pp. 167–85.

ROBINSON, W. S. 'Ecological Correlations and the Behaviour of Individuals', *American Sociological Review*, Vol. 15 (1950), pp. 351–7.

RUMMEL, R. J. 'Dimensions of Conflict Behaviour Within and Between Nations', *General Systems Yearbook*, Vol. 8 (1963), pp. 1–50.

SIMON, H. A. 'Theories of Decision-Making in Economics and Behavioural Science', *American Economic Review*, Vol. 49 (1959), pp. 253–83.

SOMMER, R. 'Further Studies of Group Ecology', *Sociometry*, Vol. 28 (1965), pp. 337–48.

STEA, D. 'Space, Territory and Human Movements', *Landscape*, Vol. 15 (1965), pp. 13–16.

WARNTZ, W. 'The Topology of a Socio-Economic Terrain and Spatial Flow', *Papers – Regional Science Association*, Vol. 17 (1966), pp. 47–61.

WOLPERT, J. 'Migration as an Adjustment to Environment Stress', *Journal of Social Issues*, Vol. 22, No. 4 (1966), pp. 92–102.

WOLPERT, J. 'The Decision Process in Spatial Context', *Annals of the Association of American Geographers*, Vol. 54 (1964), pp. 537–58.

16 Chance and landscape[1]

L. Curry
The University of Reading, U.K.

A century and a half ago the young Shelley could regard intellectual beauty as the banishing

> From all we hear and all we see,
> Doubt, chance and mutability . . .

Yet by the early twentieth century in geography we find the French school discussing the sequential ordering of events in terms of contingencies, a notion taken from the contemporary mathematical probabilists.[2] Today, in the mental atmosphere of the principle of indeterminancy it is not surprising that some geographers are seeking to use the probability calculus as well as rely on its philosophical connotations. Whether this trend is only the following of a current fad which seeks to hide our ignorance and sloth or implies living boldly in one's generation, eschewing the logic of aggregation and imprecise measurement is for the future to decide. Certainly, it is easy to sympathise with the resistance to regarding the earth's surface as governed by the mechanics of a roulette wheel and its development as a permanent floating crap game. The triumphs of nineteenth century science with its mechanistic cause-effect modes of thinking cannot be lightly set aside.

Yet apart from microphysics there are no *a priori* grounds for preferring deterministic or random formulations of physical and social behaviour. Justification lies only in *a posteriori* results: which is most convenient, which leads to insights most readily in a particular problem, which allows greater generality should a broader view be sought? The random approach has a number of advantages. The most important is that it allows a problem to be approached with the explicit admission of considerable ignorance.

In a sense, the formulation of a random process is the reverse of a deterministic one. In the latter we specify some 'causes' of certain intensity and interaction and obtain a result which will differ from reality by an 'error' term. In the former we begin, at least metaphorically, with unconstrained independent random variables and, by introducing dependencies and constraints, we achieve results of various likelihoods. Where nature shows only a single result, it is interpreted as the historical realization of a process which could just as easily have produced other results according to their attached probabilities. Where nature displays considerable fluctuation or variation and thus many results, but where there are a large number of realizations, their frequency of occurrence should agree with the derived probabilities. The more complex a process, the more likely we are to be ignorant of its working and thus the more we must rely on a probability calculus. One needs only consider the problem of the forces involved in a stone being moved down a river bed, let alone the development of a watershed, to realise that the task of setting down the equations governing this action in deterministic form will daunt the stoutest heart. In fact, we have skirted such problems altogether, preferring to eschew precision and concentrate on processes where vagueness of concept is possible or alternatively tackle extremely restricted problems about which we can be precise. Each problem of landscape has its appropriate time and space scale of study. Given our level of ignorance and a deterministic mode of thought, we cannot gain the level of precision necessary to tackle many of these problems. We must pick out problems of a particular scale which will allow us to use our mode of logic. These are problems which do not require rigorous formulation and imply considerable time or space scales of study or particular and relatively simple problems of limited dimensions such as the location of a factory. In the former class we have the historical evolution of landforms or the geographic zonality idea for associating climate, landforms, vegetation, soils and hydrology. It is thus likely that the probability approach will be used to open up new problems as much as re-examine old ones. Certainly there are many areas of concern to geography for which probabilistic thinking appears irrelevant. Descriptive and extremum problems in the spatial structure of the economy, of such wide interest today, do not fit this mould. The recounting of the historical development of particular areas needs rarely to appeal to probability reasoning except perhaps via contingency when broad enough explanatory concepts are being invoked.

In general the notion of randomness is relevant where a range of possibilities exists, implying a population of sufficient size existing either through time or over space or in the mind so that recounting these possible states of the population is of relevance to the real world. Thus, for example, it will be readily agreed that human geography has been retrospective in viewpoint with 'an express emphasis on the historical background of social behaviour'. Yet to regard society as buffeted on the advancing front of a wave rather than having the ability to see some way ahead and to partially control speed and direction seems unduly fatalistic. Society is as much an organization to face the future as a product of Pavlovian conditioning. Probability analysis emphasises choice and decisions and thus aims forwards as readily as backwards.

A probabilistic formulation of a problem can lead to results which are intuitively inconceivable to our deterministically structured minds. Thus a storage system in which mean input equals mean service rate can have no change in the amount stored. In terms of random processes, however, not only must the mean output capacity be greater than mean input to constrain storage to reasonable bounds but fluctuations in this quantity are enormous as input approaches service rate. This has obvious implications for lake levels or watertables, the characteristics of a floodplain, the operation of inventories or the heat stored in the ocean and therefore climatic change. A state of affairs can often arise via stochastic processes which has no well-defined counterpart in historical explanation: no matter what the initial state and no matter what historical course of events has occurred since, the final result is the same (at least in probabilistic terms). This 'steady state' is thus independent of time so that time will not appear in the steady state equations.

Description

The ability to describe the arrangement in time and space of features of the earth's surface in probabilistic terms is clearly a first requirement of analysis. Study of the temporal variation of the atmospheric elements has led here, although the work of Dacey[3] on two dimensional dot patterns has meant that we now have a considerable kit of tools for studying events in one and two dimensions. The fitting of standard probability series to events has had some important consequences. In the first place it has sharpened ideas on what is meant by terms such as homogeneity and heterogeneity. We can specify exactly what we mean by regular, random or clustered distributions and we realize the type and extent of dependence between events which leads to these distributions. By simply counting events per unit area or unit period and fitting a standard probability series to them, we may often infer a considerable amount of further information concerning the distances apart of nth nearest neighbours, the magnitude of variation contributed at different scales, and so on. We also obtain insights into the type of process which could have produced such distributions and are thus led into explanation.

A great deal of fundamental work still remains to be done, however, on methods of description. Court, for example, discusses the problem of the form of the probability function to be fitted to the wind rose.[4] Dacey has begun work on line patterns[5] while the work of Longuet-Higgins[6] on wave statistics has still to be exploited in geography. This account could lead into the various statistical methods being used for describing surfaces numerically, for map generalization, for regionalization or for splitting continuous areal distributions into components, but this would lead us too far afield.

Although not strictly descriptive, the use of combinatorial analysis in reducing the rather perplexing ordering displayed in city sizes, the rank size rule[7] and in the branching of streams, Horton's law of stream numbers[8] may be noted here. Given all possible arrangements of a given number of things among a given number of classes, random allocation of the things among the classes provides these empirical regularities as the most probable distribution.

Diffusion

The work of Hägerstrand provides the easiest approach technically, if not conceptually, to the study of chance on the landscape.[9] Diffusion of cultural traits has long been of major concern and the tracing of such movements through historical records, physical evidence and plausible inference is a part of the stock in trade. Yet clearly there must always be a greater or lesser element of chance in these movements which does not yield to such methods. How far is the areal spread of agricultural innovation a matter of early chance contacts preserved in the patterns of acceptance? It was Hägerstrand who conceived the probabilistic basis of such transfers of information and the manner of analysis by simulation. For him, the contorted surface of the real world could not be represented sufficiently accurately by an abstract plane so that he eschewed mathematical formulation. Instead, he provided an initial source of information and then, assuming frequencies of contacts between people by distance were described by a normal curve, he drew randomly from this series to allow transfers of information to be made. Each person contacted a sufficient number of times became an acceptor and then a further source as the process developed. Thus patterns of spread could be built up which looked like the historical record emphasising that, in this case at least, personal contact was the dominant link rather than persuasion by the mass media. A number of studies have pursued this method of analysis although nowadays the convenience of a digital computer is commonly sought[10] rather than Hägerstrand's paper and pencil. Topics have been central places in Sweden[11], urban growth[12] and liquid propane gas tanks[13] among others.

The main problem connected with this approach is the necessary comparison of the results of the simulation with the real world distribution it is designed to reproduce. Since the latter has in effect only a probabilistic existence, the results of a simulation will certainly not match it. How then are we to say that they are sufficiently alike that the rules we have used in the simulation may be reasonably taken as approximating the processes of the real world? It is here where the virtues of an analytic formulation are evident.

The most sophisticated analytic model of a stochastic diffusion process devised by a geographer is Culling's theory of erosion.[14] He refers to humid conditions, a permeable rock and relatively gentle slopes so that it is the efficiency of transport of eroded material via soil creep which is the limiting factor in denudation. The movements of a soil particle are regarded as independently and randomly directed so that application of the central limit theorem assures that the resulting probability distribution of its location after a number of individual displacements will be three-dimensionally normal about its original position. Gravity and gravitational soil moisture would provide a downward bias to this motion but because of the lower barrier to movement provided by the underlying solid rock, pore space would decrease in this direction so limiting displacements and thus removing the bias. Thus a steady state vertical distribution of particles will develop, probabilistically defined and representing a density increase and pore space decrease with depth. Density layering will be parallel to the surface and if there is no slope, no mean motion would occur although there would be a diffusion of 'marked' particles. Where there is slope, an extremely slow mean motion of particles will occur depending on the density gradient and tending to replace layering parallel to the surface with horizontal layering.

Culling has a river at the foot of the slope eroding its banks and rapidly transporting the material away so creating 'pore spaces'. In technical terms, an absorptive boundary for particles exists. Probabilities for upslope and downslope movement are now no longer equal, there being a bias in the general downslope direction. An ingenious feature of the argument is that the movement of pore spaces upslope from the river is considered; this is an inverse process to that of soil particles and follows the same laws. The course of development of various landform features is followed for given boundary conditions using mathematical results obtained in heat conduction studies.

With both Hägerstrand and Culling there is a genuine attempt to describe the small-scale process leading to large-scale results. The 'rules of the game' in simulation and the random displacements of the analytic model are open to empirical checking and perhaps modification in a way that aggregative deterministic models cannot match. Above all, we may perceive the large-scale ordering only after appreciating indeterminacy at the small scale.

For completeness, a word should be said about atmospheric diffusion since those geographers concerned with the heat and moisture balance at the earth's surface have had longest acquaintance with random displacements, here in terms of turbulence theory. However, so far as is known, no geographers have contributed to its development.

Development

Probabilistic reasoning has a new viewpoint to offer to the study of historical process in geography. If the development of landscape be a series of contingent events then in a certain sense the study of the historical record cannot reveal many of the generalities of process. Of course the development of particular places requires a particularistic historical account without the need for generality of concept. The ford, the shrine, the castle, the coalfield, the protestant ethic, the sea, appear in correct sequence to help explain Newcastle upon Tyne, and another set of factors are used to explain another town. Mr. Morris happened to be repairing bicycles at Oxford. While true and necessary there is a lack of aesthetic satisfaction about such explanations. We would certainly feel a study of land-form development inadequate which showed how, stone by stone and particle by particle, the surface of one date became that of another.

By contingency we mean that action 1 occurred but 2 might have happened just as easily; because of 1, 3 resulted rather than 4 which would have followed 2 and so on. The odd numbers describe real history and the even a possible history which did not happen. If we cannot specify this latter sequence it is ridiculous to discuss it. However, if we classify events sufficiently broadly, events which were previously regarded as different in kind now become only differences of degree. When we can specify the probabilities of occurrence of these degrees we can write a stochastic process of which the real history described above will be only one possible sample. However, because of the generality of classification adopted, it is possible to describe the histories of a thousand towns in the same terms and these represent a thousand samples. It is then possible to say something general about all these histories – they represent a thousand realizations of the common stochastic process. Clearly this is a valid procedure; whether it is a useful one depends on the question one asks of the landscape.

Why has the American manufacturing belt retained its coherence throughout the growth of the economy? How do such regions develop and retain their differential character? Momentum and inertia in this context are vaguely suggestive words rather than attempts to describe relevant processes. Take a city, representing all cities, in which the spin of a coin determines whether a manufacturing or a 'service' employee will be added to the payroll. The second employee is determined by a second toss being summed with the first and so on. We have a specialization process in which, while each increment has a random aspect, the future is being affected by the past to the extent that the past is summed up in the present, no matter how the present was arrived at. The probability distribution of the ratio of manufacturing to total employees one arrives at after a number of tosses is the U-shaped arc-sine curve implying that cities will tend to be specialized in manufacturing or service with relatively few of mixed type. One can add the same type of process going on in space with adjacent cities instead of periods affecting each other so that the same probability distribution is obtained. If any uninterrupted series of manufacturing cities is called a manufacturing zone, even if there be only one city, an extremely small number of boundaries between the two types of zones can occur. We thus obtain a marked and stable regionalization. Changes are most likely at the bounds of the zones and there is a finite probability of the whole structure changing but this is what we expect from the real world.[15]

In the previous example cities distributed in space were regarded as the sampling points of a single process. We may equally well regard time periods as samples of a process; the time perspective thus obtained is not of history unrolling but rather of a collection of periods existing simultaneously, including our own single sample. Take, for example, a stand of trees having distinct age groups visible: presumably during the life span of such trees there have been only a few occasions when natural regeneration has been allowed. We may thus speak of a probability of successful regeneration during the life-span of a tree and determine probabilities of intervals of various lengths between regenerations. In many areas there will be finite probabilities that the intervals will be longer than the lifetime of a tree plus the period of viability of seeds so that the trees will die out. With an unchanging mean climate we may nevertheless expect that vegetation will change, indeed be experiencing constant change particularly at the limits of vegetation types. It is pertinent then to regard this plant cover and all plant covers as existing at some probability level. It exists because our recent period obtained a particular sample from the climatic record whereas another but unfavourable sample might have occurred just as readily. A view of the landscape as a continuously fluctuating panorama even without exogenous intervention is thus obtained.[16] The same type of argument can be applied to climate itself. In the vegetation example, the periodicity of change is clearly related to the life-span of the trees which constitutes the length of 'memory' of the system. In the atmosphere-earth system the oceans provide an enormous memory by storing heat: because of the stirring of the water in depth very considerable quantities can be stored there with very little overall change in temperature; furthermore, these layers are stably stratified so that the heat could be held indefinitely.[17] Now, since it is inconceivable that there should be an exact balance of heat exchange between earth and atmosphere in any one year, the oceans can be regarded as gaining heat one year and losing it another via a random process. The probabilities of the relative amounts stored and the periodicities of the fluctuations may be calculated.[18] It turns out that these are of the order of those of the ice ages. We must view climate as a very wide spectrum of conditions with any particular time period spanning a relatively narrow bandwidth of conditions. Ice ages are a normal feature of the general circulation and will, with high probability, occur again. It can be helpful to say that they occur now in a region with a certain probability.

This approach can be applied to many landscape features. The landforms of the North African desert show clearly that present forms require recognition of an earlier wet phase.[19] But with the present concept of climate this should be taken as a normal state of affairs. To couch explanation in historical terms as a sequence of events may be useful but it is equally pertinent to analyse the landforms as mirroring a considerable part of the climatic spectrum.

Steady state

A feature of probabilistic processes which has been used already but not yet discussed is the notion of the steady state. Many such processes do achieve this condition. It is one without relation to any initial conditions, with fluctuations occurring constantly but with the probabilities of the various states being independent of time. Thus, for a system in a steady state there is no need to study its history in order to understand its form or the processes going on. Such an approach does not fit well the traditional trilogy of genesis, form and function as the approaches to geographical knowledge.

This approach has been expounded most clearly in relation to landforms.[20] Genesis as secular change or history is represented by the Davisian approach which Chorley represents as closed system analysis in which the potential energy due to elevation (i.e. endogenous to the system) is being degraded until, barring upsets, a peneplain is achieved. However, where the interplay of present processes and forms is of concern and much shorter time periods are relevant, the open system viewpoint is adopted. Here a continual exogenous supply of atmospheric energy, particularly the potential energy of rain, is supplied to the land and attention is focused on the present-day equilibrium of process and form. Leopold and Langbein have shown by probabilistic arguments that an exponential form of the thalweg of rivers will result as a steady state condition: the rivers referred to are those flowing through arid areas and without tributaries.[21]

Decisions and behaviour

Uncertainty is a basic fact of life for both individuals and groups of men. It matters little operationally whether this uncertainty be inherent indeterminacy or simply reflects ignorance of deterministic sets of events. Uncertainty is particularly important in connection with the future so that actions predicated on future conditions may be understood as the making of decisions within the range of possible futures. A retrospective view of society is less concerned with the decision problem since the passage of time is in a sense solving the probabilistic equations which described the future. However, it is reasonable to believe that the human landscape must to some extent mirror and be a manifestation of this uncertainty. The natural landscape must express the same principle. Three methods of adaptation appear possible although there may be others.[22] The passing of inputs through storage devices will reduce their variance while passing them through a delay device will alter their phase and allow a conforming reaction. Finally the maintenance of a varied structure will allow adaptation to changed circumstances. The first two methods have been described for a dairy farm[23] and a central place system,[24] but other examples spring to mind.

The transition from individual men making decisions to groups of men which are the main concern of human geography is not an easy one and a number of methods have been used. The most obvious approach is to postulate a 'representative man' and examine the choice matrix of his decisions; since goals must also be specified, the optimizing man that economic theory has provided has been used. Thus, for example, given the probabilities of herbage production as estimated from daily variations in weather throughout the year, how does the livestock farmer arrange his seasonal farm management programme to use these fluctuating uncertain supplies?[25] Note that this is an extremum problem in management of resources rather than in localization. It would be absurd to believe that economic man is really representative of any society but equally absurd to deny that considerable insights can be obtained by using him. He is extremely difficult to replace. Kates, for example, in a probabilistically stated problem of flood plain occupance has used the notion of 'satisficing' man – one who goes only some way toward economic rationality.[26] However, this is more of a descriptive term for the man discovered by behaviourist studies than a concept having analytic power. Indeed it is not clear whether we cannot retain the convenience of economic man and account for non-optimality by using the perceived environment rather than the objectively described environment. That is, actions may be optimal in terms of the perception of flood hazard but not in terms of the statistical forecast of floods. The relation of perception and economic calculus in uncertainty conditions will be taken up shortly.

Wolpert, although using a representative economic man, has handled the uncertainty facing him rather differently.[27] He regards risk as fluctuation on an unchanging environment, in this case that of the Swedish farmer and obtains a deterministic 'rational' allocation of these fixed resources using linear programming methods. A 'risk' component is then added to explain real world departures from calculated optimality; degree of technical knowledge obtained from a Hägerstrand diffusion model is also used. It is unlikely that this formulation has general application although the use of a 'certainty' equivalent for probabilistic situations does appear possible.

The other main line of attack to group decisions, i.e. to group behaviour conceived as a decision problem, is a 'summation man'. This 'group man' represents a convolution of the situations of many individuals and the group actions he performs have fewer constraints on them than were each individual considered individually. He thus appears more like an urn scheme of the mathematical probabilist since his choices at this more general level have a considerable random component. Some examples should make this idea clearer. Consider an area of variable rainfall in which the several crops have their yields differentially affected by differing rain totals. What proportions of land should be planted to the various crops to ensure that the minimum amount of food in any year should be as large as possible? Gould has asked this question as an approach to the understanding of land utilization in Central Ghana.[28] This situation, while having obvious affinities with the livestock farmer mentioned earlier, does not need to delve into the level of the individual farm because the problem implies a greater level of aggregation. The 'group' approach is predicated here on some egalitarian sharing either of crops for consumption or of land types for individual use. The formal procedures used for choosing an optimum combination of crops are those of the theory of games – here the village is adopting a strategy against the possible plays of nature.

Marble has used a 'summation man' in a study of traffic in a city.[29] His data consisted of a number of travel diaries kept by sample families in which all trips, including those having more than one purpose and thus allowing stopovers, were recorded. One possible tabulation of this data is on the basis of type, i.e. function, by origin and destination. A table was prepared showing for all origins that when a trip starts from, say, the work-place, the frequencies of destinations are, say, home 80 per cent, shopping 8 per cent, and so on. Although these frequencies are obtained from the summations of individuals trips, they may be regarded as the choice matrix of the summation man. By treating the frequencies as transition probabilities between states (i.e. home, shop, etc.) and by regarding home as an absorbing state (i.e. people do get home eventually) it is possible to conceive the travel pattern as generated by one summation man performing a random walk. The reason for doing this is that, by manipulation of the matrix, it is possible to arrive at a steady state condition describing the overall travel pattern although only the individual segments of many trips were provided as data. With sufficient data, locations may also be included in the analysis. It has been seen here how the summation man is acting quite randomly, constrained only by a single probability matrix – he is an unthinking urn scheme. Yet he adequately represents the sum of the highly purposeful and specialized individual travel behaviour.

Although not concerned with statistical methods here, in discussing decision-making under uncertainty it would be remiss not mention briefly the analytic role of subjective probability in the perception-decision problem. Classical statistics is not geared to making probabilistic statements about future conditions and certainly not adapted to guiding rational decisions on the basis of subjectively held beliefs of the future course of events. Yet people in their adaptation to environment act in accordance with their beliefs about the world rather than an objectively described world. They amend their beliefs with experience as a learning process and weight them with the benefits and costs to be obtained from acting on particular beliefs; some best course of action can thus be obtained. Such a scheme is a reasonable representation of the perception-decision problem: it happens to be that provided by Bayesian statistics.[30]

A revealing example may be taken from the physical planning field: assume engineering activity is to be decided on the basis of a forecast of the population of a city. The usual forecasting procedure is to manipulate the data of past populations and come up with a single value of population for a future date for all the world as if this was an academic exercise in astronomy. Given such a total, roads are built, sewers are laid and so on. Now, of course, it is absurd to act as though we know with certainty how big the city will be; it is equally foolish, although less obviously so, to regard a forecast on which decisions are made with attendant costs and benefits as a problem in pure science. We have some ideas about the city's future size but there is also uncertainty – a probability distribution is the best we can hope for. Let us keep gathering evidence and amending our beliefs; let us calculate the effects of these outcomes, keeping in mind their probability of occurrence. This seems to be a reasonable representation in formal terms of the way people operate; that it is not so for planners is due to the inappropriate statistics they use.

Conclusion

An academic discipline presumably justifies itself insofar as its individual members contribute to answering the questions we ask of nature. With regard to any particular problem or set of problems the approach of a geographer may coincide with that of a student from a neighbouring discipline. But there seems to be the obligation for geography to phrase individual explanations of events in the landscape so that these may be structured into an overall articulated view of ordering in space. Obviously no common discourse is possible for all topics studied by geographers but the formal language of probability theory does hold the greatest hope of establishing a structure of explanation to which the various parts of geography can contribute and from which they can draw. Communication can only be at a very general level but at least some co-operative building and cross-fertilization can occur.

It has been seen that there has been no real break in analytic concept throughout this essay. From the representative individual to group decisions and group behaviour, from conscious man with his choice matrix to the movement of rock particles there is no discontinuity. Very large gaps do occur in the areas of application, however, and thus all assertions as to the inclusiveness of our mathematics need to be qualified. One gap which does not seem to hold promise is that of genetic climatology. It is an unfortunate fact of history, as Leighly has pointed out,[31] that studies of the causes and of the consequences of climate have tended to be organised in separate and often antagonistic camps. The statistical description of climate has generally been pursued by the functional school so that there has been little attempt to relate these statistical studies of form with the dynamics of the atmosphere. Yet there must be complete complementarity in nature allowing pursuit of the statistics back into the dynamics and forward into the landscape. Nevertheless it can be argued that statistics are but a pale reflection of the real course of events and that the meteorologist using the perfectly general deterministic equations of physics plus boundary and initial conditions can perform his integrations and produce a closer copy of the sequence of nature. It is conceivable that a model could produce hour to hour changes all over the world and, running for the computer equivalent of years on end, accumulate climatic statistics. Such a model would imply greatly expanded understanding of the physical processes involved and since the statistical measures are but secondary rather than primary facts they do not justify separate treatment.

But does such a model allow understanding of why the statistical measures of, say, rainfall differ between stations? Here too a rigorous, quantitative, physically well-grounded theory of the atmosphere is needed. But the aim is to account for a collective rather than write a history. These statistics are primary facts as parameters of landscape and their spatial distribution should be explained directly. There are other reasons for direct probabilistic modelling of the atmosphere but that of coherence has much merit. Of course, this is not to say that such a theoretical formulation is possible: only that it does appear desirable.

To conceive the landscape as composed of elements having characteristic spectra of fluctuations in time and of differentiation in space, to depict society as an organization for facing the future and choosing among a variety of possibilities, to stress movement and development, to place uncertainty in the centre of our analysis instead of ignoring it as regrettable: all this must be welcome. That the format of the language of probabilities is equally germane to discussing both physical and human events, both spatial and temporal ordering, both retrospective and anticipatory explanations of actions, both individual and collective activities, the physical world both as exclusive of man and as the environment of man, bespeaks its versatility. A common mode of discourse thus presents itself for scientific study of the landscape, promising to promote an articulated coherent viewpoint.

References

1   Reprinted from *Northern Geographical Essays: in honour of G. Daysh,* Department of Geography, University of Newcastle upon Tyne, U.K. (1967).

2   LUKERMANN, F. 'The "calcul des probabilités" and the école française de géographie.' *Canadian Geographer,* Vol. 9 (1965), pp. 128–38.

3   DACEY, M. F. 'A compound probability law for a pattern more dispersed than random and with areal inhomogeneity.' *Economic Geography,* Vol. 42 (1966), pp. 172–9.

   DACEY, M. F. 'Modified Poisson probability law for point pattern more regular than random.' *Annals of the Association of American Geographers,* Vol. 54, 4 (1964), pp. 559–65.

   DACEY, M. F. 'Two dimensional random point patterns: a review and an interpretation.' *Papers – Regional Science Association,* Vol. 13 (1964), pp. 41–55.

   DACEY, M. F. 'Imperfections in the uniform plane.' *Michigan Inter-University Community of Mathematical Geographers,* Vol. 4 (1964).

   DACEY, M. F. 'The spacing of river towns.' *Annals of the Association of American Geographers,* Vol. 50 (1960), pp. 59–61.

   DACEY, M. F. 'Analysis of central place and point patterns by a nearest neighbour method.' *Proceedings of International Geographical Union Symposium on Urban Studies,* Lund (1960), Lund Studies in Geography, B, 24, pp. 55–76.

   DACEY, M. F. 'A note on the derivation of nearest neighbor distances.' *Journal of Regional Science,* Vol. 2 (1960), pp. 81–7.

4   COURT, A. 'Some new statistical techniques in geophysics.' *Advances in Geophysics,* Vol. 1, New York: Academic Press (1952), pp. 45–85.

5   DACEY, M. F. 'Description of linear patterns,' Garrison, W. L. and Marble, D. F. (Eds.), *Quantitative Geography,* Northwestern Studies in Geography, Chicago (1967).

6   LONGUET-HIGGINS, M. S. 'The statistical analysis of a random moving surface.' *Philosophical Transactions of the Royal Society.* A., Vol. 249, pp. 321–87.

7   CURRY, L. 'The random spatial economy: an exploration in settlement theory.' *Annals of the Association of American Geographers,* Vol. 54 (1964), pp. 138–46.

8   SHREEVE, R. L. 'Statistical law of stream numbers,' *Journal of Geology,* Vol. 74 (1966), pp. 17–37.

9   HÄGERSTRAND, T. 'The propagation of innovation waves.' *Lund Studies in Geography,* B, Vol. 4 (1952).

10 PITTS, F. R. 'Problems in computer simulation of diffusion.' *Papers – Regional Science Association,* Vol. 9 (1962).

11 MORRILL, R. L. 'The development of spatial distributions of towns in Sweden: an historical-predictive approach.' *Annals of the Association of American Geographers,* Vol. 53 (1963), pp. 1–14.

   MORRILL, R. L. 'Simulation of central place patterns over time.' *Proceedings, International Geographical Union Symposium in Urban Geography,* Lund (1960), Lund Studies in Geography, Series B, Vol. 24, Lund: Gleerup (1962), pp. 109–20.

12 GARRISON, W. L. 'Toward simulation models of urban growth and development,' *Proceedings, International Geographical Union Symposium in Urban*

Geography, Lund (1960), Lund Studies in Geography, Series B, Lund: Gleerups (1962), pp. 91–108.

13 BROWN, L. 'The diffusion of innovation: a Markov chain-type approach.' *Discussion Paper No. 3, Department of Geography, Northwestern University, Evanston, Ill.* (1964).

14 CULLING, W. E. H. 'Theory of erosion on soil-covered slopes.' *Journal of Geology*, Vol. 73 (1965), pp. 230–54.
   CULLING, W. E. H. 'Soil creep and the development of hillside slopes,' *Journal of Geography*, Vol. 71 (1963), pp. 127–61.

15 CURRY, L., op. cit. (1964).

16 CURRY, L., 'Climatic change as a random series.' *Annals of the Association of American Geographers*, Vol. 52, 1 (1962), pp. 21–31.

17 ROSSBY, C. G. 'Current problems in meteorology.' *The atmosphere and the sea in motion: Rossby memorial volume*, Bolin, B. (Ed.), New York (1959), pp. 9–50.

18 CURRY, L., op. cit. (1962).

19 CAPOT-REY, R. 'Recherches récentes et tendances nouvelles en morphologie désertique.' *La géographie française au milieu du XXe siecle*, Paris (1957), pp. 43–52.

20 CHORLEY, R. J. 'Geomorphology and general systems theory.' Theoretical papers in the hydrologic and geomorphic sciences, *Geological Survey Professional Paper* 500-B., Washington, D.C. (1962).

21 LEOPOLD, L. B. and LANGBEIN, W. B. 'The concept of entropy in landscape evolution.' *Geological Survey Professional Paper* 500-B (1962).

22 CURRY, L. 'Central places in the random spatial economy' (unpublished paper).

23 CURRY, L. 'Regional variation in the seasonal programming of livestock farms in New Zealand.' *Economic Geography*, Vol. 39 (1963), pp. 96–118.
   CURRY, L. 'Canterbury's grassland climate.' *Proceedings of the Second New Zealand Geography Conference* (1958).

24 CURRY, L. 'The geography of service centres within towns: the elements of an operational approach.' *IGU Symposium in Urban Geography*, Lund (1960), Lund Studies in Geography, B, Vol. 24, pp. 31–53.

25 CURRY, L. 'The climatic resources of intensive grassland farming: the Waikato, New Zealand.' *Geographical Review*, Vol. 52 (1962), pp. 174–94.

26 KATES, R. W. 'Hazard and choice perception in flood plain management.' *Department of Geography Research Paper* **78**, University of Chicago (1962).

27 WOLPERT, J. 'The decision process in spatial context.' *Annals of the Association of American Geographers*, Vol. 54, 4 (1964), pp. 537–58.

28 GOULD, P. R. 'Man against his environment: a game theoretic framework.' *Annals of the Association of American Geographers*, Vol. 53, 3 (1963), pp. 290–7.

29 MARBLE, D. 'Simple Markovian model of trip structure in a metropolitan region.' *Proceedings of the Western Section Regional Science Association*, Tempe, Arizona (1964).

30 CURRY, L. 'Seasonal programming and Bayesian assessment of atmospheric resources.' Human Dimensions of Weather Modification, Sewell, W. R. D. (Ed.), *Department of Geography Research Paper* **105**, University of Chicago (1966).

31 LEIGHLY, J. B. 'Climatology,' in *American Geography, Inventory and Prospect*, James, P. E. and Jones, C. F. (Eds.), Syracuse (1954).

17   Spatial theory and human behavior[1]

Gunnar Olsson and Stephen Gale
University of Michigan, U.S.A.

A significant number of spatial analysts have recently shifted their emphasis away from economic theories of location to a more behavioristic approach. It is the purpose of this paper to review this development and to suggest a general framework within which an integrated theory of human spatial behavior can be developed. More specifically, the first part of the paper will isolate the major behavioristic and spatial assumptions in classical location theory; the second section will deal with some recent attempts to relax the most limiting of these assumptions. Since a large part of this new work has employed probability models, some comments will be directed toward the concept of stochastic model-building in general. In the last part of the paper, this discussion will be extended to comments on q-way probability matrices and their use in multidimensional Markovian models.

Economics and spatial theory

It has been argued that the basic problem in economics is to study how limited material resources become allocated among different users. This allocation process can be viewed as a game between producers and consumers in which economic man behaves in a predictable manner. More precisely, the consumer is assumed to follow a strategy that maximizes his utility function, while the producer attempts to maximize his profit. In this simplified economic game, both players operate within given constraints, such as budget, demand, technology, etc., and each is further restrained by the moves of his opponent. After a period of mutual adjustment, both will find that they cannot improve their respective positions and an equilibrium situation has been reached.[2] The allocation problem can be analyzed with tools other than game theory and equilibrium models, but most available techniques are in fact based on the same premises. This indicates that Katona's and Simon's characterization of classical economics is still basically valid and that most existing theories are normative.[3, 4]

Since most spatial theories are extensions of economic theories, traditional work in regional science and theoretical geography has been directed toward the same normative, optimizing constructs as have the studies in economics. The modifications of classical economics by Lösch, Isard, and others have led to a relaxation of the presumption of a nondimensional economic world but they have not affected any of the normative-behavioristic assumptions. As a consequence, the players in the spatial economic game are blessed with the attributes of economic man, which makes it meaningful to apply the same equilibrium approach as in nonspatial economics. The extension of the classical theory into a spatial context implies, however, that the players can influence the outcome of the game not only by manipulating utility and production functions, or supply and demand curves, but also by manipulating space. In a *ceteris paribus* situation, this indicates that the producer tries to increase his profits by enlarging the size of his market area, while the consumer tries to minimize his purchasing costs by forcing the producer to come as close to him as possible.

It was in this *areal* sense that von Thünen and Lösch treated space in their models. Lösch arrived at his spatial solution by combining the notion of the demand cone with a set of special areal equilibrium conditions which stated that no parts of the total area can be left unserved, and that supply, production, and sales areas must be as small as possible.[5] Isard later extended this into a more general formulation embracing both Lösch's and von Thünen's derivations as special cases.[6] This was made possible by the use of a substitution framework in which the notion of transportation inputs played a major role. As a consequence, Isard was able not only to fuse existing location theories into one model but also to incorporate pertinent parts of production and distribution theory.

In addition to the studies of space in the areal sense, there exists a large number of quasi-spatial models in which space has been simplified and collapsed into a set of points. Some interregional equilibrium models are good examples of this approach[7] as are the transportation variant of the linear programming model, most gravity formulations, and the application of graph theory to spatial problems.[8] In other cases, such as general interregional linear programming and input-output analysis, space has been included in the form of point regions simply by expanding the traditional models;[9] symbolically. This means that new subscripts and superscripts have been introduced and the traditional equations have been repeated for each region.

ASSUMPTIONS

It has already been noted that most spatial theories are based on the same behavioristic assumptions as the nondimensional theory of the firm. Cyert and March[10] (p. 8 ff) have suggested that these premises can be reduced to two rationality assumptions; first, that firms consistently seek to maximize profits and, second, that they possess perfect knowledge. It is only if these assumptions are fulfilled that traditional spatial theory is valid, and it is only then that it is meaningful to employ it in the computation of optimality and equilibrium solutions.

In spatial theories, the maximization of profits has usually been assumed to result from the minimization of transportation costs or, in the simplified case, from the minimization of physical distance. In this respect, locational optimality can be related to the social physics principle of least effort, which Bunge recently restated as a nearness principle by which he means that interacting objects tend to place themselves as closely together as possible.[11] In effect, Bunge and the social physicists have drawn analogies with the fundamental optimality principles in physics, and, in this respect, their approach is similar to some biological studies[12] and to general systems theory in particular.[13]

It is superfluous to discuss in any detail how the goal of maximizing profits has been criticized in economics. It is enough to recall that particular attention has been drawn to the importance of non-economic motives in the decision-making process of the firm. As a result, it has often been suggested that the optimizing function should include noneconomic factors and that the objective, or purely economic, maximizing function be replaced by a subjective one. For the spatial analyst, similar reasoning implies that minimization of physical distance is only a theoretical notion intended to simplify the formulation of abstract concepts. Actors in the real world realize of course that minimizing functional or subjective distance is often more appropriate than putting interacting objects as closely together as possible in the physical sense. The use of subjective distance functions in spatial models is closely related to the use of subjective utility and profit functions in non-dimensional economics, and, in this regard, subjective distance could be one of the variables entered into those functions. A seemingly non-optimal location decision or a biased interaction field, for example, may appear optimal for the individual concerned simply because he understands how to specify the relevant subjective distance and maximizing functions.

The notion of subjective distance functions is also relevant to an understanding of Cyert and March's second proposition, i.e., that the classical rationality assumption presupposes perfect knowledge on the part of the decision-makers. What appears as an optimal location or interaction pattern for the individual may thus do so simply because he is not aware of all the possibilities. For the most part, this unawareness relates either to insufficient information channels or to the decision-maker's subjective filtering of the information he receives. This suggests a close relationship between interaction fields and information flows of the type discussed in several migration studies.[14–18] Pred has treated urban and industrial growth in similar terms and suggested that spatial economic growth is governed by a circular and cumulative process by which the creation and dissemination of inventions is facilitated.[19] Such processes create tendencies toward selected growth, and these are further enhanced by biases in the flow of information. Thorngren[20] and Wärneryd[21] have analyzed the Swedish economy in related terms and found that the mechanisms of spatial growth are best understood if existing spatial theories are complemented with the concepts of organization theory. Pred[22] has recently extended his idea on circular and cumulative causation into a detailed and excellent discussion of the behavioristic assumptions in location theory – re-emphasizing the fact that spatial growth, to a large extent, reflects the spatial patterns of information flows.

As already noted, the data submitted through biased information channels are further distorted by each individual's perception or subjective filtering of the information he receives. On a large scale, Gould[23] has demonstrated that such distortions of information content can influence people's ranking of different countries, and Wolpert[24] has noted how the perception of a social and physical environment affects an individual's decision to migrate. Finally, a number of related studies have treated the perception of natural hazards as a particular aspect of the physical environment.[25-29]

These brief comments on the perception and flow of information indicate how the subjective evaluation of spatial alternatives makes the traditional behavioristic assumptions extremely unrealistic. In addition, spatial theories and models often involve specifically spatial premises that are equally questionable. Thus, most areal spatial models depend on the notion of a homogeneous plain over which the actors in the economic game move themselves and their activities. In the case of the interregional point models, on the other hand, it is not always clear which spatial assumptions have been made; presumably, however, they all involve some measure of areal dispersion and homogeneity which is then used as a basis for determining the center of each region. If this is done on a functional basis, each region may be imagined as a punctiformally generated demand cone of the Lösch-type or as a Clark-type city[30] in which sales, interaction intensities, and population densities decrease with increasing distance from the center. Gurevich and Saushkin[31] have discussed this problem and have demonstrated that Clark's model can be extended to take non-homogeneities into account. Their contribution is important since reality is only remotely similar to the theoretical isotropic plain – a recognition that has caused many empirical geographers to refute practically all spatial theory. One of the few constructive suggestions in this dialogue between empiricists and theoreticians has been provided by Tobler[32] in his work on mathematical functions by means of which complex realities can be transformed into homogeneous plains. Conceptually, his approach is related to that of von Thünen, who deformed the theoretical pattern of concentric land-use zones by introducing a river on which goods could be transported more cheaply than over the land.

RECENT APPROACHES TO SPATIAL MODEL-BUILDING

The discussion thus far has recapitulated and criticized the basic assumption in classical spatial theories. There is general agreement that these assumptions are highly unrealistic, but there has been little work on the specification of another more rational set of premises. Furthermore, the few attempts along these lines have been primarily verbal rather than symbolic and critical rather than constructive. One reason for this may be that knowledge about spatial behavior is far from adequate. Isard and Dacey[33] came to similar conclusions in their study of individual behavior in which they pointed out that an individual's choice among alternatives varies widely with his attitudes. By restating the assumptions of rational behavior in axiomatic form they were able to isolate the few instances in which spatial behavior can comply with theoretical expectations. In situations where the knowledge is so scant, it is meaningless to employ highly simplified deterministic models with predefined cause and effect relationships. Some economists have therefore suggested that the classical *homo economicus* be replaced by a *homo stochasticus*, and it is obvious that a concomitant modification of spatial theory is highly desirable. As noted by Curry[34] in a discussion of spatial probability models, an important advantage of the probabilistic approach is precisely this explicit admission of the researcher's ignorance.

Although there are great conceptual differences between deterministic and stochastic approaches, it often happens that the outcomes from the two types of formulation are the same. As a consequence, the insights into a process provided by one model can help in the interpretation and development of another logically more attractive one. For the spatial analyst, it is particularly fortunate that this applies to one of his most widely used models. The basically deterministic gravity formulation can thus be related to several existing stochastic models of human interaction, and it can, in fact, itself be transformed into a model of that type.[35] Typically, these interaction models consider two sets of probabilities, one set specifying the number of opportunities already reached and accepted by a trip-maker, and the second set specifying the proportion of trip-makers going beyond a certain distance to reach an acceptable opportunity. Consequently, the probabilities themselves relate human behavior to the spatial unevenness of opportunities and therefore indirectly to the spatial homogeneity assumption.

The probabilities in some cell-counting models can be interpreted in essentially the same manner. Medvedkov,[36] for example, has employed the entropy concept of information theory as a technique for quantifying the agreement between empirical and theoretical settlement distributions. More specifically, he suggested that Christaller's hexagonal central place pattern be regarded as the signal, and the random noise be attributed to deviations from the isotropic conditions. Medvedkov thus viewed the spatial distribution of places as a result of a set of 'regulated accidents' in which chance or randomness has affected the outcome from the otherwise deterministic central place theory. A similar approach has been taken by Dacey[37-41] who has conceived of urban places as points which topography or institutional factors have dislocated from their theoretical positions on the hexagonal lattice. A variant of the cell-counting technique was used to verify that the distribution of places in almost homogeneous areas follows a double Poisson distribution, although the locations in inhomogeneous areas are better described by a double negative binomial. These and other distribution functions have recently been discussed by Olsson.[42] Hudson[43] has used them in his analysis of rural settlement and Rogers[44] has employed them in studies of shops within cities. Finally, Harvey[45] has furnished an excellent discussion of cell count analysis in diffusion studies, pointing out, in particular, how the technique offers a way in which time and space can be treated in the same model.

Even though the relationships between distribution functions and physical urn models make it possible to draw conclusions about generating processes, the great value of cell-counting analysis is as a sophisticated technique for spatial description. Implicit in this approach to description is the idea that the real world represents only one of many possible realizations of a specified stochastic process. If this realization has had a small likelihood of occurrence, it in itself becomes difficult to understand without reference to the underlying probability matrix from which chance happened to pick an unusually small figure. What may initially appear to be a unique case can thus be fitted into more general constructs. This demonstrates that even basically descriptive probability models lend themselves to causal inference by which large scale aggregate regularities can be related to small scale processes. Employing a stochastic approach, one may therefore determine the degree to which reality deviates from the homogeneous plain and also, in theoretical terms, account for this deviation. If the focus is on the behavior of real world people, as compared with the behavior assumed in traditional spatial theory, the possibilities for such causal inferences become indispensable.

An interesting and somewhat different attempt along these lines has been made by Curry,[46] who claims that the time factor is the key to understanding the spatial distribution of central place activities. In particular, he has suggested that the spatial pattern of service outlets is a result of consumer behavior and its changes over time. Employing a set of Poisson distributions with the mean varying between different goods, Curry used this approach to determine not only the number of purchases per week but also the size of various service areas. Most importantly, he later extended these arguments into a general theory of consumer and retailer behavior[47] in which interesting analogies were drawn from turbulence and communications theories. As a consequence, the use of spectral analysis was suggested as an appropriate technique both for determining the convertability between the time and space domains and for detecting noise due to uncertainties faced by consumers and suppliers.

Spectral analysis offers a new and challenging approach to traditional spatial theory, primarily because it provides a sophisticated technique for describing and analyzing an agglomerated spatial economy in terms of its harmonics. The generation of such an economy can be related to the flow of information as Pred has proposed,[19, 22] but it may also be understood in terms of the stochastic interplay between the forces of distance and specialization as suggested by Curry.[48] More specifically, the generation of an undulating spatial economic surface can be related to the idea of contingency in probability theory; this, by definition, means that one particular outcome of a stochastic process can change the probability for similar events occurring in spatial and temporal proximity. In one of Curry's examples,[48] such an accumulating process gave rise to high industrial and areal specialization as in the American manufacturing belt; in another case,[49] it provided an explanation for the occurrence of ice ages.

Suggested framework for spatial behavioral analysis

The models reviewed in the preceding section are all promising and attractive, even though Curry's explicit recognition of chance and uncertainty makes his approach more appealing than most of the others. No matter how thought-provoking a particular formulation may be, however, it is always worth recalling that several quite different models sometimes give rise to almost identical results.[50] Attention will therefore be drawn to a somewhat different, but very general framework within which spatial theory and human behavior may be studied.

Keeping in mind that precise knowledge about spatial behavior is lacking, a deterministic approach seems indiscriminate. Further, recalling that human behavior is affected by a large number of variables, it appears appropriate to consider explicitly the complexity or multidimensionality of the underlying data. Finally, a model of behavioral processes should treat spatial and temporal sequences concomitantly.

TRADITIONAL MARKOV MODELS

If these are the three central criteria which the ideal model of spatial behavior must meet, then it is possible that a Markovian formulation can be extended to provide a meaningful analytic framework. To facilitate a more detailed elaboration of this point the basic properties of Markovian models will be briefly summarized.

The general Markov process[51-53] is a sequential model characterized by a set of state conditions, S, where

$$(1) \qquad S = \{s_1, \ldots, s_m, \ldots\}$$

and by a set of sequence conditions, T, where

$$(2) \qquad T = \{t_1, \ldots, t_n, \ldots\}.$$

In addition, the Markov process embodies a probability criterion, termed the Markov property, defined as

$$(3) \qquad P(t_n = s_i | t_{n-1}, \ldots, t_{n-j} = s) = P(t_n = s_1 | t_{n-1} = s_j)$$

which says that the probability that any state s_i occurs on sequence t_n is independent of all other states in the sequence except that one which occurred on t_{n-1}. In operational terms, the Markov process is a sequence of matrix operations of the form

$$(4) \qquad {}^nP \cdot {}^nV = {}^{n+1}V'$$

where P is a matrix of transition probabilities, and V is a vector of observations. The superscript in (4) represents the element t_n from the set of sequences

T, where time is the implicit variable in the model. The cells of the matrix P, $\{p_{ij}\}$, are the probabilities of moving from state i to state j. Functionally the $\{p_{ij}\}$ can be denoted as

$$(5) \qquad {}^{n+1}p_{ij} = f({}^{n}p_{ij}, {}^{n}v_j, n)$$

which means that the probability matrix in a Markov process is related to the value of the probability matrix at the preceding sequence, the observed state v_j at the preceding sequence, and the place of that particular sequence in the set T. Given a process in a particular state s_i at sequence t_n, no additional information is therefore needed to characterize completely the distribution of the process at any time $t > t_n$.

Whereas the general class of Markov process models has not found wide application in the study of social processes, Markov chains have been used to a considerable extent. Since the Markov chain forms a subset of the Markov process, it is characterized by the same basic properties; the only difference is that a Markov chain assumes that the probability that any state s_i occurs on a particular sequence t is independent of the position of that sequence in the set T. The condition for a Markov chain can therefore be written as

$$(6) \qquad P(t_n = s_i | t_{n-1}, \ldots, t_{n-j} = s) = P(t_m = s_i | t_{m-1}, \ldots, t_{m-j} = s)$$

which implies that a Markov chain is a Markov process with the added condition of stationarity. Operationally, this means that the matrix P is invariant over time such that

$$(7) \qquad {}^{n}P_{ij} = {}^{0}P_{ij}$$

where ${}^{0}P_{ij}$ represents the initial value of the transition probability matrix at t_0. Since each combination of V and P thus results in a unique Markov chain, simple matrix operations can be used to determine the steady state or equilibrium conditions for the model.

There exists a number of applications of Markovian models to spatial studies. Brown[54] for instance, employed a chain model to describe the spread of liquid propane tanks, and Rogers[44] used a similar formulation in his analysis of interregional migration flows. However, migration behavior does not usually conform to the stationarity assumption in the Markov chain model, and some writers have consequently suggested that Markov process models be used instead.[55, 56] As the transition probability matrices in the latter can change over time, they are well suited for handling problems involving search procedures and learning processes of the type encountered not only in migration but also in marketing behavior.[57, 58] In addition, Markovian models have been applied in a number of other spatial studies[59-62] with subjects ranging from city size distributions to the movement of points in two dimensions.

A VARIATION

The brief summary of Markovian models and their use in spatial research indicates at least three points worthy of further attention. Firstly, the traditional models employ ordinary matrix structures which limit the number of analyzed variables to one. Secondly, the ordinary rules for matrix multiplication imply a linearity assumption such that the element j in ^{n+1}V in (4) is equally dependent on all of the entries of the ith row of ^{n}P and the jth column of ^{n}V. Thirdly, the traditional model is limited in its temporal scope because it specifies the probability of any state s_i occurring on sequence t_n to be independent of all states in the sequence except the immediately preceding one.

It is almost a truism that models are used in scientific investigations because they limit the number of variables employed in the analysis. With respect to spatial behavior, however, present knowledge is too scant to suggest which of the many possible variables should be employed. To limit oneself to the use of the one (or sometimes two) variables whose interactions can be portrayed in ordinary matrix structures would therefore imply severely limiting *ceteris paribus* assumptions. The answer to this problem comes in a logical extension of the matrix form such that the ordinary two-dimensional matrix

$$
p = \begin{matrix} p_{11} & \cdots & p_{1m_2} \\ \vdots & & \vdots \\ p_{m_1 1} & \cdots & p_{m_1 m_2} \end{matrix}
$$

be expanded into three dimensions, where

$$
P' = \begin{matrix} & p_{11m_3} & \cdots & p_{1m_2m_3} \\ & & \ddots & & \ddots \\ p_{111} & \cdots & p_{1m_21} & \\ \vdots & & \vdots & \\ & p_{m_1 1m_3} & \cdots & p_{m_1 m_2 m_3} \\ & & & \\ p_{m_1 11} & \cdots & p_{m_1 m_2 1} \end{matrix}
$$

or, in the more general case, into a matrix of q dimensions. Following Oldenburger's notation,[63] such a structure can be called a q-way matrix represented as

(8)
$$
P'' = \{ p_{m_1, \; m_2, \; \cdots, \; m_q} \},
$$

i.e., as a vector space with any number of axes, q, where q is a finite positive integer.

To appreciate the significance of q-way matrices, it may be helpful to recall how Clark[59] tested the usefulness of Markov chain models as a tool for analyzing rental housing distributions in urban areas. As noted by Clark himself, the limitations of the two-dimensional transition probability matrix allowed him to trace only the movement of a tract within the rental classes; the spatial pattern of these movements could not be considered explicitly. This means that the traditional Markov chain approach enabled him to derive a gross measure of change in one variable at the expense of neglecting the associated spatial process. Map analysis could shed light on his problem.

but it could be treated more directly by means of a four-way matrix, allowing the indexing not only of the probabilities of moving within rental classes but also within space.

To turn to the second point raised above – i.e., the operator in Markov models – it should be recalled that ordinary matrix multiplication can be functionally denoted as

(9) $$V' = f(V,P)$$

where f is a linear transformation specified as

(10) $$v'_{ij} = \sum_k p_{ik} \cdot v_{kj}.$$

Historically, this operator was derived for the solution of sets of simultaneous linear equations, but its inherent property of linearity could become a limiting factor when applied to models of social processes. As seen from (9) and (10), the function f is solely responsible for the linearity of the transformation; the matrices V, P, V' serve merely as indexing or representational mechanisms. The obvious generalization of the sequential operator therefore involves a redefinition of f into a function f', which can be either linear or nonlinear, continuous or discontinuous.

Much work has yet to be done on the specification of the function f', but a similar approach has recently been used by Tobler,[64] who suggested a two-way linear transformational model of the form

(11) $$V' = P \cdot V \cdot P^T$$

where P^T is the transpose of the transition probability matrix P. Since the multiplication in Tobler's case is still defined as a linear operator, the premultiplication of V by P and the postmultiplication of V by P operates respectively on the rows and the columns of V. Thus V' is a linear combination of both the associated rows and columns of the probability matrix, such that

(12) $$v''_{ij} = \sum_k p_{ik} v_{kj}$$
$$v'_{ij} = \sum_k v''_{ik} p^T_{kj}$$

where V'' is employed solely for computational simplicity. Tobler has also discussed the relationship between (12) and the idea of a local operator by which the interaction between juxtaposed elements can be established. For the two-way matrix, the local operator can be defined as

(13) $$v'_{ij} = f[(v_{i-1,j-1}, v_{i-1,j}, v_{i-1,j+1}, v_{i,j-1}, v_{i,j}, v_{i,j+1}, v_{i+1,j-1}, v_{i+1,-j},$$
$$v_{i+1,j+1}), (p_{i-1,j-1}, p_{i-1,j}, p_{j-1,j+1}, p_{i,j-1}, p_{i,j}, p_{i,j+1},$$
$$p_{i+1,j-1}, p_{i+1,j}, p_{i+1,j+1})]$$

from which generalization to the case of q-way matrices follows immediately.

The concept of a local operator is closely related to the notion of neighborhood effects in spatial interaction and diffusion. It seems possible, therefore, that detailed knowledge of systematic variations in the b-value of the gravity model may help to determine precisely multiplication rules that would yield the desired spatial contingency effects. Presently, it is less clear how the operator should be defined for those variables having no spatial dimension.

Finally, to turn to the third point raised above, it should be recalled that the transition probability matrices are usually determined from the formulation

$$(14) \qquad P = [p_{ij}] = \frac{a_{ij}}{\sum\limits_{k} a_{ij}}$$

where the $\{a_{ij}\}$ are the observed frequencies of transitions from state i to state j. The results of this calculation apply to ordinary matrix structures, but, once again, they can easily be extended to q-way matrices, although the interpretation of each cell of the probability matrix is then altered; in the two-way matrix, each p_{ij} represents the many-to-one mapping of a complex process, while the values in the q-way matrix specify simple conditional probabilities. In both cases, however, the probability statements are empirical and consequently conceived of in an empirical or frequentist manner. In the case of behavioral modeling, such a conception could be somewhat limiting and it may be more attractive to employ the subjectivist view instead.

The specification of the matrix P also bears on another problem related to Markov models. Thus, the Markov property states that the outcome of each trial depends solely on the outcome of the immediately preceding one which makes the length of each trial period crucial. If it is desirable to keep the Markov property unchanged, the obvious solution to this difficulty is to determine the probabilities over short time periods. If, on the other hand, it is deemed more meaningful to change the Markov property itself, the transition probabilities can be made functions of any number of previous states, or symbolically

$$(15) \quad P(t_n = s_i | t_{n-1}, \ldots, t_{n-j} = s) = P(t_n = s_i | t_{n-1}, \ldots, t_{n-a} = s')$$

where a is any positive finite integer $\leq n$.

Summary and conclusions

The first part of this paper discussed the well-known relationships between classical economics and spatial theory, and it was noted that most spatially-oriented studies have been based on the same behavioristic assumptions as the non-dimensional theory of the firm. To overcome the limitations imposed by these assumptions, it was pointed out that subjective distance functions must be introduced as a complement to the economists' subjective profit-maximizing functions.

After this introduction, the paper turned to a discussion of the notion of subjective functions and the manner in which they might be entered into operational spatial models. Since the knowledge of spatial behavior is limited, it was argued that the use of simplified deterministic models with predefined cause and effect relationships would be meaningless. Therefore, as a preliminary to the search for more appropriate constructs, the use of probabilistic models in spatial research was reviewed. Particular attention was drawn to Curry's studies, in which stochastic process models have been employed as a device for relating observed large-scale regularities to small-scale behavior.

Although much still remains to be done, most of the models discussed in the second part are promising. Recalling, however, that two very different formulations may sometimes produce the same outcome, the third part of the paper draws attention to an alternative approach. More specifically, it was suggested that the usefulness of Markovian models in spatial research should be investigated in more detail, even though a subsequent review of existing studies indicated that traditional Markov formulations may impose several limitations. As an example, it was noted that the use of ordinary two-way matrix structures in effect limits the number of variables to one necessitating far-reaching *ceteris paribus* assumptions. It was suggested that the easiest way to avoid this difficulty was to extend the ordinary matrices into several dimensions. This, however, raised the question of how to define appropriate multiplication rules, and it was noted that the operator can be specified in a large number of ways from which the researcher can then choose those most relevant to his specific problem. With respect to spatial studies, it was argued that the condition of a linear sequential operator be modified in order to account for neighborhood and contingency effects.

The suggestions concerning q-way matrices and local operators are admittedly very general, and much work remains before they can be made operational and subsequently employed in testable models. In this respect, the discussion has been typical of the inductive stage in research. Experiments with a set of specific models of the optimality type have thus indicated the respects in which spatial theories are least satisfactory, but these experiments have not shown how existing models can be reformulated to take the noted deficiencies into account. In such situations, it is often necessary to sacrifice simplicity for accuracy and increase the number of variables until the mechanisms of the studied process are fully understood. Once this understanding has been reached, the number of variables may then be reduced to achieve the economy of thought which is a primary characteristic of a good model. It is possible that the notion of q-way matrices and multidimensional Markov formulations constitute a framework for the analysis of spatial behavior in general and for the derivation of simplified but realistic models in particular.

References

1   The support of the Swedish Council for Social Science Research is gratefully acknowledged. Reprinted from *Papers and Proceedings of the Regional Science Association,* Vol. 15 (1967), University of Pennsylvania, U.S.A.

2   KUENNE, ROBERT E. *The Theory of General Economic Equilibrium,* Princeton, New Jersey: Princeton University Press (1963).

3   KATONA, GEORGE *Psychological Analysis of Economic Behavior,* New York City: McGraw-Hill Book Company, Inc. (1951).

4   SIMON, HERBERT A. *Models of Man,* New York City: John Wiley & Sons, Inc. (1957).

5   LÖSCH, AUGUST *The Economics of Location,* Woglom-Stolper translation, New Haven, Connecticut: Yale University Press (1954).

6   ISARD, WALTER *Location and Space Economy,* New York City: John Wiley & Sons, Inc., (1956).

7   ISARD, WALTER and OSTROFF, DAVID 'General Interregional Equilibrium,' *Journal of Regional Science,* Vol. 2 (1960), pp. 67–74.

8   NYSTUEN, JOHN D. and DACEY, MICHAEL F. 'A Graph Theory Interpretation of Nodal Regions,' *Papers and Proceedings of the Regional Science Association,* Vol. 7 (1961), pp. 29–42.

9   ISARD, WALTER *Methods of Regional Analysis: An Introduction to Regional Science,* New York City: John Wiley & Sons, Inc. (1960).

10 CYERT, RICHARD M. and MARCH, JAMES G. *A Behavioral Theory of the Firm,* Englewood Cliffs, New Jersey: Prentice Hall, Inc. (1963).

11 BUNGE, WILLIAM *Theoretical Geography,* Lund: Gleerups Förlag (1966).

12 ROSEN, ROBERT *Optimality Principles in Biology,* London: Butterworths (1967).

13 VON BERTALANFFY, LUDWIG 'General System Theory: A Critical Review,' *General Systems,* Vol. 7 (1962), pp. 1–20.

14 HÄGERSTRAND, TORSTEN *Innovationsforloppet ur korologisk synpunkt,* Lund: Gleerups Förlag (1953).

15 HÄGERSTRAND, TORSTEN 'Migration and Area' in David Hannerberg *et. al.* (Eds.), *Migration in Sweden,* Lund: Gleerups Förlag (1957).

16 LÖVGREN, ESSE 'The Geographical Mobility of Labor,' *Geografiska Annaler,* Vol. 38 (1956), pp. 344–94.

17 NELSON, PHILLIP 'Migration, Real Income and Information,' *Journal of Regional Science,* Vol. 1 (1959), pp. 43–74.

18 WOLPERT, JULIAN 'Behavioral Aspects of the Decision to Migrate,' *Papers of the Regional Science Association,* Vol. 15 (1965), pp. 159–69.

19 PRED, ALLAN *The Spatial Dynamics of Urban-Industrial Growth 1800–1914: Interpretive and Theoretical Essays.* Cambridge, Massachusetts: The M.I.T. Press (1966).

20 THORNGREN, BERTIL 'Regional External Economics'. *Stockholm School of Economics:* The Economic Research Institute, mimeograph (1967).

21 WÄRNERYD, OLOF 'Urban Regions as Spatial Systems.' Paper presented at the second Polish-Scandinavian Regional Science Meeting, Copenhagen (1967).

22 PRED, ALLAN *Behavior and Location. Foundations for a Geographic and Dynamic Location Theory, Part I,* Lund: Gleerups Förlag (1967).

23 GOULD, PETER R. 'On Mental Maps,' *Michigan Inter-University Community of Mathematical Geographers,* Discussion Paper No. 9 (1966).

24 WOLPERT, JULIAN 'Migration as an Adjustment to Environmental Stress,' *Journal of Social Issues,* Vol. 22 (1966), pp. 92–102.

25 BURTON, IAN and KATES, ROBERT W. 'Perception of Natural Hazards in Resources Management,' *Natural Resources Journal,* Vol. 3 (1964), pp. 412–41.

26 KATES, ROBERT W. 'Hazard and Choice Perception in Flood Plain Management', *University of Chicago: Department of Geography,* Research Paper No. 78 (1962).

27 LOWENTHAL, DAVID (Ed.), 'Environmental Perception and Behavior', *University of Chicago: Department of Geography,* Research Paper No. 109 (1967).

28 SAARINEN, THOMAS F. 'Perception of the Drought Hazard on the Great Plains',

University of Chicago: Department of Geography, Research Paper No. 106 (1966).

29 WHITE, GILBERT F. 'Choice of Adjustment to Floods.' *University of Chicago: Department of Geography,* Research Paper No. 93 (1964).

30 CLARK, COLIN 'Urban Population Densities,' *Journal of the Royal Statistical Society,* Series A, Vol. 114, pp. 490–6.

31 GUREVICH, B. L. and SAUSKIN, YU G. 'The Mathematical Method in Geography,' *Soviet Geography,* Vol. 7 (1966), pp. 3–35.

32 TOBLER, WALDO R. 'Geographic Data and Map Projections,' *Geographical Review,* Vol. 53 (1963), pp. 59–78.

33 ISARD, WALTER and DACEY, MICHAEL F. 'On the Projection of Individual Behavior in Regional Analysis, I and II,' *Journal of Regional Science,* Vol. 4 (1962), pp. 51–96.

34 CURRY, LESLIE 'Chance and Landscape' in House, J. W. (Ed.), *Northern Geographical Essays in Honour of G. W. J. Daysh.* Newcastle upon Tyne University: Department of Geography Publications (1967).

35 OLSSON, GUNNAR 'Central Place Theory, Spatial Interaction, and Stochastic Processes,' *Papers of the Regional Science Association,* XVIII (1967), pp. 13–45.

36 MEDVEDKOV, YURIY V. 'The Regular Component in Settlement Patterns Shown on a Map,' *Soviet Geography,* Vol. 8 (1967), pp. 150–68.

37 DACEY, MICHAEL F. 'Modified Probability Law for Point Patterns More Regular than Random,' *Annals of the Association of American Geographers,* Vol. 54 (1964), pp. 559–65.

38 DACEY, MICHAEL F. 'Order Distance in an Inhomogeneous Random Point Pattern,' *Canadian Geographer,* Vol. 9 (1965), pp. 144–53.

39 DACEY, MICHAEL F. 'A Probability Model for Central Place Locations,' *Annals of the Association of American Geographers,* Vol. 56 (1966), pp. 549–68.

40 DACEY, MICHAEL F. 'A Compound Probability Law for a Pattern More Dispersed than Random and with Areal Inhomogeneity,' *Economic Geography,* Vol. 42 (1966), pp. 172–9.

41 DACEY, MICHAEL F. 'A County Seat Model for the Areal Pattern of an Urban System,' *Geographical Review,* Vol. 56 (1966), pp. 527–42.

42 OLSSON, GUNNAR 'Lokaliseringsteori och stokastiska processer,' in Tor Fr. Rasmussen (Ed.), *Forelesninger i regionale analysematoder,* Oslo: Norsk Institutt for By-og Regionforskning (1967).

43 HUDSON, JOHN G. 'Theoretical Settlement Location', unpublished doctoral dissertation, Department of Geography, University of Iowa (1967).

44 ROGERS, ANDREI 'A Markovian Policy Model of Interregional Migration,' *Papers of the Regional Science Association,* Vol. 17 (1966), pp. 205–24.

45 HARVEY, DAVID 'Geographic Processes and the Analysis of Point Patterns: Testing Models of Diffusion by Quadrat Sampling,' *Transactions of the Institute of British Geographers,* Vol. 40 (1966), pp. 81–95.

46 CURRY, LESLIE, 'The Geography of Service Centers within Towns: The Elements of an Operational Approach' in Knut Norborg (Ed.), *Proceedings of the IGU Symposium in Urban Geography, Lund* (1960), Lund: Gleerups Förlag (1962).

47 CURRY, LESLIE 'Central Places in the Random Spatial Economy,' *Journal of Regional Science,* Vol. 7 (Supplement) (1967), pp. 217–38.

48 CURRY, LESLIE 'The Random Economy: An Exploration in Settlement Theory,' *Annals of the Association of American Geographers,* Vol. 54 (1964), pp. 138–46.

49 CURRY, LESLIE 'Climatic Change as a Random Series,' *Annals of the Association of American Geographers,* Vol. 52 (1962), pp. 21–31.

50   CHORLEY, RICHARD J. 'Geography and Analogue Theory,' *Annals of the Association of American Geographers,* Vol. 54 (1964), pp. 127–37.

51   BHARUCHA-REID, A. T. *Elements of the Theory of Markov Processes and Their Applications,* New York City: McGraw-Hill Book Company, Inc. (1960).

52   KEMENY, JOHN G. and SNELL, J. LAURIE *Finite Markov Chains,* Princeton, New Jersey: D. Van Nostrand Company, Inc. (1960).

53   PARZEN, EMANUEL *Stochastic Processes,* San Francisco: Holden-Day (1962).

54   BROWN, LAWRENCE A. 'The Diffusion of Innovation: A Markov Chain-Type Approach.' Northwestern University: *Department of Geography, Discussion Paper* No. 8 (1963).

55   MCGINNIS, ROBERT, MYERS, GEORGE, C. and PILGER, JOHN E. 'Internal Migration as a Stochastic Process. Paper presented at the meeting of the International Statistical Institute, Ottawa, (August 1963).

56   MCGINNIS, ROBERT and PILGER, JOHN E. 'On a Model for Temporal Analysis,' Paper presented at the meeting of the American Sociological Association, Los Angeles (August 1963).

57   GOLLEDGE, REGINALD G. 'A Conceptual Framework of a Market Decision Process,' University of Iowa: *Department of Geography, Discussion Paper* No. 4 (1967).

58   HERNITER, JEROME D. and HOWARD, RONALD A. 'Stochastic Marketing Models' in Hertz, David B. (Ed.), *Progress in Operations Research,* Vol. 2, New York City: John Wiley & Sons, Inc. (1964).

59   CLARK, W. A. V. 'Markov Chain Analysis in Geography: An Application to the Movement of Rental Housing Areas,' *Annals of the Association of American Geographers,* Vol. 55 (1965), pp. 351–9.

60   FUGUITT, GLENN W. 'The Growth and Decline of Small Towns as a Probability Process,' *American Sociological Review,* Vol. 30 (1965), pp. 403–11.

61   HUDSON, JOHN C. *Maps and Spatial Processes Describable with Markov Chains,* University of North Dakota: Department of Geography, mimeograph (1966).

62   MARBLE, DUANE F. 'Simple Markovian Model of Trip Structure in a Metropolitan Region,' *Proceedings of the Western Section of the Regional Science Association* (1964).

63   OLDENBURGER, R. 'Composition and Rank of *n*-way Matrices and Multilinear Forms,' *Annals of Mathematics,* Vol. 25 (July 1934), pp. 622–57.

64   TOBLER, WALDO R. 'Of Maps and Matrices,' *Journal of Regional Science,* Vol. 7, No. 2 (Supplement) (1967), pp. 275–80.

Selected Bibliography for further reading

FELLER, WILLIAM *An Introduction to Probability Theory and Its Applications, Volume I,* New York City: John Wiley & Sons Inc. (1957).

ROGERS, ANDREI 'A Stochastic Analysis of the Spatial Clustering of Retail Establishments,' *Journal of the American Statistical Association,* Vol. 60 (1965), pp. 1094–1103.

18 Behavioral aspects of the decision to migrate[1]

Julian Wolpert
University of Pennsylvania, U.S.A.

During the decade 1950–60, there were sufficient changes from previous patterns of migration streams in the United States to warrant some re-examination and re-evaluation of model-building attempts in migration analysis. It must be admitted that the gravity model and its elaborations appear to lose explanatory power with each successive census. When flows are disaggregated, the need becomes greater selectively to determine unique weights for areas and unique distance functions for subgroups of in- and out-migrants. The Stouffer model of 'competing migrants'[2] provided a rather poor prediction of migration streams for the 1955–60 period. Perhaps the most successful of spatial interaction models, which does take into consideration the spatial arrangement of places of origin and destination, is sufficiently rooted in the 1935–40 depression-period movements as to present serious deficiencies when applied to recent streams. Plots of migration distances defy the persistence of the most tenacious of curve fitters.

The defenders of the wage theory of economic determinism find some validity for their constructs, so long as net, and not gross, migration figures are used and regional disaggregation does not proceed below the state level, thereby neglecting much of the intrastate heterogeneity.[3–7]

The extremely scanty empirical evidence of the 'friends and relatives effect' in directing migration has given birth to a generation of models which, although offering the solace of a behavioral approach, provide little explanation of the actual process involved.[8, 9] Perhaps the most serious gap occurs in the transition from micro- to macro-model and in the selection of appropriate surrogates for testing. Here, the inadequacy of published data in the United States appears to have its most telling effect. Though almost every conceivable method of combining existing data into useful indicators has been tried, explanation through surrogates hardly provides an analysis which is independent of the bias which is introduced.

A good deal of useful information has come from the analysis of migration differentials by categories of occupation, income, race, and, especially, age,[4, 10–12] However, predictive models have not been designed to include these findings and to consider the interdependence of these characteristics in migration behavior. Demonstrating the potential usefulness of the migration differential approach is one of the objectives of this paper.

A composite of interesting ideas about migration behavior has been incorporated within Price's ambitious proposed simulation model.[13] On the basis of selected characteristics of individuals and of places of origin and destination, migration probabilities are generated reflecting empirically observed regularities. As far as is known, the model has not become operational – the task for simulating United States migration would overtax the most modern computer. The only successful attempt in this direction has been Morrill's study of the emerging town development in south central Sweden.[14]

The use of Monte Carlo simulation models in migration analysis does offer a viable and promising approach, especially considering the rather persistent tendencies for critical elements or parameters to remain stable over time. Thus, although the streams show considerable variation over time, and the characteristics of the population and of places continuously change, stability persists in migration behavior.

To illustrate this observation it may be noted that Bogue, Shryock, and Hoermann,[4] in their analysis of the 1935–40 migration streams, summarize with the following statements that could as well be applied to the 1955–60 streams:

1   Basic shifts in the regional and territorial balance of the economy guided the direction and flow of migration streams.
2   The two factors that seem to contribute most to the mobility of the population are above average educational training and employment in white collar occupations.
3   Any theory of economic determinism in migration is inclined to be incomplete.

It appears, therefore, that understanding and prediction of migration streams require determining of the constants in migration behavior and distinguishing these from the variables with respect to population composition and place characteristics which evolve differentially over time.

As indicated, attempts at model-building in migration research have largely focused on variables and surrogates such as distance and ecological characteristics of places exerting 'push and pull' forces[4, 15] to the exclusion of behavioral parameters of the migrants. The model suggested here is of doubtful usefulness as an exact predictive tool. It borrows much of its concepts and terminology from the behavioral theorists, because of the intuitive relevance of their findings to the analysis of mobility. Verification will be only partial because of the general absence in this country of migrational histories. Instead, greater reliance will be placed upon evidence from a variety of sources and special studies. The framework of the analysis must be classified as descriptive or behavioral and partially dynamic.

Clearly, the focus must remain with the process of internal migration, i.e., a change of residence which extends beyond a territorial boundary. Some attempt will be made, however, to relate this process of 'long distance' movement to the more general topic of mobility which encompasses not only shifts within areal divisions but also movement between jobs and social categories. This larger zone of investigation is referred to as the 'mover-stayer' problem.[16]

Central concepts of migration behavior

The central concepts of migration behavior with which we shall be concerned are: 1. the notion of place utility, 2. the field theory approach to search behavior, and 3. the life-cycle approach to threshold formation.

Before translating these concepts into an operational format within a proposed model, some attempt will be made to trace their relevance to migrational decisions.

PLACE UTILITY

Population migration is an expression of interaction over space but differs in certain essential characteristics from other channels of interaction, mainly in terms of the commodity which is being transported. Other flows, such as those of mail, goods, telephone calls, and capital also reflect connectivity between places, but, in migration, the agent which is being transported is itself active and generates its own flow. The origin and destination points take on significance only in the framework in which they are perceived by the active agents.

A degree of disengagement and upheaval is associated with population movements; thus, households are not as readily mobile as other phenomena subject to flow behavior. Yet it would be unrealistic to assume that sedentariness reflects an equilibrium position for a population. Migrational flows are always present, but normally the reaction is lagged and the decision to migrate is non-programmed. Thus, migration is viewed as a form of individual or group adaptation to perceived changes in environment, a recognition of marginality with respect to a stationary position, and a flow reflecting an appraisal by a potential migrant of his present site as opposed to a number of other potential sites. Other forms of adaptation are perhaps more common than change of residence and job. The individual may adjust to the changing conditions at his site and postpone, perhaps permanently, the decision to migrate. Migration is not, therefore, merely a direct response or reaction to the objective economic circumstances which might be incorporated, for example, within a normative transportation model.

In designing the framework for a model of the migration decision, it would be useful at the outset to enumerate certain basic descriptive principles which have been observed to have some general applicability and regularity in decision behavior. To a significant degree these principles have their origin in the studies of organizational theorists.

We begin with the concept of 'intendedly rational' man[17] who, although limited to finite ability to perceive, calculate, and predict and to an otherwise imperfect knowledge of environment, still differentiates between alternative courses of action according to their relative utility or expected utility. Man responds to the perception of unequal utility, i.e., if utility is measured broadly enough to encompass the friction of adaptation and change.

The individual has a threshold of net utility or an aspiration level that adjusts itself on the basis of experience.[17-22] This subjectively determined threshold is a weighted composite of a set of yardsticks for achievement in the specific realms in which he participates. His contributions, or inputs, into the economic and social systems in terms of effort, time, and concern are rewarded by actual and expected attainments. The threshold functions as an evaluative mechanism for distinguishing, in a binary sense, between success or failure, or between positive or negative net utilities. The process is self-adjusting because aspirations tend to adjust to the attainable. Satisfaction leads to slack which may induce a lower level of attainment.[18] Dissatisfaction acts as a stimulus to search behavior.

Without too great a degree of artificiality, these concepts of 'bounded rationality'[17] may be transferred to the mover-stayer decision environment and a spatial context. It is necessary only to introduce a place subscript for the measures of utility. *Place utility,* then, refers to the net composite of utilities which are derived from the individual's integration at some position in space. The threshold reference point is also a relevant criterion for evaluating the individual's place utility. According to the model, the threshold will be some function of his experience or attainments at a particular place and the attainments of his peers. Thus, place utility may be expressed as a positive or negative quantity, expressing respectively the individual's satisfaction or dissatisfaction with respect to that place. He derives a measure of utility from the past or expected future rewards at his stationary position.

Quite different is the utility associated with the other points which are considered as potential destinations. The utility with respect to these alternative sites consists largely of anticipated utility and optimism which lacks the re-enforcement of past rewards. This is precisely why the stream of information which is so important in long-distance migration – information about prospects must somehow compensate for the absence of personal experience.

All moves are purposeful, for an evaluation process has preceded them, but some are more beneficial in an *ex post* sense, because of the objective quality of search behavior, the completeness of the information stream and the mating of anticipated with realized utility. If migrations may be classified as either successes or failures in a relative sense, then clearly the efficiency of the search process and the ability to forecast accurately the consequences of the move are essential elements.

Assuming intendedly rational behavior, then the generation of population migration may be considered to be the result of a decision process which aims at altering the future in some way and which recognizes differences in utility associated with different places. The individual will tend to locate himself at a place whose characteristics possess or promise a relatively higher level of utility than in other places which are conspicuous to him. Thus, the flow of population reflects a subjective place-utility evaluation by individuals. Streams of migration may not be expected to be optimal because of incomplete knowledge and relocation lag but neither may we expect that individuals purposefully move in response to the prospect of lower expected utility.

The process of migration is conceived in the model as: 1. proceeding from sets of stimuli perceived with varying degrees of imperfection, and 2. involving responses in a stayer-mover framework.

The stayers are considered to be lagged movers postponing the decision to migrate for periods of time extending up to an entire lifetime. Thus, the mover-stayer dichotomy may be reduced to the single dimension of time – when to move.

Distinction must clearly be made between the objective stimuli which are relevant for the mover-stayer decision and the stimuli which are perceived by individuals and to which there is some reaction. The stimuli which are instrumental in generating response originate in the individual's action space which is that part of the limited environment with which the individual has contact.[19] Thus, the perceived state of the environment is the action space within which individuals select to remain or, on the other hand, from which to withdraw in exchange for a modified environment.

FIELD THEORY APPROACH TO SEARCH BEHAVIOR

Though the individual theoretically has access to a very broad environmental range of local, regional, national, and international information coverage, typically only some rather limited portion of the environment is relevant and applicable for his decision behavior. This immediate subjective environment or action space is the set of place utilities which the individual perceives and to which he responds. This notion of the action space is similar to Lewin's concept of life space – the universe of space and time in which the person conceives that he can or might move about.[19] Some correspondence may exist with the actual external environment, but there may also be a radical degree of deviation. The life space is a surface over which the organism can locomote and is dependent upon the needs, drives, or goals of the organism and upon its perceptual apparatus.[23] Our concern is with man in terms of his efficiency or effectiveness as an information collecting and assimilating organism and thus with his ability to produce an efficient and unbiased estimation or evaluation of the objective environment. It is suggested that the subjective action space is perceived by the individual through a sampling process whose parameters are determined by the individual's needs, drives, and abilities. There may not be a conscious and formal sampling design in operation, but, nevertheless, a sampling process is inherently involved in man's acquisition of knowledge about his environment.

Both sampling and non-sampling errors may be expected in the individual's perceived action space – a spatial bias induced by man's greater degree of expected contact and interaction in his more immediate environment, as well as sampling errors introduced because of man's finite ability to perceive and his limited exposure and observation. The simple organism which Simon describes has vision which permits it to see, at any moment, a circular portion of the surface about the point in which it is standing and to distinguish merely between the presence or absence of food within the circle.[23]

The degree to which the individual's action-space accurately represents the physically objective world in its totality is a variable function of characteristics of both man and the variability of the environment. Of primary emphasis here are the consequences of man's fixity to a specific location – the spatial particularism of the action space to which he responds.

What is conspicuous to the individual at any given time includes primarily information about elements in his close proximity. Representing the information bits as points, the resulting sampling design most closely resembles a cluster in the immediate vicinity of the stationary position. The individual may be considered at the stationary position within the cluster of alternative places, each of which may be represented by a point on a plane. The consequences of this clustered distribution of alternatives within the immediate vicinity of the individual is a spatially biased information set, or a mover-stayer decision based upon knowledge of only a small portion of the plane.

Cluster sampling may be expected to exhibit significantly greater sampling bias for a given number of observations than random sampling; its most important advantage is in the reduction of the effort or cost in the collection of information. In the absence of a homogeneous surface, however, the difference in cost may be more than outweighed by the loss in representativeness of a given cluster.

The local environment of the individual may not, of course, be confined purely to his immediate surroundings. The action space may vary in terms of number and intensity of contacts from the limited environmental realm of the infant to the extensive action space within which diplomats, for example, operate. The degree of contact may perhaps be measured by the rate of receipt or perception of information bits.[24] Mass communications and travel, communication with friends and relatives, for example, integrate the individual into a more comprehensive spatial setting but one which is, nevertheless, still biased spatially. Mass communications media typically have coverage which is limited to the service area of the media's transmission center. Here a hierarchy of nodal centers exists in terms of the extent of service area and range of coverage. Thus the amount of transmission and expected perception of information by individuals is some function of the relative position of places within the network of communication channels. The resident in the area of a primary node has an additional advantage resulting from his greater exposure to information covering a relatively more extensive area of choice. His range of contact and interaction is broader, and the likelihood of an unbiased and representative action space is greater.

THE LIFE CYCLE APPROACH TO THRESHOLD FORMATION

Another significant determinant of the nature and extent of the individual's action space (i.e. the number and arrangement of points in the cluster) consists of a set of factors which may be grouped under the heading of the 'life cycle.' Illustrative of this approach is Hägerstrand's analysis of population as a flow through a system of stations.[25] Lifelines represent individuals moving between stations. The cycle of life almost inevitably gives rise to distinct movement behavior from birth, education, and search for a niche involving prime or replacement movements. Richard Meier also has examined this notion of the expanding action space of the individual from birth through maturity.[26, 24] The action space expands as a function of information input – and growth depends on organization of the environment so that exploration becomes more efficient. Associated with the evolution of the individual's action space through time is a complex of other institutional and social forces which introduce early differentiation. Differences in sex, race, formal education, family income, and status are likely to find their expression early in shaping the area of movement and choice. Although the action space is unique for each individual, still there is likely to be a good deal of convergence into a limited number of broad classes. The congruity and interdependence of the effects of race, family income, education, and occupation are likely to result in subgroups of individuals with rather homogeneous action spaces.

In Lewin's concept, behavior is a function of the life space, which in turn is a function of the person and the environment.[19] The behavior-influencing aspects of the external (physical and social) environment are represented through the life space. Similarly, but in a more limited fashion, the action space may be considered to include the range of choice or the individual's area of movement which is defined by both his personal attributes and environment. Most prominent among the determinants of the alternatives in this action space which are conspicuous to the individual is his position on one of divergent life cycles and location in terms of the communication networks linking his position to other places. His accumulated needs, drives, and abilities define his aspirations – the communication channels carry information about the alternative ways of satisfying these aspirations. To illustrate this structure in terms of the simple organism, we may turn to Simon's model of adaptive behavior.[23] The organism he describes has only the simple needs of food-getting and resting. The third kind of activity of which it is capable is exploration for food by locomotion within the life space where heaps of food are located at scattered points. In the schema, exploration and adaptive response to clues are necessary for survival; random behavior leads to extinction. The chances of survival, i.e., the ability to satisfy needs are dependent upon two parameters describing the organism (its storage capacity and its range of vision) and two parameters describing the environment (its richness in food and in paths). Of course, with respect to the human organism, aspirations require the fulfilment of many needs, and thresholds are higher. Exploratory search is aided by clues provided by the external environment through communication channels which extend the range of vision.

OTHER BEHAVIORAL PARAMETERS

The discussion was intended to develop the concept of action space as a spatial parameter in the mover-stayer decision. Thus the action space of the individual includes not only his present position but a finite number of alternative sites which are made conspicuous to him through a combination of his search effort and the transmission of communications. The action space refers, in our mover-stayer framework, to a set of places for which expected utilities have been defined by the individual. A utility is attached to his own place and relatively higher or lower utility has been assigned to the alternative sites. The variables here are the absolute number of alternative sites and their spatial pattern or arrangement with respect to his site. The sites may consist of alternative dwellings within a single block, alternative suburbs in a metropolitan area, or alternative metropolitan areas. The alternatives may not all present themselves simultaneously but may appear sequentially over time.

There are other components of behavioral theories which are relevant in the analysis of migration, especially with respect to the problem of uncertainty avoidance. We have already mentioned the sequential attention to goals and the sequential consideration of alternatives.[18] The order in which the environment is searched determines to a substantial extent the decisions that will be made. In addition, observations appear to confirm that alternatives which *minimize uncertainty* are preferred and that the decision maker *negotiates for an environment of relative certainty*.[18] Evidence shows also that there is a tendency to *postpone decisions* and to rely upon the *feedback of information*, i.e., policies are reactive rather than anticipatory.[18] Uncertainty is also reduced by imitating the successful procedures followed by others.[18]

The composite of these attempts to reduce uncertainty may be reflected in a lagged response. A lapse of time intervenes in a cause and effect relationship – an instantaneous human response may not be expected. As with other stimulus-response models, events are paired sequentially through a process of observation and inference into actions and reactions, e.g., unemployment and out-migration. As developed in economics, a lag implies a delayed, but rational, human response to an external event. Similarly, with respect to migration, responses may be measured in terms of elasticity which is in turn conditional upon factors such as complementarity and substitutability. A time dimension may be added to measures of elasticity, and the result is a specific or a distributed lag – a response surface reflecting the need for re-enforcement of the perception of the permanance of change.

Framework of a proposed operating model

The model which is proposed attempts to translate into an operational framework the central concepts with which we have been concerned: the notion of place utility, the field theory approach to search behavior, and the life cycle approach to threshold formation.

The model is designed to relate aggregate behavior in terms of migration differentials into measures of place utility relevant for individuals. The objective is a prediction of the composition of in- and out-migrants and their choice of destination, i.e., by incorporating the stable elements which are involved in the changes in composition of population of places.

Inputs into the system are the following set of matrices:

1   Matrix A, defining the migration differentials associated with the division of the population by life cycles and by age, represented respectively by the rows and columns.

2   Matrix B, representing the distribution of a place's population within the life cycle and age categories.

3   Matrices C, D, E, and F, representing respectively the gross in-, out-, and net-migration and 'migration efficiency'[27] for each of the cell categories corresponding to Matrix B.

The rates for the A matrix are determined on an aggregate basis for the United States population by means of the 'one in a thousand' 1960 census sample. These rates are then applied to the B matrix entries for specific places to predict the expected out-migration rates of profile groups at these places. The differences between the expected rates and those observed in the C, D, E, and F matrix tabulations are then used to provide a measure of the relative utility of specific places for the given profile groups which may be specified as a place utility matrix. The net migrations, whether positive or negative for the given cell, represent the consensus of cell members of the utility which the place offers relative to other places which they perceive. The migration efficiency measures not only the relative transitoriness of specific subgroups of the population but also the role of the specific place as a transitional stepping stone or station for certain groups.

There is an additional matrix, Matrix G, representing the parameters of search behavior which are characteristic of the subgroup populations. These are specified in terms of the number of alternatives which are perceived and the degree of clustering of these alternatives in space. The destination of the out-migrants predicted by means of the G matrix entries are tested against the observed migration flows in order to derive measures of distance and directional bias.

The concepts of place utility, life cycle, and search behavior are integrated, therefore, within the classification of the population into subgroups. Preliminary testing has revealed a significant degree of homogeneity of migrational behavior by subgroup populations in terms of differential rate of migration, distance, and direction of movement. The classification procedure, involving the use of multivariate analysis, is designed to provide a set of profile or core groups whose attributes may be represented by prototype individuals. The differential migration rates of Matrix A are assumed, therefore, to be parameters in the migration system, at least for the purposes of short-term forecasting. Individuals move along each row as they grow older and, to some extent, move in either direction along age columns as socio-economic status changes over time, but the migration rates for the cells remain relatively constant.

Similarly, the utility to the population subgroups of the specific places of origin and destination shift over the long-term but remain relatively constant in the short-run. For long-term forecasting, exogenous measures of economic trends in specific places would be necessary inputs.

References

1   The support of the Population Council and of the Regional Science Research Institute is gratefully acknowledged.
     Reprinted from *Papers and Proceedings of the Regional Science Association,* Vol. 15 (1965), University of Pennsylvania, U.S.A.

2   STOUFFER, SAMUEL A. 'Intervening Opportunities and Competing Migrants,' *Journal of Regional Science,* Vol. 2 (1960), pp. 1–26.

3   BLANCO, CICELY 'Prospective Unemployment and Interstate Population Movements,' *Review of Economics and Statistics,* Vol. 46 (1964), pp. 221–2.

4   BOGUE, DONALD J., SHRYOCK, HENRY S., HOERMANN, SIEGFRIED, 'Streams of Migration between Subregions,' *Scripps Foundation Studies in Population Distribution* (1957), No. 5, Oxford, Ohio.

5   BUNTING, ROBERT L. 'A Test of the Theory of Geographic Mobility,' *Industrial and Labor Relations Review,* Vol. 15 (1961), pp. 76–82.

6   RAIMON, ROBERT L. 'Interstate Migration and Wage Theory,' *Review of Economics and Statistics,* Vol. 44 (1962), pp. 428–38.

7   SJAASTAD, LARRY A. 'The Relationship Between Migration and Income in the United States,' *Papers of the Regional Science Association,* Vol. 6 (1960), pp. 37–64.

8   KERR, CLARK 'Migration to the Seattle Labor Market Area, 1940–42,' *University of Washington Publications in the Social Sciences,* 11 (1942), pp. 129–88.

9   NELSON, PHILLIP, 'Migration, Real Income and Information,' *Journal of Regional Science,* Vol. 1 (1959), pp. 43–74.

10   ELDRIDGE, HOPE T. and THOMAS, DOROTHY SWAINE, *Population Redistribution and Economic Growth, United States 1870–1950,* Philadelphia: American Philosophical Society (1964).

11   THOMAS, DOROTHY S. 'Age and Economic Differentials in Internal Migration in the United States: Structure and Distance,' *Proceedings, International Population Conference,* Vienna (1959), pp. 714–21.

12   WILBER, GEORGE L. 'Migration Expectancy in the United States,' *Journal of the American Statistical Association,* Vol. 58 (1963), pp. 444–53.

13   PRICE, D. O. 'A Mathematical Model of Migration Suitable for Simulation on an Electronic Computer,' *Proceedings, International Population Conference,* Vienna (1959), pp. 665–73.

14 MORRILL, RICHARD L. 'The Development of Models of Migration,' *Entretiens de Monaco en Sciences Humaines* (1962).

15 BURFORD, ROGER L. 'An Index of Distance as Related to Internal Migration,' *Southern Economic Journal*, Vol. 29 (1962), pp. 77–81.

16 GOODMAN, LEO 'Statistical Methods for the Mover-Stayer Model,' *Journal of the American Statistical Association*, Vol. 56 (1961), pp. 841–68.

17 SIMON, HERBERT A. 'Economics and Psychology,' in Koch, Sigmund (Ed.), *Psychology: A Study of a Science*, 6, New York: McGraw-Hill (1963).

18 CYERT, R. M. and MARSH, J. G. *A Behavioral Theory of the Firm*, Englewood Cliffs, N.J.: Prentice-Hall (1963).

19 LEWIN, KURT *Field Theory in Social Science*, New York: Harper and Row (1951).

20 MCGUIRE, JOSEPH W. *Theories of Business Behavior*, Englewood Cliffs, N.J.: Prentice-Hall (1964).

21 SIEGEL, S. 'Level of Aspiration and Decision-Making,' *Psychological Review*, Vol. 64 (1957), pp. 253–63.

22 STARBUCK, WILLIAM H. 'Level of Aspiration Theory and Economic Behavior,' *Behavioral Science*, Vol. 8 (1963), pp. 128–36.

23 SIMON, HERBERT A. 'Rational Choice and the Structure of the Environment,' *Psychological Review*, Vol. 63 (1956), pp. 129–38.

24 MEIER, RICHARD L. *A Communications Theory of Urban Growth*, Cambridge: Massachusetts Institute of Technology Press (1962).

25 HÄGERSTRAND, TORSTEN 'Geographical Measurements of Migration,' *Entretiens de Monaco en Sciences Humaines* (1962).

26 MEIER, RICHARD L. 'Measuring Social and Cultural Change in Urban Regions, *Journal of the American Institute of Planners*, Vol. 25 (1959), pp. 180–90.

27 SHRYOCK, H. S. Jr. 'The Efficiency of Internal Migration in the United States,' *Proceedings, International Population Conference*, Vienna (1959), pp. 685–94. (Migrational efficiency refers to the ratio of net migration to total gross migration.)

Selected bibliography for further reading

AJO, REINO 'New Aspects of Geographic and Social Patterns of Net Migration Rate,' *Svensk Geografisk Årsbok* (1954), Lund, Sweden.

BOGUE, DONALD J. 'Internal Migration' in Hauser, P. M. and Duncan, O. D. (Eds.), *The Study of Population*, Chicago: The University of Chicago Press (1959).

BUNTING, ROBERT L. 'Labor Mobility in the Crescent,' in Chapin, F. Stuart, Jr., and Weiss, Shirley F. (Eds.), *Urban Growth Dynamics in a Regional Cluster of Cities*, New York: Wiley (1962).

ELDRIDGE, HOPE T. 'A Cohort Approach to the Analysis of Migration Differentials,' *Demography*, Vol. 1 (1964), pp. 212–19.

MC GINNIS, ROBERT and PILGER, JOHN E. 'On a Model for Temporal Analysis' and 'Internal Migration as a Stochastic Process,' (mimeo). Department of Sociology, Cornell University (1963).

MORRILL, RICHARD L. and FORREST PITTS, R. 'Marriage, Migration and the Mean Information Field,' *Annals of the Association of American Geographers*, Vol. 57 (1967), pp. 401–22.

OLIVER, F. R. 'Inter-Regional Migration and Unemployment, 1951–61,' *Journal of the Royal Statistical Society*, Series A, 127 (1964), pp. 42–75.

ROSEN, HOWARD 'Projected Occupational Structure and Population Distribution,' *Labor Mobility and Population in Agriculture*. Ames: Iowa State University Press (1961).

ROSSI, PETER *Why Families Move,* Illinois: The Free Press of Glencoe (1955).

SJAASTAD, LARRY A. 'Occupational Structure and Migration Patterns,' *Labor Mobility and Population in Agriculture,* Ames: Iowa State University Press (1961).

TAEUBER, KARL E. and TAEUBER, ALMA F. 'White Migration and Socio-Economic Differences between Cities and Suburbs,' *American Sociological Review,* Vol. 29 (1964), pp. 718–24.

TER HEIDE, H. 'Migration Models and Their Significance for Population Forecasts,' *Milbank Memorial Fund Quarterly,* Vol. 41 (1963), pp. 56–76.

THOMAS, DOROTHY S. 'Age and Economic Differentials in Interstate Migration,' *Population Index* (1958), pp. 313–24.

WEBBER, MELVIN M. 'The Urban Place and the Nonplace Urban Realm,' *Explorations into Urban Structure,* Philadelphia: University of Pennsylvania Press (1964).

19 Search, learning, and the market decision process[1]

Reginald G. Golledge and Lawrence A. Brown
Ohio State University and University of Iowa, U.S.A.

Current theories concerning the spatial behavior of producers and consumers are cross-sectional in approach and emphasize equilibrium situations in which stereotyped behavior is assumed. For example, the market decision process is generally conceptualized in a deterministic framework with individuals being allocated once-and-for-all to discrete nodes, and with areas of nodal dominance being separated by finite linear boundaries. Such a conceptualization runs contrary both to empirical evidence of market behavior, and to psychological evidence concerning the probabilistic nature of human behavior.[2]

In this paper an attempt is made to build a conceptual framework for the market decision process. A basic hypothesis is that equilibrium states of behavior such as those summarized in Löschian-type models are the outcomes of a learning process, and the models in which decisions are narrowed down to a single market represent but one of a family of equilibrium solutions which can be produced by the learning process. It is proposed to formalize the conceptual framework in terms of a learning model.

The activities of search and learning

In the present day world it is axiomatic that individuals by themselves cannot produce everything they desire and hence must trade for many commodities. Although there are potentially many courses of action open to each individual, some of these exert only a restricted influence because of intervening opportunities or because of diseconomies associated with patronizing specific centres. Within any given spatial range therefore, individuals have a finite number of feasible alternative markets at which they can satisfy their wants and needs.

In the course of satisfying such desires, individuals will test a number of possible combinations of likely markets (i.e., they will adopt a number of different marketing strategies). From their experience of the results they will 'learn' which decision process (or which strategies) give them the greatest rewards (or payoffs), and they will tend to retain 'satisfactory' responses and delete 'unsatisfactory' responses. This process is likely to be continued as the search for the 'most satisfactory' pattern of responses is carried out.

Thus, few choice-decisions are made without some preliminary search activity. Except for random choice situations search activity is primarily problem-oriented. Most searching is therefore motivated in some way, and is continued until a solution to the problem situation is achieved. The *degree* of search activity in a system of responses depends largely on the force of the initiating motivation.

Cyert and March[3] claim that there are a few simple rules that provide an initial basis for search activity. These are:

1   search in the neighborhood of the problem situation
2   search in the neighborhood of the current alternative
3   if the neighborhood search process does not provide an adequate solution, then successively use 'more distant' search procedures.

Such rules give an unmistakable spatial bias to the search process.[4] This spatial bias may complement search biases resulting from such things as special skills or training, aspirations held prior to search, or incomplete information about the problem situation.

Not all spatial activities, however, are search activities. For example, while the *initial* casting around for a suitable market can be interpreted as a somewhat random event that might be typified as a Bernoulli trial,[5] later events may be quite well oriented and organized. During search activity, the sequences of successful and unsuccessful outcomes on each trial can be represented as an array of 1's (for success) and 0's (for failure), for example,

1 1 1 0 0 0 1 1 0 1 0 1 1 0 0 1 1 0 1 0

search activity

1 1 1 1 1 1 1 1 1 1

learned activity

Here a path to a market is chosen on the provisional try and may be repeatedly chosen on a few subsequent trials while an individual accumulates information about the system in which he finds himself. This temporary stability may be followed by exploratory behavior (or search activity) with mixed outcomes of success and failure. Search will continue until the individual's aim – be it least cost, minimized distance, maximized aesthetic value, or the like – appears to be accomplished. Once a decision has been made concerning the nature of the most favorable response, search activity is reduced and is replaced by some regular or habitual pattern of responses.

The principal characteristics of *habitual* response patterns are invariability, repetition, and persistence. Each patterned response is said by Hull[6] to have a varying degree of habit strength, and each takes a different period of time to become established. It is significant that habitual response patterns can include either single or multiple responses. In terms of the market decision process, habitual responses may include the patronage of one or many market centers.

The exhibition of such habitual response patterns is evidence that a problem situation has been mastered and that future responses can be predicted with a high degree of accuracy. This regularity is particularly important to the geographer if the response pattern stabilizes the individual's movements in space. In terms of the aggregate market decision problem, for any given population at any given time it may be deduced that only part of the population is in a search phase and that the remainder is in a habitual response phase, but that all elements of the population are aspiring towards the development of the habitual response phase. Such an aim may or may not be achieved in the short run, and it may or may not result in the choice of a single market center.

As an example of this procedure let us consider the actions of a producer desirous of marketing a product. For purposes of clarity we adopt a theoretical point of view. It can be assumed that, at successive time intervals, this producer has products to sell and that at each time interval he does in fact sell all products then available. Assume further that the producer is new in an area at the time of his first sale and consequently has but scant or no knowledge of the relative advantages of alternative marketing outlets. At first, then, he may select markets either at random or because of some bias due to imperfect knowledge of the competitive situation. The outcome of his initial response is a reward which gives him a measure of satisfaction for performing his productive service. As the producer has no previous knowledge of the variability (or range) of rewards open to him, he would probably be reluctant to assess the magnitude of the initial reward as 'large' or 'small.' At this stage his only criterion of measurement would appear to be a dichotomous classification into 'positive' (i.e., returns over and above the cost of production) and 'negative' (i.e., returns less than costs of production). When the next marketing opportunity occurs, the producer finds himself in a slightly different situation from previously. He has some 'experience' of possible market conditions and of the satisfaction he can get from making a specific response. However, there is still a degree of risk involved in that the same conditions may not govern the market, and there is uncertainty as to whether the response-reward situation already experienced is the 'best' alternative available.[7]

Thus the producer with some market-experience has three alternatives open to him:

a   to repeat the behavior of the previous trial
b   to make a different response excluding his previous responses
c   to make a different response including at least part of the previous responses.

Obviously, given a situation which undergoes no major change over time, successive responses would be conditioned largely by the results of previous responses and by any extraneous information gained about the system. Of particular importance would be the magnitude of the rewards obtained on each trial. Should the value of several or all response-rewards be similar, then for all practical purposes the producer either would become indifferent as to which response he made, or his final decision would turn on 'fringe-benefits' factors such as personal relationships or services provided by the market.

Thus by a trial and error procedure which may be seen as a learning process, the producer limits his range of marketing behavior until some equilibrium pattern emerges. The actual process involves collating information from previous behavior and from outside sources in order to limit the area of trial and error choice. In this way the number of alternatives considered at successive intervals may become smaller and smaller as information mounts. The equilibrium position involves either reliance on a single market (the Löschian solution) or reliance upon a group of markets which are patronized in the long run according to their ability to generate returns. In either of these equilibrium cases, stereotype behavior responses occur.[8] Such a process is similar to the learning process described by Hull,[9] Spence,[10] and Bush and Mosteller.[11]

By drawing on the probability concepts that have been applied to learning theory, it can be argued that at any given moment of time, and for any given trial, the aggregate responses of a population can be predicted with a certain degree of accuracy (i.e., there will be a certain proportion of right guesses and a certain proportion of wrong guesses of the responses selected by the population). We can therefore justifiably claim that, at any given time, a given population has a certain probability of making a selected marketing response, and that the magnitude of the probability is conditioned by a set of parameters, some increasing the probability of a given response (the most rewarding), and others decreasing the probability of alternative responses (the least rewarding).

A stochastic model of the market decision process

The most geographically relevant learning theories appear to be contiguity theories (Guthrie,[12] Hull[13]) and sign theories (Tolman[14]). Contiguity theories and their modifications are often stated in the form of Markov chains. This type of expression will also be used in this paper.

Stated simply, a Markov chain consists of n discrete 'states of being,' a set of phenomena which occupy each state i at time t with probability s_{it}, $(i = 1, \ldots, n)$, and behavior characteristics such that movement from state i to state j within a single discrete time interval occurs with probability P_{ij} $(i = 1, \ldots, n; j = 1, \ldots, n)$. Furthermore, since all phenomena at any time t are distributed only among the n discrete states, $\sum_{i=1}^{n} s_{it} = 1$. Also, since phenomena in state i at time t can only move to each of the remaining $n - 1$ states or remain in the same state by time $t + 1$, then $\sum_{j=1}^{n} P_{ij} = 1$. This system can be expressed mathematically by a 'states' vector $S(t)$, which consists of elements s_{it}, and a transition matrix P which consists of elements P_{ij}. The states vector for time $t + 1$ is derived by multiplication: i.e.,

$$S(t + 1) = S(t)*P$$

which is equivalent to

$$[s_{1,\,t+1} s_{2,\,t+1} \cdots s_{n,\,t+1}] = [s_{1,\,t} s_{2,\,t} \cdots s_{n,\,t}] \begin{bmatrix} P_{11} P_{12} & \cdots & P_{1n} \\ P_{21} P_{22} & \cdots & P_{2n} \\ \cdot & & \cdot \\ \cdot & & \cdot \\ \cdot & & \cdot \\ P_{n1} P_{n2} & \cdots & P_{nn} \end{bmatrix}$$

Similarly,

$$S(t + 2) = S(t + 1) \qquad * P$$

$$\cdot \qquad\qquad\qquad \cdot$$
$$\cdot \qquad\qquad\qquad \cdot$$
$$\cdot \qquad\qquad\qquad \cdot$$

$$S(t + m) = S(t + (m - 1)) * P$$

Thus, given an initial states vector for a base time period 0 and a transition matrix, a states vector for every succeeding time period may be derived. In the context of the market decision problem, each state would represent a market that might be chosen at any particular time t, and the transition probability P_{ij} would represent the probability of choosing market j at time $t + 1$, given that the choice for time t was market i.

This model represents a basic form of the market decision process. It expresses the idea that the outcome of each decision period is determined solely by the outcome of the immediately preceding time period, and that the probability of going from any one market decision (outcome) to any other decision (within the span of a single time interval) is constant over time. Thus, a *constant* stochastic operator is applied to the states vectors, and the mathematical process is readily discernible as a Markov chain. Using such a process an equilibrium state of behavior is achieved when the Markov chain reaches a steady state.

The requirement of a constant stochastic operator, however, limits the scope of the proposed model because as conceptualized, the conditional probability of choosing a particular market *need not be constant over time*. This is particularly true if we modify assumptions concerning the constancy of factors affecting rewards and the limitation of information sources to personal experience. If it is hypothesized that the realm of knowledge about a market system changes as learning proceeds, it follows that an organism views the choice situation in a different light at each time. His viewpoint will be conditioned by his accumulated experience with the market system – which is reflected in his actions at the preceding stage – but his perception of the choice system open to him is also tempered both by the rate of extinction of previously unrewarded responses, and by his rate of approach to fully learned behavior. The effect of both these factors may be to alter the transition probabilities for each state. Conceivably there is a large set of possible transition probability matrices that *could* be applied to the existing states vector at any time. The selection of the appropriate matrix operator will depend on the way the subject organism views (or perceives) the market situation at each state, which in turn is influenced by the amount of learning he has achieved. It appears, therefore, that some variation of the simple Markov model initially proposed would be desirable.

Bush and Mosteller[15] have developed three types of learning models which do appear suitable for describing the market decision process. The essential difference between these and the Markov chain model, described above, is the use of a transition matrix which varies over time. Thus, we now have $P(t)$ instead of P. For simplicity, consider the case of only two markets and let transition probabilities be

$$P_{11}(t) = 1 - b(t)$$
$$P_{12}(t) = b(t)$$
$$P_{21}(t) = a(t)$$
$$P_{22}(t) = 1 - a(t)$$

This is possible since

$$P_{11}(t) + P_{12}(t) = P_{21}(t) + P_{22}(t) = 1$$

In this context, a and b represent the probabilities of switching markets in time $t + 1$: from market 1 to market 2 is represented by $b(t)$; from market 2 to market 1 is represented by $a(t)$. Similarly, $1 - b(t)$ and $1 - a(t)$ represent the probabilities of using the same market in time $t + 1$ as in time t: $1 - b(t)$ for market 1; $1 - a(t)$ for market 2.

Given these representations of transition probabilities, the probability of choosing market 1 at time $t + 1$ is

(1) $$s_1(t + 1) = s_1(t)*[1 - b(t)] + s_2(t)*a(t)$$

The probability of choosing market 2 at time $t + 1$ is

(2) $$s_2(t + 1) = s_1(t)*b(t) + s_2(t)*[1 - a(t)]$$

Now, letting

$$\alpha(t) = 1 - a(t) - b(t)$$

and employing the relationship that

$$s_2(t) = 1 - s_1(t),$$

then

(3) $$s_1(t + 1) = s_1(t)*\alpha(t) + a(t).$$

Similarly, it may be shown that

(4) $$s_2(t + 1) = s_2(t)*\alpha(t) + b(t)$$

The bounds of α, a, and b are

$$-1 \leq \alpha(t) \leq 1,$$
$$0 \leq a(t) \leq 1,$$
$$0 \leq b(t) \leq 1.$$

Equations (3) and (4) are essentially of the type $Y = bX + a$ and are called the *slope-intercept form* by Bush and Mosteller. In these equations, the value of $\alpha(t)$ may be seen either as reflecting behavior of the consumer in searching for a suitable market between time t and time $t + 1$, or as the likelihood of using the same market in time $t + 1$ as in time t. As $\alpha(t) \Rightarrow 1$, the probability of remaining in the same state (i.e., patronizing the same market) at times t and $t + 1$ approaches 1. As $\alpha(t) \Rightarrow -1$, the probability of *changing* markets from time t to time $t + 1$ tends toward 1. Thus, to the degree that $\alpha(t)$ tends toward 1, the individual approaches an asymptotic behavioral pattern which may become stereotyped; to the degree that $\alpha(t)$ tends toward -1, it can be inferred that the individual is in a search phase of behavior. Consequently an $\alpha(t)$ value changing from -1 to $+1$ over time would indicate a transition from search to fully learned or stereotyped behavior.

A second variation of the Markov model is called the *gain-loss form*. This can be derived from equations (1) and (2) as follows. Again, employ the relationship that

$$s_2(t) = 1 - s_1(t)$$

Then, from (1),

(5) $$s_1(t + 1) = s_1(t)*[1 - b(t)] + [1 - s_1(t)]*a(t)$$
$$= s_1(t) + a(t)[1 - s_1(t)] - b(t)s_1(t)$$

and from (2), using similar procedures,

(6) $$s_2(t + 1) = s_2(t) + b(t)[1 - s_2(t)] - a(t)s_2(t)$$

These difference equations assert that the action of patronizing a particular market at time $t + 1$ may result either by patronizing that same market at time t and continuing to behave similarly, or by patronizing the other market at time t and changing the pattern of behavior. Clearly, the temporal trend of transition probabilities is also important in determining market behavior at time $t + 1$.

A third variation of the model is the *fixed-point form*. This is obtained by manipulation of the slope-intercept form. To obtain this, we introduce another operator, $\lambda(t)$.
Let

$$\lambda_1(t) = \frac{a(t)}{a(t) + b(t)}$$

$$\lambda_2(t) = \frac{b(t)}{a(t) + d(t)}$$

Then,

$$a(t) = [1 - \alpha(t)]\lambda_1(t)$$

and

$$b(t) = [1 - \alpha(t)]\lambda_2(t).$$

As a consequence, (3) may be restated as

(7) $$s_1(t + 1) = s_1(t)\alpha(t) + [1 - \alpha(t)]\lambda_1(t)$$

Similarly, (4) may be restated as

(8) $s_2(t + 1) = s_2(t)\alpha(t) + [1 - \alpha(t)]\lambda_2(t)$

In this case $a(t) + b(t) \Rightarrow 2$ if the probability of switching markets between time t and time $t + 1$ tends toward 1, and $a(t) + b(t) \Rightarrow 0$ if the probability of remaining in the same market from time t to time $t + 1$ (i.e., not switching) tends toward 1. Thus, if a consumer is employing market 2 in time t and assuming that the transition probabilities do not change over time, $\lambda_1 = 1$ implies that the consumer will change to market 1 and continue to employ market 1; $\lambda_1 = \frac{1}{2}$ implies that the consumer will change to market 1 in time $t + 1$ but move back to market 2 in time $t + 1$ and continue oscillation in all future time periods; $\lambda_1 = 0$ implies that the consumer will continue to employ market 2 in all future time periods. The implications of various values of λ_2 are similar, except that λ_2 refers to the behavior of a consumer initially using market 1 as it is relevant to employing market 2 in the future. The fixed point format is favored by Bush and Mosteller as a model which has considerable scope, and it has already been used with success by Haines[16] in discussing the development of a buying pattern for new goods introduced into stores, and by Sprowls and Asimow[17] in their analysis of customer behavior for the TASK corporation. Either it can be used in situations where only two alternatives are considered, or it can be transformed to account for more complex situations.

Extensions of the model

Elaboration of the model presented above may profitably focus upon two aspects of the market decision process – (1) the specific probability of changing from one market to another during any particular time interval, as indicated by the elements in the transition probability matrix $P(t)$; and (2) the search behavior associated with learning a particular response pattern or settling upon the patronization of a particular set of markets. These aspects are related and are not mutually exclusive.

In formulating a model of the market decision process, it has been suggested that there is a set of transition probability matrices, each one of which is applicable to a particular time interval. Each element $P_{ij}(t)$ of any single such matrix (i.e. the probability that a person utilizing market i in time t will come to utilize market j in time $t + 1$), is an index representing the combination of all factors influencing a particular decision about where to buy or sell. As an example, the value assumed by any particular transition probability may be seen as a function of:

1   The probability that the reward at market j in time $t + 1$ is greater than rewards at each of the other markets in time $t + 1$ (including market i) which is defined here as $r_{ij}(t)$;

2   The probability that a person patronizing market i at time t will be

informed of the advantages of market j by time $t + 1$, which is defined here as $d_{ij}(t)$;

3   The probability that a person patronizing market i at time t will be induced to patronize market j in time $t + 1$ as a result of supplementary rewards such as benefits from multiple shopping, which is defined here as $m_{ij}(t)$.

These and other factors work together to define each $P_{ij}(t)$, i.e.,

(9) $$P_{ij}(t) = f(r_{ij}(t), d_{ij}(t), m_{ij}(t) \ldots)$$

The form of (9) suggests that the dependency relationship may be treated in a regression-type format. Such an approach, however, does not illuminate the dynamic processes underlying the value of each of the elements $P_{ij}(t)$, $r_{ij}(t)$, $d_{ij}(t)$, $m_{ij}(t)$, etc., and knowledge of these is important for a full understanding of the market decision process. A more satisfactory approach, then, would be the employment of sub-models to account both for the values of $r_{ij}(t)$, $d_{ij}(t)$, $m_{ij}(t)$, etc., and the interactions among these variables. Combining the sub-models would produce particular values of $P_{ij}(t)$.[18] For example, Hägerstrand,[19] Rapoport,[20] and Brown[21] have proposed models of the process of information flow which underlies the values of $d_{ij}(t)$. Marble[22] and Nystuen[23] have contributed empirical research and a model to account for the effects of the advantages of multiple shopping and other supplementary rewards upon market choice. Such models could be used to estimate $m_{ij}(t)$. However, the probability that the reward at market j in time $t + 1$ is greater than that at market i in time t ($r_{ij}(t + 1)$) has been given much less attention by geographers and others. Evidently, however, two factors are important – (1) the policies of distributors in each market which determine prices in the market, and (2) the relative accessibility of a market to persons making the market decision. Brown[24] has provided a format which may be useful for treating these two factors.

Clearly, any one or all of the factors underlying the transition probabilities will have an influence upon patterns of search for a market and the rate with which the decision maker settles upon a particular set of markets. This influence is reflected in the alpha-value ($\alpha(t)$), which has been termed the learning parameter in the model presented above. In particular, a large alpha-value would indicate that the decision-maker will quickly terminate search activity and will come to a stable or settled behavior pattern in which he trades regularly at a single set of markets. Such a situation might be possible where previous experience, latent learning, or urgency of action are factors which influence decisions made on each trial. Large alpha-values can occur either when the few trials needed to approach a stable state are spaced close together in time or are separated by long periods of time. A small alpha-value on the other hand would indicate that a long period of searching-type behavior occurs in which the decision-maker will try many markets before settling on a few. This might occur either when numerous alternative courses of action exist, or where search is accompanied by an extended period of information gathering. However, if behavior under a small alpha-value is coupled with numerous trials per time interval, the real time taken to reach stable behavior may be comparable to that characterizing market behavior under a large alpha-value with fewer trials per time interval. This aspect of the market decision process leads one to consider the possibility of differentiating alpha-values for the same individual according to the type of good being purchased or sold. Thus, since low order or convenience type goods such as groceries are purchased often, they may be characterized by a low alpha-value but many trials. The purchase or sale of high order or shopping type goods, on the other hand, which constitutes a major monetary decision, might be characterized by a high alpha-value and few trials, but with a long period of information-gathering preceding any actual trial at a market (i.e., a purchase or sale). This further suggests that an information-gathering parameter similar to the alpha-parameter should be added to the model, so as to reflect the readiness of the decision-maker to attempt a market trial. For convenience type goods this parameter might very quickly go to a value of one, thus inhibiting trial behavior only slightly, whereas it would approach one slowly for shopping-type goods, thus acting as a dampener upon the model's predictions of actual trial behavior.[25]

Another important factor related to the size of the alpha-value for any given situation is the degree to which a transfer of responses made in similar situations is possible. This constitutes the act of *generalization* which is an integral part of the learning process. The opportunity to generalize is likely to be influenced by the degree of satisfaction elicited by a response. For example, a rural producer may achieve a satisfactory return from a sale of goods to a farm co-operative. This success prompts the producer to include the co-operative in his search for a source of machinery, or fertilizer, or the like. The probability of generalization may be enhanced by the advantages of multiple patronage of different parts of the co-op and by dividends received, and the result is a limitation of the search for outlets for the producer's other products and a minimization of the search for satisfactory supply sources for his needs. In effect generalization allows the individual to slip quickly into a learned response pattern that often becomes habitual. With continued reinforcement in the form of satisfactory outcomes for trips undertaken, such patterns develop a high degree of habit strength and are very difficult to extinguish. Once this habit pattern has evolved, trip patterns through space are regularized, and the market decision problem virtually ceases to exist as a problem per se. If the habits can be identified, choices made at future decision points can then be predicted with a high degree of accuracy. Thus, where generalization is possible, we expect a large alpha-value, whatever the type of good being traded. It might be ventured, however, that generalization is more likely on convenience-type goods, since these are traded more often and by more people.

To illustrate the pattern of search that might occur during the trial period, whatever its length, consider a situation in which an 'optimal' response pattern was unknowingly adopted early in the search process. Because of our assumptions of incomplete knowledge, it might be expected that on future trials one of the following will happen.

Case (1): Search will be terminated.

Case (2): Continued search will take place with the 'optimal' being the basic point of reference for comparison of rewards and ultimately the 'optimal' pattern will be accepted as the equilibrium response pattern.

Case (3): The more favorable elements of the optimal pattern will be continued and search will be limited to a minor part of the response pattern – this may mean addition and deletion of the smaller elements in the chosen response vector, and perhaps also some modification of the larger responses.

In the first of the three cases equilibrium is attained and settled behavior may result immediately (i.e., the individual may regard himself as a satisficer rather than an optimiser and may stop search activities as soon as a response elicits a sufficiently 'satisfactory' reward).

In Case (2), search is continued but rewards are compared with a fixed point that is part of the individual's experience. Recursiveness will be involved unless a change in market situations forces the adoption of a new fixed point for comparative purposes.

The third case involves the most 'rational' behavior. At each time rewards are checked against existing knowledge of past rewards and search is confined to exploring markets for only a small part of the total product. At each stage adjustments are made to all elements of the states-vector and if search activities find an alternative more suitable than any already held, all elements are revised in accordance with the new information. This approach is more characteristic of the optimiser than the satisficer. However, by retaining some exploratory elements in the behavior vector (S in the model above), the case-three market decision may be regarded as tending towards a dynamic equilibrium situation, rather than a steady state. As further illustration of the third approach consider the states vectors.

$$\text{Time } t_1 = \begin{bmatrix} .50 & \text{Centre} & \text{A} \\ .25 & ,, & \text{B} \\ .15 & ,, & \text{C} \\ .07 & ,, & \text{D} \\ .03 & ,, & \text{E} \end{bmatrix} \quad \text{Time } t_2 = \begin{bmatrix} .51 & \text{Centre} & \text{A} \\ .24 & ,, & \text{B} \\ .15 & ,, & \text{C} \\ .08 & ,, & \text{N} \\ .02 & ,, & \text{G} \end{bmatrix}$$

$$\text{Time } t_3 = \begin{bmatrix} .48 & \text{Centre} & \text{A} \\ .24 & ,, & \text{N} \\ .24 & ,, & \text{B} \\ .04 & ,, & \text{C} \end{bmatrix}$$

At time t_1 three major and two minor patronages can be discerned. By time t_2, however, the decision-maker has come to retain A, B, and C in his response pattern, while deleting D and E. Also, t_2 features the addition of two new responses, N and G, which are tested only slightly at first. Should N's attraction be found high, it will be retained as an alternative in future decision situations, and the entire behavioral pattern may be reappraised in light of the satisfactions gained through N. Thus, a new probable response pattern is generated for t_3, with G completely deleted and C occupying a very minor position. In the future perhaps only one new alternative (replacing C) would be examined at any given time. This presumes that a reasonable measure of 'satisfaction' has been gained and that the 'need' for continued search among alternatives has decreased.

This example and the discussion above suggests the complexity of the market decision process. Of particular importance is the role of learning and trial behavior which leads to a stable pattern of market behavior. What criteria control decisions in the trial phase of the market decision process and how do these affect spatial behavior? What is the role of pre-trial information in the market decision process? What type of information is most influential on the decision-maker? And how does the decision-maker go about gathering adequate pre-trial information? How are stable market behavior patterns affected by system shocks such as a drastic change in prices at one market or a change in the transportation system (e.g. a new freeway) which suddenly makes one market more accessible than others? And finally, how does the role of all these factors and others change according to the type of good being purchased or sold? These and other relevant questions are suggested by the model presented above, but testing of that model must await answers to some of these questions. Apparently, this will come about largely through empirical research, some of which is currently being carried on by the authors.

References

1   Reprinted from *Geografisker Annaler,* Vol. 49B (1967), Stockholm, Sweden.
2   GOLLEDGE, R. G. 'Conceptualizing the Market Decision Process,' *Journal of Regional Science,* Vol. 7 (December 1967).
3   CYERT, R. M. and MARSH, J. G. *A Behavioral Theory of the Firm,* Englewood Cliffs, New Jersey: Prentice Hall (1963).
4   This has also been pointed out by GOULD, PETER L., 'A Bibliography of Space Searching Procedures,' *Research Note, Department of Geography, Pennsylvania State University* (1965).
5   For amplification see PARZEN, E. *Stochastic Processes,* San Francisco: Holden-Day (1964), and Feller, W. *An Introduction to Probability Theory and Its Applications,* Vol. 1, New York: John Wiley & Sons Inc. (1950).
6   HULL, C. L. *A Behavior System,* Science Editions, New York: John Wiley & Sons, Inc. (1962).
7   It is a feature of the market decision process that risk and uncertainty occur to some extent throughout the entire process and help produce a dynamic rather than a static equilibrium as the end product.
8   An alternative suggestion is that such a stage corresponds to 'complete learning'; for further information on this see Hilgard, E., *Theories of Learning,* New York: Appleton-Century-Crofts (1956), and Bush, R. R. and Mosteller, F. *Stochastic Models for Learning,* New York: John Wiley & Sons Inc. (1955).
9   HULL, C. L., op. cit.

10 SPENCE, K. W. 'Theoretical Interpretations of Learning,' in Stevens, S. S. (Ed.), *Handbook of Experimental Psychology,* New York: John Wiley & Sons (1951).

11 BUSH, R. R. and MOSTELLER, F. *Stochastic Models for Learning,* New York: John Wiley & Sons Inc. (1955).

12 GUTHRIE, E. R. *The Psychology of Learning,* New York: Harper (1952).

13 HULL, C. L., op. cit.

14 TOLMAN, E. C., RITCHIE, B. F. and KALISH, D. 'Studies in Spatial Learning II, Place Learning versus Response Learning,' *Journal of Experimental Psychology,* Vol. 36 (1946), pp. 221–9.

15 BUSH, R. R. and MOSTELLER, F., op. cit.

16 HAINES, G. H. 'A Theory of Market Behavior After Innovation,' *Management Science,* Vol. 10 (1964), pp. 634–55.

17 SPROWLS, R. C. and ASIMOW, M. 'A Model of Customer Behavior for the TASK Manufacturing Corporation,' *Management Science,* (1961), pp. 251–4.

18 A review and bibliography of models relevant to this task can be found in BROWN, LAWRENCE A. 'Diffusion Processes and Location: A Conceptual Framework and Bibliography'. *Bibliography Series, No. 4, Regional Science Institute* (1968).

19 HÅGERSTRAND, TORSTEN 'A Monte Carlo Approach to Diffusion,' *Archives Européennes de Sociologie (European Journal of Sociology),* 6 (1965), pp. 43–67.

HÅGERSTRAND, TORSTEN 'Quantitative Techniques for the Analysis of the Spread of Information and Technology,' in Anderson, C. Arnold and Bowman, Mary Jean (Eds.), *Education and Economic Development,* Chicago: Aldine (1965), pp. 244–80.

HÅGERSTRAND, TORSTEN 'Aspects of the Spatial Structure of Social Communication and the Diffusion of Information,' *Papers of the Regional Science Association,* Vol. 16 (1966), pp. 27–42.

20 RAPOPORT, ANTOL 'Nets with Distance Bias,' *Bulletin of Mathematical Biophysics,* Vol. 13 (1951), pp. 85–92.

RAPOPORT, ANTOL 'Spread of Information Through a Population with Socio-Structural Bias: I. Assumption of Transitivity,' *Bulletin of Mathematical Biophysics,* Vol. 15 (1953), pp. 523–34.

RAPOPORT, ANTOL 'Spread of Information Through a Population with Socio-Structural Bias: II. Various Models with Partial Transitivity,' *Bulletin of Mathematical Biophysics,* Vol. 15 (1953), pp. 535–47.

RAPOPORT, ANTOL 'Spread of Information Through a Population with Socio-Structural Bias: III. Suggested Experimental Procedures,' *Bulletin of Mathematical Biophysics,* Vol. 16 (1954), pp. 75–82.

RAPOPORT, ANTOL 'The Diffusion Problem in Mass Behavior,' *General Systems Yearbook,* Vol. 1 (1956), pp. 48–55.

RAPOPORT, ANTOL 'Contributions to the Theory of Random and Biased Nets,' *Bulletin of Mathematical Biophysics,* Vol. 19 (1957), pp. 257–77.

21 BROWN, LAWRENCE A. *Diffusion Dynamics: A Review and Revision of the Quantitative Theory of the Spatial Diffusion of Innovation,* Lund Studies in Geography, No. 29, Lund, Sweden: Gleerup (1968).

22 MARBLE, DUANE F. 'A Theoretical Exploration of Individual Travel Behavior,' in Garrison, W. L. and Marble, D. F. (Eds.), *Quantitative Geography Part I: Economic and Cultural Topics,* Department of Geography, Northwestern University, Studies in Geography, Illinois: Evanston (1967), pp. 33–53.

23 NYSTUEN, JOHN D. 'A Theory and Simulation of Urban Travel,' in Garrison, W. L. and Marble, D. F. (Eds.), *Quantitative Geography Part I: Economic and Cultural Topics,* Department of Geography, Northwestern University, Studies in Geography, Illinois: Evanston (1967), pp. 54–83.

24 BROWN, LAWRENCE A., op. cit.

25 The exact nature of this parameter has not as yet been determined.

20 Interdependency of spatial structure and spatial behavior: a general field theory formulation[1]

Brian J. L. Berry
Center for Urban Studies, University of Chicago, U.S.A.

Spatial behavior – the movement of people, goods, messages, ideas – takes place in a field of forces. This field links individuals and areas of varying characteristics and locations in a system of potential interaction. The basic postulate of general field theory is that actual behavior is patterned upon the potentials for interaction (potential energies) that exist within the force field, and that the force field is, in turn, a product of the system of characteristics and relative locations of the individuals and areas. A second postulate is that general properties of this system show persistence through time, although the characteristics of particular individuals or areas may change relatively quickly, leading to shifts in potentials for certain interactions and, ultimately, in revealed spatial behavior. It follows that, at any point in time, behavior, potential energies in the potential field and the spatial system of individuals and their characteristics should be isomorphic. Through time, on the other hand, a succession of changes in the characteristics of individuals and areas can accumulate enough deviations from general system properties that a revolutionary shift in spatial structure occurs, thereby transforming the bases of potential interaction and spatial behavior.[2] Thus, the isomorphism between spatial behavior and spatial structure at any given point in time has generality only within the context of a particular structural paradigm.

This paper does not concern itself with the translation of one paradigm into another. Rather, the purpose is to explore in a general way the bases of spatial structure, translation of these into potential fields, identification of the bases of spatial behavior, and establishment of the interdependencies between structure and behavior through the medium of potentials. Every step has a mathematical equivalent in linear algebra and vector geometry. Each involves multivariate analysis, and, for the most critical step, that of the establishment of the nature and strength of interdependencies, multivariate tests of significance are available. First, we introduce the general concepts in sequence, and then we provide an example of application in a study of Indian commodity flows.[3]

In general concept, we owe much to Lewin's social field theory[4] and to the approaches to conflict behavior developed by Rummel.[5, 6]

Bases of spatial structure

There is clearly an infinity of characteristics of individuals and areas that can be observed and recorded, both quantitative and qualitative. But, like symptoms of diseases, they are interrelated and interdependent, accumulating to syndromes that stem from the underlying disease mechanism. We make an important first step in analysis when we accept this fact – that the infinity of characteristics in fact represents a limited number of underlying *bases* of spatial structure. If we can identify these bases and index the individuals or areas on them, we will not only have achieved considerable intellectual economy but also will have defined spatial structure unambiguously; in the language of general field theory, we will have delineated the total situation within which behavior takes place.

The methodology is increasingly well-known and relatively simple. Components or factor analysis, proceeding from a matrix of correlations among characteristics produced from a data matrix of measurements of individuals or areas on the characteristics, will produce the bases. Interpretation of the resulting output – factor loadings, communalities, eigenvalues and factor scores – in light of relevant theory and previous empirical work will suggest what underlying themes the bases represent. Thus, in studies of urban ecology, many research workers have now confirmed an underlying three-factor structure comprising socio-economic rank (social status), stage-in-life cycle (age- and family-structure), and segregation as the bases of residential sifting within cities. Or, again, cross-national studies of economic-demographic development record three principal ways by which countries may be ordered: level of per capita wealth, energy inputs, etc. ('economic development'), size of country (aggregate wealth, scale, and 'power'), and demographic well-being ('population pressure)', etc. Other examples of replicated bases could be cited from each of the social sciences, but these two should suffice right now.

Each basis summarizes the structural pattern in a group of correlated characteristics. The bases are, in turn, mutually independent. This lack of correlation means orthogonality in a geometrical sense. Thus, if there were only two bases, the individuals or areas could be arrayed in a two-dimensional Euclidean space using their index values (factor scores) on the bases. The variance of the observations in this space would span all the structural differentiation in the original larger set of characteristics. If there are more bases, the structural space will have more dimensions, but the important property of spanning the total situation – defining spatial structure – remains.

HOMOGENEOUS TYPES AND UNIFORM REGIONS

As an aside, distances between observations in this Euclidean space are inverse measures of their similarity. Thus, grouping algorithms using the distances can be used to subdivide the space into the nested subsets of a hierarchy of homogeneous types of observations or uniform regions of areas. Again, the methodology is now well-known.[3]

Interaction potentials

An over-all distance separates each possible pair of observations in the structural space – the hypotenuse, in the Euclidean metric. But, from the point of view of interaction potentials, it is useful to consider the set of distances, one for each basis, that separates each pair of observations (dyad). Each distance is a relative measure, the interval between members of the pair on an axis. For the set of $(n^2 - n)$ dyads created from n observations, a complete matrix of $(n^2 - n) \times r$ distances can thus be prepared, where there are r bases. The intervals will be symmetric for observations $i - j$ and $j - i$, but a sign indicating which is the larger will generally be inserted. The basic field theory theorem is that these distances index interaction potentials. Thus, in the ecological example, pairs of areas within cities can be coded according to the distances separating them in social space (social distance), according to stage of life cycle, and according to degree of segregation.

Bases of spatial behavior

A dyadic format will also enable bases to be derived for a variety of kinds of movements, for example, shipments of many kinds of commodities among a set of areas. The initial data source is usually three-dimensional – an $n \times n$ commodity flow matrix for each commodity. Each of these matrices becomes a column vector in the dyadic formulation, however. Factoring thus proceeds from an $(n^2 - n) \times c$ data matrix (c commodities) and produces as bases t types of spatial behavior:

FUNCTIONAL TYPES AND ORGANIZATIONAL REGIONS

Again as an aside, the factor space in the behavioral case has dyads located in it. Use of numerical taxonomy in connection with distances among dyads will produce sets and subsets of the space corresponding to hierarchies of functional (interacting) types or organizational regions (nodal regions are a particular form of these). Again, there is no longer any mystery about the methodology,[3] but, once again, the ability to provide an objective regionalization is a useful offshoot.

Interdependency of behavior and structure

With dyads as observations, we can now write for each type of behavior that the location of a dyad on that type is a function of the interaction potentials (distances) in the structural space

$$b_{ij,t} = \sum r^\alpha r^{d_{ij,r}}$$

and, more generally, where there are k forms of independency

$$\sum \beta_{tk}b_{ij,t} = \sum \alpha_{rk}d_{ij,r}$$

Estimates of parameters and tests of significance are provided by canonical analysis. This multivariate procedure is not known as well as regression or factor analysis, but it constitutes an ideal format for investigation of inter-dependencies in multivariate data. For each form of interdependency, it operates like generalized regression of a set of dependent variables on a battery of predictors. But, like factor analysis, it provides k independent sets of parameters β and α for k patterns of independency among types of behavior and interaction potentials. Any good text on multivariate analysis describes the method clearly.

GENERAL TYPES AND REGIONS

Since for any k, $\beta_{tk}b_{ij,t}$ and $\alpha_{rk}d_{ij,r}$ are isomorphic, in all k identical dimensions are created by each side of the general field theory equation. Therefore, dyads have identical locations in each space. Application of numerical taxonomy to either side will therefore produce a *general* regionalization combining both structural and behavioral elements, something which heretofore has been considered technically and conceptually impossible.

An Indian case study

We have applied the entire procedure to the case of India. The Indian sub-continent is an interesting example because it affords a range of types of economic organization (Fig. 16), distributed spatially in a pattern closely re-sembling the map of population potential (Fig. 17) with commercialized economics in the urban-industrial cores and isolated tribal economies lying at the peripheries between core regions. For the country's 325 districts, we had available from the Census of India and other official sources 166 variables relating to labor force characteristics, social and economic charac-teristics of the population, nature of the settlement pattern, production of a variety of crops and products, and measures of relative location (potentials). In addition, the Inland Trade Accounts provide each year 63 36 × 36 com-modity flow matrices – for 63 commodities and 36 trade blocks: the states of India, plus major cities such as Calcutta, Bombay, Madras, and Delhi. Factoring and taxonomic analysis of the matrix of total commodity flows among the trade blocks showed the country's principal functional regions to correspond to initial impressions of the spatial structure, Fig. 18 (compare with Figs. 16 and 17), and so we had reasonable expectations as to the out-come of the analysis. Inspection of individual commodity flow maps (Figs. 19–30) suggested a dominant metropolitan orientation of many movements, with a substantial intermetropolitan component in the modern sectors, as well as east-west differences stemming from both environmentally-related

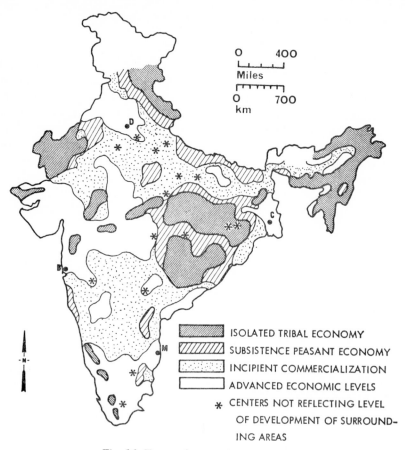

Fig. 16. *Types of economic organization.*

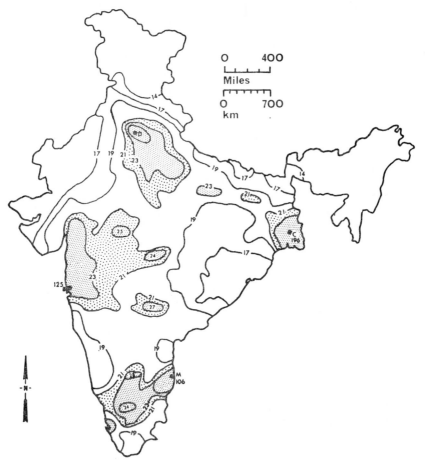

Fig. 17. Urban population potentials.

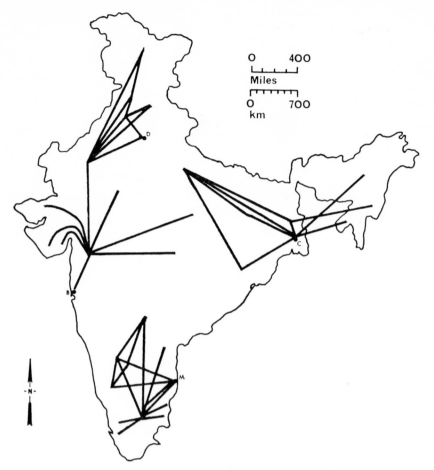

Fig. 18. Functional regions based on total quantity shipped.

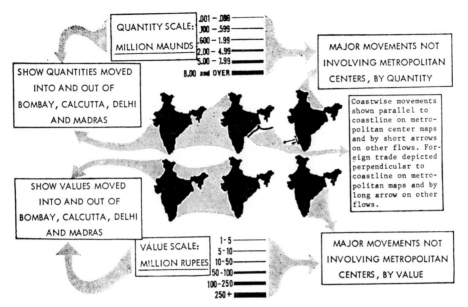

Fig. 19. Legend for flow maps.

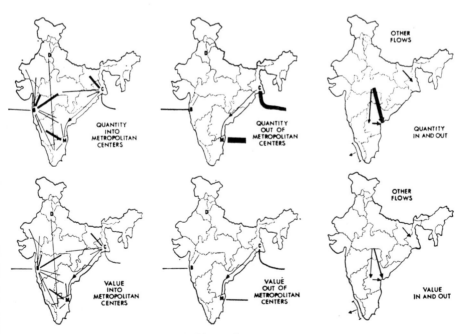

Fig. 20. Flows of manganese ore.

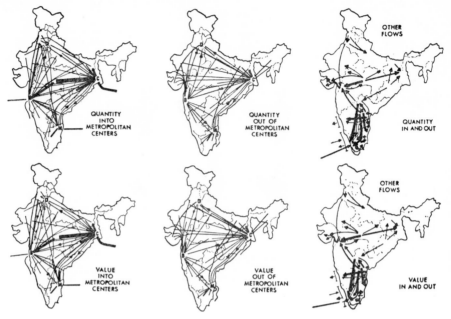

Fig. 21. Flows of unhusked rice.

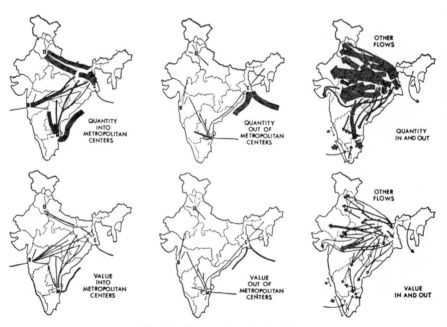

Fig. 22. Flows of coal and coke.

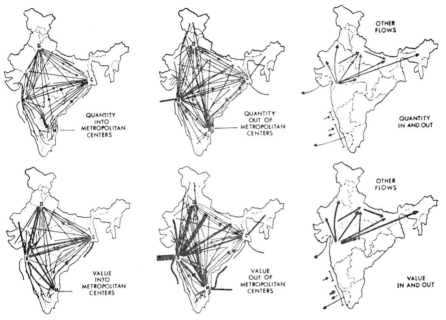

Fig. 23. Flows of Indian cotton piece goods.

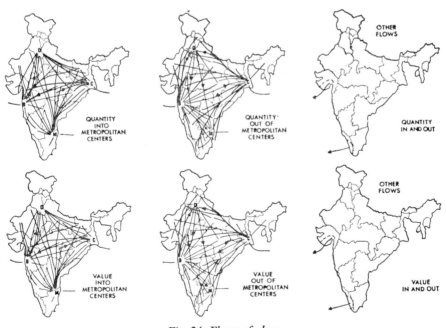

Fig. 24. Flows of glass.

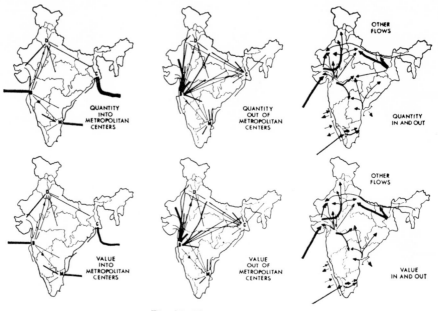

Fig. 25. Flows of kerosene.

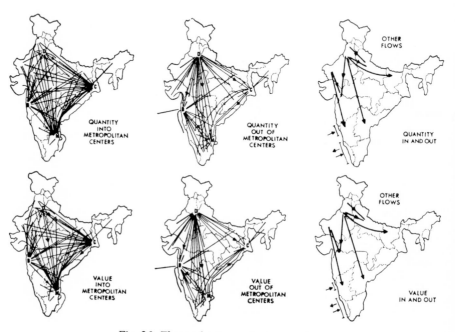

Fig. 26. Flows of gram and gram products.

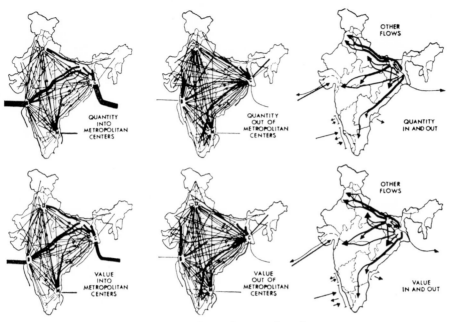

Fig. 27. Flows of iron and steel.

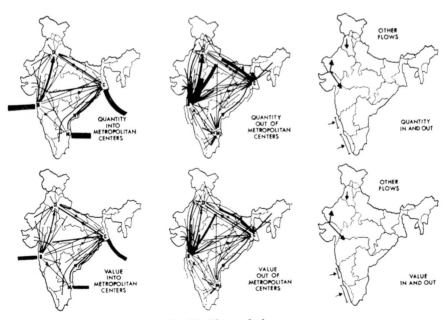

Fig. 28. Flows of wheat.

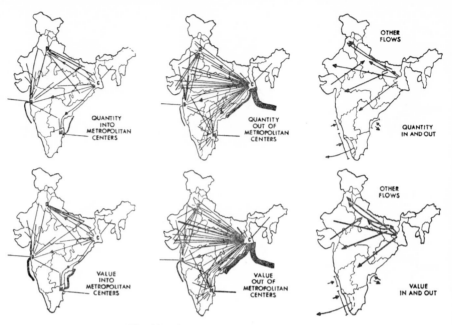

Fig. 29. Flows of gunny bags and cloth.

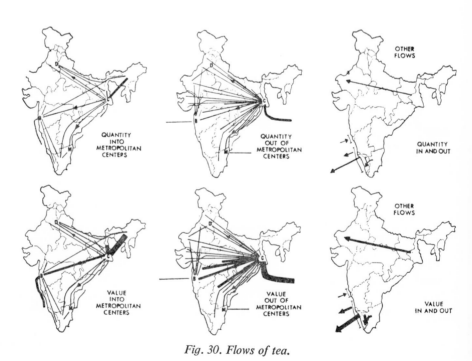

Fig. 30. Flows of tea.

agricultural specializations (dryland crops in the north-west, for example) and from the developing coal-steel complex of West Bengal-Bihar-Orissa in the east.

SPATIAL STRUCTURE

Several structural analyses of the 166 variables were undertaken, of spatial and agricultural variables, of labor forces and socio-economic variables, and of production variables. These enabled us to select a subset of 92 variables for a final combined analysis (our 7094 factor analytic program had at that time a maximum capacity of 100 variables).

Results are shown in Table 1, and Figs. 31 and 32 map two of the sets of factor scores. Briefly, the first factor identifies *urban-manufacturing regions* (Fig. 31). Urban and manufacturing potentials have high loadings, as do production data for a wide variety of industries located in metropolitan centers, cities highly accessible to these metropolises, or the few large provincial capitals located outside India's main core regions.

Factor two picks out *intensive irrigated agricultures*, factor three the principal east-west differences in agricultures, the fourth focuses upon access to India's population in the *Gangetic Valley*, the fifth the *dryland agricultural* specialities of the north-west, and the remaining factors point out different regional specializations in mineral production.

INTERACTION POTENTIALS

Since the flow data are available only for trade blocks and the structural analysis used 325 districts as observations, districts within the same trade block were treated as a swarm of points in the nine-dimensional structural space. To derive interaction potentials the centroids of each of the 36 swarms were located. Projections of the centroids on to the axes yielded scores, and differences between scores yielded nine vectors of distances for 36^2-36 trade block dyads.

SPATIAL BEHAVIOR

The original commodity flow data were rearranged in a $(36^2-36) \times 63$ matrix, and these data were analyzed to derive the bases of the behavior space. A twelve-factor structure resulted (Table 2). The first factor identified the principal regional connections focusing on Bombay, Madras, and Calcutta, the country's metropolitan port hinterlands – compare Fig. 33 with Figs. 18, 20, 21 and 25. Flows moving to and from the north-west emerge in factor two, whereas factor three picks out those originating in the east (Figs. 34, 35, etc.). Only the first five bases index more than the movement of a specific crop or product, however. Therefore, in the next step only these first five factors were used.

TABLE 1

PRINCIPAL FACTOR LOADINGS: STRUCTURAL ANALYSIS

Variable	Loading	Variable	Loading
Factor One (19.9 per cent Common Variance)		Rural Houses	0.89
		Urban Houses	0.46
Potentials-Urban	0.66	Urban Towns	0.59
Potential Manufacturing	0.71		
Potentials-Urban		Factor Three (7.6 per cent)	
Manufacturing	0.69	Production of	
Production of		Jowar	−0.79
Vegetable Canning	0.51	Bajra	−0.53
Flour Mills	0.51	Groundnut Seed	−0.64
Edible Oils	0.42	Sesamum Seed	−0.49
Hydrogenated Oils	0.53	Cotton Fiber	−0.70
Rubber	0.70	Jute Fiber	0.44
Chemicals, Heavy	0.60	Edible Oils	−0.47
Chemicals, Light	0.64	Cotton Gins	−0.74
Pottery	0.57	Longitude	0.66
Nonferrous Metals	0.56	Proportion of Population	
Metal Products	0.75	Urban	−0.42
Electrical Machinery	0.69		
Transport Equipment	0.68	Factor Four (6.4 per cent)	
Miscellaneous		Potentials	
Manufacturing	0.72	Total	−0.88
Services	0.60	Urban	−0.67
Slaughtering	0.56	Rural	−0.82
Bakery	0.60	Literate	−0.72
Beverages	0.52	Potentials	
Textiles, Wool	0.67	Manufacturing	−0.59
Piece Goods	0.66	Urban Manufacturing	−0.60
Wood Manufacturing	0.47	Production of	
Furniture	0.58	Gram	−0.45
Pulp	0.54	Gur	−0.59
Printing	0.54		
Leather	0.41	Factor Five (5.8 per cent)	
Cotton Mills	0.62	Production of	
Woollen Mills	0.48	Maize	0.60
Boots and Shoes	0.51	Wheat	0.78
Glass	0.58	Gram	0.66
Urban Population	0.68	Latitude	−0.69
Proportion of Population			
Rural	−0.57	Factor Six (5.0 per cent)	
Proportion of Population		Production of	
Urban	0.62	Iron Ore (Quantity)	0.78
Urban Houses	0.71	Iron Ore (Value)	0.78
		Manganese Ore (Q)	0.62
Factor Two (9.6 per cent)		Manganese Ore (V)	0.60
Production of		Nonferrous Ore (Q)	0.70
Rice	0.70	Nonferrous Ore (V)	0.71
All Grains	0.78	Other Minerals (Q)	0.43
Gur	0.42	Other Minerals (V)	0.42
Sugar	0.49		
Tobacco	0.49	Factor Seven (4.5 per cent)	
Rice Mills	0.53	Production of	
Total Population	0.72	Mica (Q)	0.90
Urban Population	0.46	Mica (V)	0.91
Rural Population	0.90	Coke Ovens	0.42
Proportion of Population		Mica Factories	0.70
Rural	0.47		

TABLE 1 (*continued*)

Variable	Loading	Variable	Loading
Factor Eight (4.5 per cent)		Factor Nine (4.4 per cent)	
Production of		Production of	
Limestone (Q)	−0.88	Salt (Q)	0.59
Limestone (V)	−0.87	Salt (V)	0.59
Clay and Quartz (Q)	−0.52	Coal (Q)	−0.69
Clay and Quartz (V)	−0.50	Coal (V)	−0.69

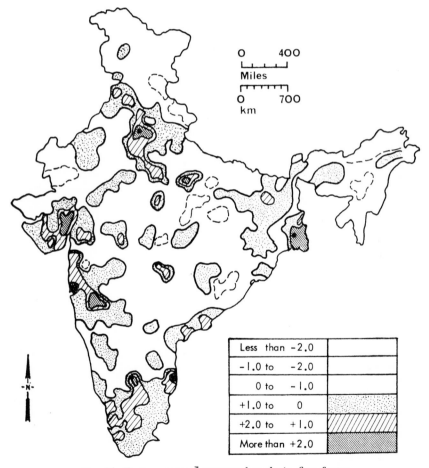

Less than −2.0	
−1.0 to −2.0	
0 to −1.0	
+1.0 to 0	
+2.0 to +1.0	
More than +2.0	

Fig. 31. Factor scores: structural analysis, first factor.

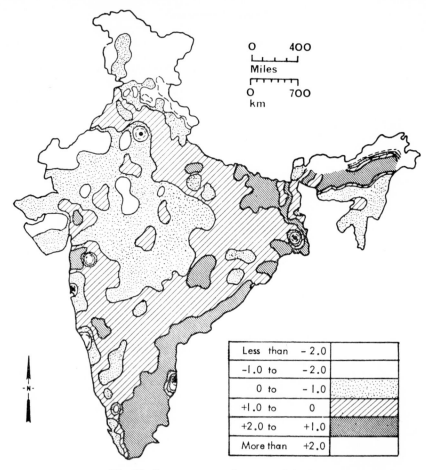

Less than	− 2.0	
−1.0 to	− 2.0	
0 to	− 1.0	
+1.0 to	0	
+2.0 to	+1.0	
More than	+2.0	

Fig. 32. Scores on second structural factor.

TABLE 2

PRINCIPAL FACTOR LOADINGS: BEHAVIORAL ANALYSIS

Commodity	Loading	Commodity	Loading
Factor One (19.7 per cent Common Variance)		Factor Four (10.6 per cent)	
		Bones	0.44
Cement	0.49	Dyes	0.69
Cotton Yarns–Indian	0.60	Rice, Unhusked	0.62
Cotton Yarns–Foreign	0.70	Manganese Ore	0.47
Piece Goods–Indian	0.62	Center Seed	0.43
Glass	0.67	Til	0.51
Rice, Husked	0.40	Teak	0.60
Wheat	0.48		
Kerosene	0.54	Factor Five (9.8 per cent)	
Castor Oil	0.57	Jowar	0.49
Coconut Oil	0.66	Bajra	0.65
Groundnut Oil	0.73	Millet	0.65
Other Vegetable Oils	0.54	Cotton Seed	0.53
Groundnut Seed	0.51	Khandsari Sugar	0.71
Ghee	0.45		
Salt	0.61	Factor Six (8.9 per cent)	
Raw Tobacco	0.55	Cattle	−0.40
		Sheep	−0.49
Factor Two (16.7 per cent)		Other Animals	−0.49
Cattle	0.53	Hemp	−0.44
Bones	0.48	Jute, Loose	−0.69
Gram	0.80	Jute, Baled	−0.77
Other Pulses	0.73	Molasses	−0.44
Jowar	0.51		
Bajra	0.51	Factor Seven (4.5 per cent)	
Hemp	0.62	Horses	0.47
Other Oil Cakes	0.51	Foreign Cotton Piece Goods	0.68
Other Vegetable Oils	0.45		
Groundnut Seed	0.44	Factor Eight (4.7 per cent)	
Linseed	0.42	Raw Cotton–Indian	0.76
Rape Seed	0.62	Raw Cotton–Foreign	0.87
Til	0.46		
Gur	0.46	Factor Nine (3.4 per cent)	
		Foreign Cotton Yarns	0.74
Factor Three (12.9 per cent)			
Coffee	0.65	Factor Ten (3.7 per cent)	
Gram	0.71	Raw Wool	−0.78
Jute, Loose	0.62		
Gunny Bags	0.67	Factor Eleven (3.1 per cent)	
Iron Forms	0.75	Raw Rubber	−0.73
Sugar	0.46		
Tea	0.42	Factor Twelve (2.9 per cent)	
		Lac	−0.56

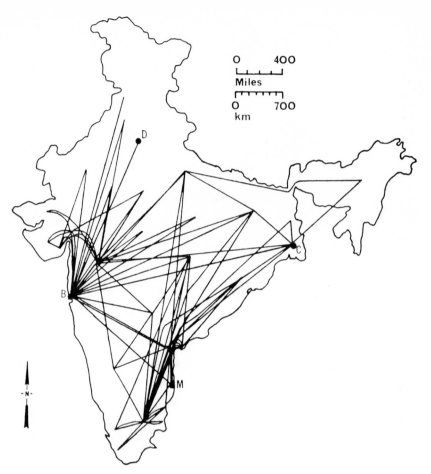

Fig. 33. Scores exceeding 2.0 on first behavioral factor.

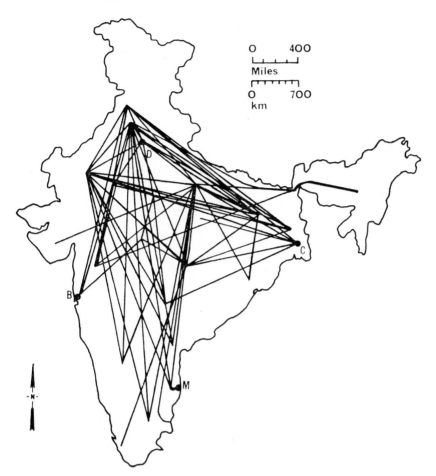

Fig. 34. Scores exceeding 2.0 on second behavioral factor.

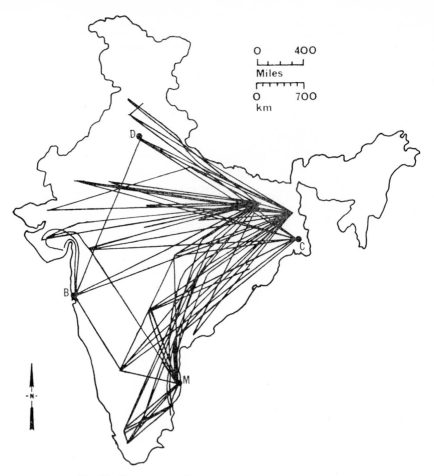

Fig. 35. Scores exceeding 2.0 on third behavioral factor.

TABLE 3

CANONICAL VECTORS OF THE GENERAL FIELD FORMULATION

		I	II	III	IV
Dyadic Factors	1	0.64	0.05	0.23	0.25
	2	0.19	0.33	−0.68	0.18
	3	0.59	−0.14	0.42	−0.68
	4	0.38	0.57	0.53	0.45
	5	−0.21	0.72	0.10	−0.47
Structural Factors	1	0.41	0.32	0.65	−0.39
	2	0.26	−0.40	0.43	0.45
	3	−0.12	−0.53	−0.09	−0.55
	4	−0.37	−0.40	0.35	0.01
	5	−0.35	−0.32	0.65	0.50
	6	0.35	−0.22	0.23	0.21
	9	0.28	0.05	−0.12	−0.32
Canonical Correlation		0.35	0.23	0.12	0.08
Chi-Square		106.30	34.82	13.58	3.88
Degrees of Freedom		24	15	8	3
Probability		.001	.01	.10	.25
Wilks Lambda for set		0.805	(Chi-Square 279.62)		

INTERDEPENDENCIES

There are at most three significant patterns of interdependency between types of spatial behavior and bases of spatial structure (Table 3), recalling also that we have available the loadings of each dyad on each of the k dimensions:

1   Exchange of urban-manufacturing specialities of one region for agricultural specialities of other regions, taking place through an intermetropolitan network of interregional connections. This is the fabric that holds separate parts of the economy together in a national system, linking areas of high potential.

2   Intraregional redistribution of regional products, assembly of specialities for export to other regions or overseas, or distribution of imports, within the four principal metropolitan-centered regions in patterns that mirror the spatial organization of the economy.

3   Direct transfers of agricultural specialities between agricultural regions and manufacturing specialities among manufacturing regions.

The results confirm Friedmann's generalizations in his 'general theory of polarized development'[7] and show great similarities to some of Linnemann's results[8] in the study of international trade.

Conclusions

A general field theory, made operational through multivariate analysis, very obviously provides a powerful way of analyzing spatial structure, spatial behavior and their interdependencies. The essential link is provided by dyadic representation of interaction potentials in structural space, thus drawing into the theory the social physicists' preoccupation with potentials and providing a multivariate generalization of it.

With interdependencies established, the next step is obviously to look for generating mechanisms, and the first steps of Friedmann[7] and Pred[9] should be noted. A further avenue of exploration is that of system change and diffusion processes, and here the interaction potentials provide a relevant field for operation of diffusion mechanisms.[10]

Most generally, of course, the field theory provides a general format for concept-building and testing in any situation where connections are sought between interactional and structural data.

References

1   Reprinted from *Papers and Proceedings of the Regional Science Association,* Vol. 21 (1968), pp. 205–27.

2   KUHN, THOMAS S. *The Structure of Scientific Revolutions,* Chicago: International Encyclopedia of Unified Science (1957).

3   BERRY, BRIAN J. L. 'Essays on Commodity Flows and the Spatial Structure of the Indian Economy', Chicago: *Department of Geography Research Paper* No. 111, University of Chicago (1966).

4   LEWIN, KURT *Field Theory in Social Science,* New York City: Harper Torchbooks (1964).

5   RUMMEL, RUDOLPH J. 'A Field Theory of Social Action with Application to Conflict within Nations,' *General Systems,* Vol. 10 (1965), pp. 183–211.

6   RUMMEL, RUDOLPH J. 'A Social Field Theory of Foreign Conflict,' *Peace Research Society Papers,* Vol. 9 (1966), pp. 131–50.

7   FRIEDMANN, JOHN *A General Theory of Polarized Development,* Santiago: Ford Foundation Program of Urban and Regional Development in Chile (1967).

8   LINNEMANN, H. *An Econometric Study of International Trade Flows,* Amsterdam: North Holland Publishing Company (1966).

9   PRED, ALLAN *The Spatial Dynamics of U.S. Urban-Industrial Growth, 1840–1900,* Cambridge: The M.I.T. Press (1966).

10   BOON, FRANCOISE *A Simple Model for the Diffusion in an Innovation Urban System,* Chicago: Center for Urban Studies, University of Chicago (1967).